Selected Titles in This Series

PIONEERS OF REPRESENTATION THEORY:
Frobenius, Burnside, Schur, and Brauer

History of Mathematics
Volume 15

PIONEERS OF REPRESENTATION THEORY:
Frobenius, Burnside, Schur, and Brauer

Charles W. Curtis

American Mathematical Society
London Mathematical Society

1991 *Mathematics Subject Classification.* Primary 01A55, 01A60, 20C15, 20C20;
Secondary 01A70, 16G10, 20G05.

A list of photograph credits and archival sources is included at the beginning of this volume.

Library of Congress Cataloging-in-Publication Data
Curtis, Charles W.
   Pioneers of representation theory : Frobenius, Burnside, Schur, and Brauer / Charles W. Curtis.
      p. cm. — (History of mathematics, ISSN 0899-2428 ; v. 15)
   Includes bibliographical references and index.
   ISBN 0-8218-9002-6 (alk. paper)
   1. Representations of groups—History. 2. Finite groups—History. I. Title. II. Series.
QA176.C87   1999
   512′.2—dc21   99-14983
                                                                                            CIP

To the memory of
Irving Reiner

Contents

Preface

This book contains an account of the history of the representation theory of finite groups, presented through an analysis of the published work of the four principal contributors to the theory in its formative stages: Ferdinand Georg Frobenius, William Burnside, Issai Schur, and Richard Brauer. The impetus for the project comes from my collaboration with the late Irving Reiner on a series of expository books on representation theory, and a gradually awakening interest on the part of both of us in the history of the subject.

The articles surveyed in the book were published mainly in the last two decades of the nineteenth century, and in the first half of the twentieth century. An introductory chapter contains an outline of some research on algebra and number theory involving characters of finite abelian groups, which appeared earlier in the nineteenth century, and was part of the tradition inherited by Frobenius and his successors. Later chapters contain reports on the main papers of the principals, and related work done at about the same time by others: Jacques Deruyts, Theodor Molien, Élie Cartan, Alfred Young, Alfred Loewy, Heinrich Maschke, Emmy Noether, Emil Artin, Tadasi Nakayama, Hsio-Fu Tuan, and Michio Suzuki, in more or less chronological order.

For the most important papers, enough of the mathematics is included to make it possible for a reader to follow the thread of the argument in detail. While many of the proofs of the main theorems are given today using new methods, the first proofs, obtained a century ago, are still interesting, and complete the historical picture by showing exactly how the creators of the theory used the mathematics that was available to them at the time. While the steps taken from one point to the next in the presentation of the mathematics are set down much as they appear in the original papers, I have taken the liberty of rewriting the arguments in modern mathematical terminology, and have sometimes given modern versions of parts of them.

The mathematical parts of the book are intended to be accessible to students as well as professional mathematicians and others with a basic knowledge of algebra. Background material needed to understand the papers is summarized in Chapter I, §1, and throughout the book as needed.

By concentrating on the mathematics in the original papers, I have not given as complete a survey of the historical setting of the mathematical discoveries as a reader might desire. Fortunately a full historical account of the development of the theory of representations of finite groups and associative algebras has been given in a series of articles by Thomas Hawkins, and frequent references to them are included.

The creators of representation theory all had long and productive mathematical careers, and made important contributions to other areas of mathematics besides the representation theory of finite groups. Each of them deserves a more complete mathematical biography than it has been possible for me to include here. To this end, however, sections containing biographical sketches, excerpts from correspondence, together with information about their teaching and relationships with their contemporaries, have been included in some of the chapters. Many quotations from the work under discussion have been inserted; they contain statements of some of the main results, remarks about motivation for the research, and comments on its relation to other contemporary investigations. All four of the principals are counterexamples to the myth that first-rate mathematics is done only by persons in their youth. Now that the collected works of Frobenius, Schur, and Brauer have been published, hopefully soon to be followed by the publication of the collected papers of Burnside, the full range of their mathematical work is readily available.

There are several reasons for undertaking an historical study of this particular branch of mathematics. We have passed the centennial of the publication of the first papers on characters and representations of finite groups, by Frobenius, in 1896-97. The principal contributors at the early stages, Frobenius, Burnside, Schur, and Brauer, each made discoveries which had a profound influence on the direction taken in research in algebra and number theory throughout most of the twentieth century. Their work deserves to be better known, along with something about their personal circumstances. The subject is flourishing today, with applications to other parts of mathematics as well as physics and chemistry. Its current vigor reinforces Schur's enthusiastic statement about it, in his inaugural lecture following his election to the Berlin Academy of Sciences in 1922: "What fascinated me so extraordinarily in these investigations was the fact that here, in the midst of a standstill that prevailed in other areas in the theory of forms, there arose a new and fertile chapter of algebra which is also important for geometry and analysis and which is distinguished by great beauty and perfection." The origins of this well-established area of modern mathematics, however, bear little resemblance to its content and methodology at the present time, so that it is a matter of historical interest, and mathematical interest as well, to follow the early steps in the development of the subject from this point of view.

We include a few remarks that are intended to guide the reader towards an historical overview, based on the survey to follow, of the first half century of the representation theory of finite groups. The prehistory, encompassing developments in nineteenth century mathematics which led to group representation theory, is summarized in Chapter I, Chapter V (on the work of Jacques Deruyts), and in the articles of Hawkins [172], [173], and [174]. For the history itself, we make two suggestions.

The first suggestion is to follow the changes in approach to the subject at different stages in the development, and the reasons for them. First one has Frobenius's theory of characters and his factorization of the group determinant (see Chapter II). This was followed by the classification of matrix representations of finite groups, first by Burnside (see Chapter III), and then by Schur, using a different method (see Chapter IV). Finally there was the absorption of the representation theory of finite groups over arbitrary fields in the theory of representations of nonsemisimple algebras, by Brauer, Noether, and Nakayama (Chapters VI and VII).

The second suggestion is to follow the problems, conjectures, and applications considered by Frobenius, Burnside, Schur, and Brauer. It will be seen that substantial parts of the theoretical development came about as a result of an attempt to solve a problem, or to make an application. We cite a few examples. Frobenius invented characters in connection with the problem of factoring the group determinant. He applied his theory of induced characters to the calculation of characters of the symmetric group, and to determining the structure of Frobenius groups (see Chapter II). Burnside raised the problem of existence of nonabelian finite simple groups of a given order, and in particular whether a nonabelian simple group of odd order can exist, or whether a nonabelian simple group whose order contains two distinct prime factors can exist. He solved the second problem with his $p^a q^b$-Theorem, while the first problem was solved more than fifty years later by Walter Feit and John Thompson (see Chapter III). Maschke and Burnside considered the splitting field problem, and solved it in special cases. Schur settled it for solvable groups using his theory of the Schur index (see Chapter IV). Schur classified the polynomial representations of the general linear group, using, among other things, Frobenius's work on the characters of the symmetric group (see Chapter V). Brauer, in collaboration with Hasse and Noether, proved Dickson's conjecture concerning division algebras over algebraic number fields (see Chapter VI). Brauer proved Artin's conjecture concerning L-series with general group characters, as a consequence of what is now called the Brauer Induction Theorem. He found several other applications of the Induction Theorem, including a solution of the splitting field problem in the general case.

The importance of challenging problems for the health of a branch of mathematics has long been recognized. In his lecture *Mathematical Problems*,[1] David Hilbert said, "As long as a branch of science offers an abundance of problems, so long is it alive; a lack of problems foreshadows extinction or the cessation of independent development." The representation theory of finite groups has always had an abundance of challenging problems. The fact that Frobenius, Burnside, Schur, and Brauer not only raised many of them, but had the strength and resourcefulness to solve some of them, is a measure of their greatness.

Charles W. Curtis

Eugene, Oregon, December, 1998

---

[1]Bull. Amer. Math. Soc. **8** (1902), 437-479, translated by Mary Winston Newson from the original, which appeared in the Göttinger Nachrichten, 1900, 253-297.

Acknowledgments

First of all, I want to thank the Institute of Theoretical Science, University of Oregon, for providing me with an office where most of the writing was done, and the members of the Institute for their friendly interest in the project.

Special thanks go to Walter Ledermann, for an interview in 1992 about his student days in Berlin, for sharing with me his notes on lectures of Schur, for assistance with translation of German text on several occasions, and for translating the Noether-Brauer letters, from the Archives of the Bryn Mawr College Library, which appear in Chapter VI, §2 and §3.

I am indebted to the Institute for Experimental Mathematics, University of Essen, for hospitality during the Fall of 1992, and to the Alexander von Humboldt Foundation for financial support at that time. While in Essen, I had an opportunity to discuss the project with Gerhard Michler, the director of the Institute, and with Christine Bessenrodt, a member of the Institute, who introduced me to Hannalore Bernhardt, at Humboldt University, Berlin. Dr. Bernhardt arranged appointments for me at the Archives of Humboldt University, and the Archives of the Berlin-Brandenburg Academy of Science, to study documents relating to Frobenius and Schur. During that time, I also profited from discussions with Urs Stammbach concerning Frobenius and Schur. Jan Saxl and Peter Neumann arranged visits to Cambridge and Oxford in December, 1992. In Cambridge, Dr. Saxl arranged meetings with librarians at Pembroke College and St. John's College, where I obtained copies of the Burnside-Baker letters discussed in Chapter III, and other material on Burnside. I had learned of the existence of the Burnside-Baker correspondence from Louis Solomon. At Oxford, I had a useful interview with Peter Neumann on Burnside and his work.

I have also had the benefit of suggestions and comments concerning the manuscript from Jonathan L. Alperin, Dean Alvis, Harold Edwards, Walter Feit, James A. Green, Thomas Hawkins, Jens C. Jantzen, Gerald Janusz, Tsit-Yuen Lam, Karen Parshall, Gerhard Röhrle, Klaus Roggenkamp, George Seligman, and Katsuhiro Uno. Jens Jantzen also transcribed the handwritten documents, quoted in Chapter II, supporting Frobenius's election to the Berlin Academy of Science, and provided me with other linguistic assistance with German text.

I am indebted to Bhama Srinivasan for calling my attention to the letters from Noether to Brauer mentioned above, and to Eiichi Bannai, who arranged the acquisition of the photographs of the Brauers in Japan and the photograph of Suzuki, Higman, and Feit.

Photo Credits

The American Mathematical Society gratefully acknowledges the kindness of these institutions in granting the following photographic permissions:

Archives of the Institute for Advanced Study.

Photo of Hermann Weyl; p. 190; Courtesy of the Archives of the Institute for Advanced Study, Princeton, NJ

Bryn Mawr College Archives.

Photo of Emmy Noether; p. 211; Courtesy of Bryn Mawr College Archives, Mariam Coffin Canaday Library, Bryn Mawr, PA

Institute Mittag-Leffler.

Photo of Ferdinand Georg Frobenius; cover and p. 36 ; Courtesy of Institute Mittag-Leffler, Djursholm, Sweden

Photo of William Burnside; cover and p. 89; Courtesy of Institute Mittag-Leffler, Djursholm, Sweden

Mathematisches Forschungsinstitut Oberwolfach.

Photo of Issai Schur; cover and p. 130; Courtesy of Mathematisches Forschungsinstitut Oberwolfach, Oberwolfach-Walke, Germany

Nagoya Mathematical Journal.

Photo of Tadasi Nakayama; p. 246; Courtesy of Nagoya Mathematical Journal, Graduate School of Mathematics, Nagoya University, Nagoya, Japan

The American Mathematical Society gratefully acknowledges the kindness of these individuals in granting the following permissions:

Charles W. Curtis.

Photos of the University of Berlin, pp. 39 and 133; Courtesy of Charles W. Curtis

Mina Hagie and Eiichi Bannai.

Photo of Professors Michio Suzuki, D. G. Higman, and W. Feit, p. 262; Courtesy of Mina Hagie and Eiichi Bannai

Walter Ledermann.

Photocopy of Cover for Issai Schur lecture notes, ETH Zürich, p. 136; Courtesy of Walter Ledermann

K. Shiratani and Eiichi Bannai.

Photos of the Brauers' visit to Japan, pp. 207–208; Courtesy of K. Shiratani and Eiichi Bannai

The American Mathematical Society holds the copyright to all other photographs in this volume.

Photo of Richard Brauer; cover and p. 204

Photo of John Thompson; p. 257

Photo of John Von Neumann; p. 192

Archival Sources

The American Mathematical Society gratefully acknowledges the kindness of these institutions in granting the following permissions to quote material from their archives:

The Berlin Academy of Sciences [Berlin-Brandenburgische Akademie der Wissenschaften] and the University of Berlin for the use of two letters of support for Ferdinand Georg Frobenius in Chapter II, Section 1, pp. 37–38 , and also for the material concerning Issai Schur's academic history in Chapter IV, Section 1, pp. 132, 137.

The Pembroke College Library, Cambridge, U.K., and The Master and Fellows of St. John's College, Cambridge, U.K., for the letters between Burnside and Baker quoted in Chapter III, Section 1, pp. 91–96, and the use of the clipping containing Burnside's obituary in Chapter III, Section 1, p. 96.

The Mariam Coffin Canaday Library of Bryn Mawr College, Bryn Mawr, PA, for the use of Emmy Noether's letters quoted in Chapter VI, Sections 2 and 3.

CHAPTER I

Some 19th-Century Algebra and Number Theory

The year 1997 was the centennial of two events in the history of mathematics: the publication of the first paper on representations of finite groups by Ferdinand Georg Frobenius (1849-1917), and the appearance of the first treatise in English on the theory of finite groups, by William Burnside (1852-1927). Burnside soon developed his own approach to representations of finite groups, and in the next few years, Frobenius and Burnside, working independently, explored the new subject and its applications to finite group theory. They were soon joined in this enterprise by Issai Schur (1875-1941), and some years later, by Richard Brauer (1901-1977). Their pioneering research is the subject of this book; it contains an account of the early history of representation theory, through an analysis of the published work of the principals and others with whom their work was intertwined: Emil Artin, Élie Cartan, Richard Dedekind, Jacques Deruyts, Alfred Loewy, Heinrich Maschke, Theodor Molien, Tadasi Nakayama, Emmy Noether, Michio Suzuki, Hsio-Fu Tuan, and Alfred Young. The book also contains biographical sketches, and a presentation of enough of the mathematics to enable a reader to learn parts of the subject as they were developed in the beginning.

The representation theory of finite groups was created in a series of papers by Frobenius, published in 1896 and 1897. These contained powerful new ideas; still, they built, to some extent, on the work of his predecessors. One source was the use, at least implicitly, of characters of finite abelian groups by some of the great 19th-century algebraists and number theorists. Another source was the emergence of the structure theory of finite groups, beginning with the brief outline of some of the main ideas left by Galois before his untimely death in 1832, and continued by Jordan, Sylow, and Frobenius and Burnside themselves.[1] A third was research on finite dimensional associative algebras by Weierstrass, Dedekind, Molien, and Cartan, and by Killing and Cartan on Lie algebras.[2]

The first chapter contains a focused survey of the mathematical tradition from which Frobenius drew much of his inspiration, and a sketch of some of the results involving characters of finite abelian groups to which Frobenius referred in the introduction to one of the papers in the 1896-1897 series (*Über Gruppencharaktere*, [**131**], 1896). As Frobenius himself remarked in the introduction to the paper, his comments followed closely information he received from Dedekind in a letter ([**104**], XLV) dated 8 July, 1896, at a time when he was in the midst of developing his new theory and writing it up. There Dedekind had written:

---

[1]No attempt is made here to give a systematic account of the history of the group concept; for this see Wussing [**290**].

[2]Some of the research on associative algebras and Lie algebras in which Frobenius and Burnside were involved is discussed in Chapters II and III; see also Hawkins [**173**], and Coleman [**94**].

I add some remarks on characters of abelian groups A. The oldest example of their application may well be the resolvent of Lagrange (for cyclic A). The Legendre symbol (generalized by Jacobi) should also be mentioned. The signs R, N used by Gauss (Art. 131) are less promising than the definite introduction of the roots of unity ± 1 by Legendre, and it appears (Art. 230) that he understood by a *character* of a class of forms or of a genus a relation, not a number, that indeed connects (Art. 246-248) the composition of genera to the corresponding composition of characters, but not as multiplication of numbers. The conversion of Gauss's genus-characters to numbers was carried out by Dirichlet (Recherche sur diverses applications etc. §3) through the use of Legendre symbols. Further, in his article on arithmetic progressions, Dirichlet used all characters ψ (-without using this name) of the abelian group $G(m)$, consisting of the $\varphi(m)$ classes of numbers relatively prime to m, and also all characters of the group of form-classes (in the proof of the representation of infinitely many prime numbers by a quadratic form).[3]

As this letter makes clear, Dedekind traced the roots of character theory back to the work of Lagrange in the eighteenth century and through ideas of Legendre, Gauss, and Dirichlet in the nineteenth. Our survey will set the stage for our analysis in Chapter II of Frobenius's development of the theory of group characters; it includes historical sketches and accounts of the mathematical topics mentioned in Dedekind's letter, and others related to them. The mathematical discussions in this chapter, and later in the book, are given in the language of current mathematical discourse, and should be understandable to students and professional mathematicians without any special preparation beyond the background outlined in the first section.

1. Introduction

In this section, we review finite abelian groups and their characters, with some examples and historical remarks, and the Galois theory of fields.

Familiarity with the ideas and notation introduced in this section will be needed for the proofs of theorems presented later.

We first recall that a *finite group* G of *order* $n = |G|$ is a finite set $\{x, y, \dots\}$ of cardinality n, together with a rule of multiplication, which assigns to each ordered pair of elements x, y a uniquely determined product $xy \in G$. The product operation is assumed to satisfy the associative law:

$$x(yz) = (xy)z, \quad \text{for all } x, y, z \in G,$$

and for each pair y, z, the equations

$$sy = z \text{ and } yt = z$$

have unique solutions s and t, respectively. The group is said to be *commutative*, or *abelian*, after Niels Henrik Abel (1802-1829), if the multiplication also satisfies the commutative law

$$xy = yx, \quad \text{for all } x, y \in G.$$

---

[3]Dedekind, R., *Gesammelte Mathematische Werke*, Friedr. Vieweg & Sohn, Braunschweig, 1931. Used with permission.

A standard first exercise in group theory is to show that in any group G, there exists a unique element 1, called the *identity element*, with the property that $x1 = 1x = x$ for all $x \in G$.

Important examples of finite abelian groups are the additive and multiplicative groups of residue classes modulo n, denoted by $\mathbb{Z}/n\mathbb{Z}$, and $G(n)$, respectively. A residue class $\bar{a}$ modulo n is, by definition, the set of all integers s such that $s \equiv a$ (mod n), in other words the set of all integers $a + kn, k \in \mathbb{Z}$. The group $\mathbb{Z}/n\mathbb{Z}$ consists of all residue classes modulo n, with the group operation (this time called addition) defined by

$$\bar{a} + \bar{b} = \overline{a+b},$$

for all residue classes $\bar{a}$ and $\bar{b}$; it has order n. The group $G(n)$ consists of those residue classes $\bar{a}$ such that a is relatively prime to n, with the group operation

$$\bar{a}\bar{b} = \overline{ab},$$

for all residue classes $\bar{a}$ and $\bar{b}$ in $G(n)$. It has order $\varphi(n)$, where $\varphi(n)$ is the Euler function, which counts the number of integers between 1 and n which are relatively prime to n.

The proofs of the group properties for the abelian groups $\mathbb{Z}/n\mathbb{Z}$ and $G(n)$ are the very beginnings of group theory. For example, the result that if a and b are relatively prime to n, then there exists an integer x relatively prime to n such that $ax \equiv b$ (mod n), can be traced to a Sanskrit astronomical work of the fifth century A.D., and was rediscovered successively by Bachet, Fermat, Wallis, and Euler.[4] This fact, and the other group properties, are all carefully stated and proved by Gauss in §I of the *Disquisitiones Arithmeticae*. There one also finds proofs of Fermat's Theorem, that for a prime p,

$$\bar{a}^{p-1} = 1$$

for each element $\bar{a}$ of $G(p)$, and Euler's Theorem, that

$$\bar{a}^{\varphi(n)} = 1,$$

for all elements $\bar{a}$ in $G(n)$, along with references to their occurrence in the work of Fermat and Euler.

The set of residue classes modulo p, for a prime p, is a field, denoted by $\mathbb{Z}_p$, whose additive and multiplicative groups are $\mathbb{Z}/p\mathbb{Z}$ and $G(p)$, respectively. We require the basic fact, stated and proved[5] by Euler in 1773, that the multiplicative group $G(p)$ is cyclic, with $\varphi(p-1)$ generators. Integers a, relatively prime to p, for which the corresponding residue class $\bar{a}$ is a generator of $G(p)$, are called *primitive roots* modulo p.

A *character* χ of a finite abelian group A is by definition a function from A to the set of nonzero complex numbers such that

$$\chi(ab) = \chi(a)\chi(b) \quad \text{for all} \quad a, b \in A.$$

The character 1 (or 1_A), which takes the value 1 at each element of A, is called the *trivial* character (or sometimes the *principal* character). Incidentally, the mathematical term *character* was first introduced by Gauss, beginning with §230 of the *Disquisitiones Arithmeticae* [**155**], to assign numerical information to classes of

---

[4]See Weil [**283**], Chap. III, §5, for the history related to this and other properties of the groups $G(n)$, along with bibliographical references.

[5]See Gauss [**155**], §§55, 56.

binary quadratic forms, in order to separate classes of forms with the same determinant into different genera. A sketch of his ideas, and the subsequent interpretation of them by Dirichlet and Dedekind, will be given in §8.

Examples of nontrivial characters of the additive group $\mathbb{Z}/n\mathbb{Z}$ are the functions χ defined by

$$\chi(\bar{a}) = \omega^a,$$

for a complex nth root of unity $\omega \neq 1$, and $a \in \mathbb{Z}$. It is easily proved that all characters of the group $\mathbb{Z}/n\mathbb{Z}$ have this form.

As Dedekind mentioned in his letter quoted in the introduction to this chapter, the theory of quadratic residues provided one of the first examples of a character of a finite abelian group, in this case the multiplicative group $G(p)$, for a prime p. First of all, an integer a, not divisible by p, is called a *quadratic residue* modulo p if and only if the congruence $x^2 \equiv a \pmod{p}$ has a solution. Legendre introduced the notation

$$\left(\frac{a}{p}\right) = 1 \ \text{ or } \ -1$$

to indicate that a is, or is not, a quadratic residue modulo p. We shall use instead the notation (a/p) for Legendre's symbol. It has the property that

$$(ab/p) = (a/p)(b/p),$$

for all integers a and b not divisible by p, and hence defines a character of the multiplicative group $G(p)$.

A basic result is the following theorem, which describes what are known as the *Orthogonality Relations for Characters of Finite Abelian Groups.*

THEOREM 1.1. *Let χ and ψ be characters of the finite abelian group A of order $|A|$. Then*

$$\sum_{x \in A} \chi(x)\psi(x^{-1}) = |A| \ \text{ or } \ 0,$$

according as χ is, or is not, equal to ψ. In particular,

$$\sum_{x \in A} \chi(x) = |A| \ \text{ or } \ 0$$

according as χ is, or is not, the trivial character.

For the proof, assume first that $\chi = \psi$; then $\chi(x)\psi(x^{-1}) = 1$ for all elements x in A, and the result follows. Now assume that $\chi \neq \psi$, and let y be an element of A such that $\chi(y)\psi(y^{-1}) \neq 1$. Then

$$\chi(y)\psi(y^{-1}) \sum_{x \in A} \chi(x)\psi(x^{-1}) = \sum_{x \in A} \chi(xy)\psi((xy)^{-1}) = \sum_{x \in A} \chi(x)\psi(x^{-1}),$$

so

$$(\chi(y)\psi(y^{-1}) - 1) \sum_{x \in A} \chi(x)\psi(x^{-1}) = 0,$$

and it follows that

$$\sum_{x} \chi(x)\psi(x^{-1}) = 0.$$

This completes the proof of the theorem.

The set of characters of A forms an abelian group itself, called the *dual group* of A, and denoted by $\hat{A}$. The operation of multiplication of two characters φ and ψ is given by $\varphi\psi(x) = \varphi(x)\psi(x)$. It is then easy to see that, for each element $x \in A$, the map which assigns to each character ψ the element $\psi(x)$ is a character of $\hat{A}$, and that all characters of $\hat{A}$ are obtained in this way. Upon applying the previous theorem to $\hat{A}$, one obtains:

COROLLARY 1.1. *Let A be a finite abelian group of order $|A|$ and let $x \in A$. Then*

$$\sum_{\psi \in \hat{A}} \psi(x) = |A| \text{ or } 0,$$

according as $x = 1$ or $x \neq 1$.

Dedekind's letter, and the introduction to *Über Gruppencharaktere*, cited the work of Lagrange on the resolvent of a polynomial equation, published in 1770, as the first application of characters of finite abelian groups. Lagrange's theory of the resolvent, and a related investigation by Gauss, will be taken up later in this chapter. Their work was a forerunner of Galois theory; in our presentation of it, however, it will be convenient to state the results in the language of the modern Galois theory of fields, which we review briefly here.

For our purposes, it is only necessary to consider fields which occur as subfields of the field of complex numbers $\mathbb{C}$. A subfield F of $\mathbb{C}$ is simply a subset containing nonzero elements, which is closed under the operations of addition, subtraction, multiplication, and division (by a nonzero element). It is called a *finite extension* of a subfield $E \subseteq F$ if the dimension of F, viewed as a vector space over E, is finite. For a finite extension F of E, the dimension of the vector space F over E is denoted by $(F : E)$, and called the *degree* of F over E.

A field F containing a field E is said to be generated over E by a set of elements $\omega_1, \omega_2, \ldots$, and denoted by $F = E(\omega_1, \omega_2, \ldots)$, if F is the uniquely determined minimal subfield of $\mathbb{C}$ containing E and the elements $\omega_1, \omega_2, \ldots$. If $\omega_1, \ldots, \omega_s$ are roots of a monic[6] polynomial equation with coefficients in E, of the form

$$p(x) = x^n + a_1 x^{n-1} + \cdots + a_n = 0, \quad \text{with} \quad a_i \in E,$$

then $F = E(\omega_1, \ldots, \omega_s)$ is a finite extension of E.

The finite extension $F = E(\omega_1, \ldots, \omega_n)$ generated by all the roots $\omega_1, \ldots, \omega_n$ of a single polynomial equation $p(x) = 0$, as above, is called the *splitting field* of the polynomial $p(x)$. The polynomial $p(x)$ can be factored into linear factors

$$p(x) = (x - \omega_1) \cdots (x - \omega_n)$$

in the polynomial ring $F[x]$.

An extension field $F = E(\omega)$ generated by one element is a finite extension of E if and only if ω is a root of a polynomial equation $p(x) = 0$ as above. In case $(E(\omega) : E)$ is finite, there is a uniquely determined polynomial $m(x) \in E[x]$ of minimal degree, with leading coefficient 1, such that $m(\omega) = 0$; this polynomial is called the *minimal polynomial* of ω.

Finite extensions of the rational field $\mathbb{Q}$ are called *algebraic number fields*, and their elements are called *algebraic numbers*. An algebraic number α is called an *algebraic integer* whenever its minimal polynomial has coefficients in the ring of rational integers $\mathbb{Z}$. The set of all algebraic integers in an algebraic number field E forms a ring, denoted by alg.int. (E); its quotient field is the field E. It follows

---

[6] *monic* means: leading coefficient $= 1$.

easily from the definition that the only algebraic integers that belong to the rational field $\mathbb{Q}$ are in fact rational integers:

$$\text{alg.int. } (\mathbb{C}) \cap \mathbb{Q} = \mathbb{Z}.$$

Let F be a finite extension field of E. An *automorphism* σ *of* F *over* E is an automorphism of the field F, whose restriction to the subfield E is the identity, that is $\sigma(\xi) = \xi$ for all elements $\xi \in E$. The set of all automorphisms, over E, of a finite extension F of E, is a finite group, denoted by $\mathcal{G}$; the multiplication of two elements σ, τ of $\mathcal{G}$ is defined by $\sigma\tau(\eta) = \sigma(\tau(\eta))$ for $\eta \in F$. A finite extension field F of E is called a *Galois extension* of E, and $\mathcal{G}$ is called the *Galois group* of F over E, if the set of elements of F which are fixed by all automorphisms $\sigma \in \mathcal{G}$ coincides with the field E. To put it another way, the elements of E are automatically fixed by all automorphisms $\sigma \in \mathcal{G}$; to say that F is a Galois extension of E means that there are no elements of F outside E with this property. The images $\sigma(\xi), \sigma \in \mathcal{G}$, of an element $\xi \in F$ by elements of the Galois group are called *conjugates* of ξ.

We can now state two of the main results of classical Galois theory.

THEOREM 1.2. *A finite extension field F of E is a Galois extension of E if and only if F is the splitting field of some polynomial $p(x) \in E[x]$.*

THEOREM 1.3. *Let F be a finite Galois extension field of E, with Galois group $\mathcal{G}$. Then the following statements hold.*

(i) The order of the Galois group $\mathcal{G}$ is equal to the degree $(F : E)$.

(ii) There is a one-to-one correspondence from the set of all subgroups $\{\mathcal{H}\}$ of $\mathcal{G}$ to the set of all fields K such that $E \subseteq K \subseteq F$. The field K corresponding to the subgroup $\mathcal{H}$ is the set of elements in F fixed by all the automorphisms belonging to $\mathcal{H}$. The subgroup $\mathcal{H}$ of $\mathcal{G}$ corresponding to a field K such that $E \subseteq K \subseteq F$ is the set of all automorphisms $\sigma \in \mathcal{G}$ which act as the identity map on K.

(iii) Let $\mathcal{H}$ be a subgroup of $\mathcal{G}$, and K the subfield of F corresponding to it, as in (ii). Then F is a Galois extension of K, with Galois group $\mathcal{H}$. On the other hand, K is a Galois extension of E if and only if $\mathcal{H}$ coincides with all its conjugates $\sigma\mathcal{H}\sigma^{-1}$, for $\sigma \in \mathcal{G}$; when this occurs, $\mathcal{H}$ is a normal subgroup of $\mathcal{G}$, and the Galois group of K over E is the factor group $\mathcal{G}/\mathcal{H}$.

While Galois indicated statements and ideas involved in the proofs of these theorems in his letter to his friend Chevalier written on the eve of his death [**154**], it was many years before the results were widely understood in the mathematical community. The subject became a standard item in courses on abstract algebra following the publication of van der Waerden's book [**276**] in 1930. A particularly clear and elementary account was given by Artin [**10**]. For proofs of the theorems, and other results from Galois theory used later, we refer the reader to Artin's booklet, van der Waerden, or any textbook on modern algebra.

2. Ruler and Compass Constructions and Cyclotomic Fields

In the rest of this chapter, we shall follow the plan described in the introduction, by illustrating some of the ways characters of abelian groups were used, at least implicitly, in 19th century algebra and number theory, beginning with the work of Gauss and Lagrange on polynomial equations. In modern terms, they considered fields and finite extensions of them, especially cyclotomic fields, and the polynomial equations defining them, anticipating by several years the general theory of Galois.

We start with Gauss's exploration of the subject, in Part VII of the *Disquisitiones Arithmeticae* [**155**], published in 1801 when he was 24. It will be useful to give a fairly thorough survey of the contents of Part VII, as it contains other things of importance as background for representation theory.

Part VII begins with the theory of regular polygons, and the problem of constructing them with a ruler and compass. From Euler's formula

$$e^{i\theta} = \cos\theta + i\sin\theta$$

it follows that the complex roots of the equation $x^m - 1 = 0$, viewed as points in the plane, are the vertices of a regular m-gon inscribed in the unit circle, and are given by

$$\cos(2\pi k/m) + i(\sin 2\pi k/m), \ k = 0, \dots, m-1.$$

The ruler and compass constructibility of the regular m-gon is equivalent to the question of whether $\cos(2\pi k/m)$ and $\sin(2\pi k/m)$ can be obtained by taking a succession of square roots and rational expressions involving them, starting from rational numbers.

Several years earlier, Gauss had proved the ruler and compass constructibility of the regular 17-gon, and announced his achievement in 1796 in the first entry in his mathematical diary [**158**]. As the story goes, it was this discovery that sealed his decision to devote himself to mathematics [**62**].

The ruler and compass constructibility of the regular m-gon had been established by the Greek geometers in case $m = 3, 5, 6 = 2 \cdot 3, 15 = 3 \cdot 5$ and in other situations derived from these. Gauss showed that for integers $a, b, c, \dots$ without common factors, the constructibility of the polygon with m sides with $m = a \cdot b \cdot c \cdots$ follows from the constructibility of the regular a-gon, b-gon, etc. . He went on to explain what was required to construct the regular polygon with p^λ sides from a construction of the polygon with p sides, for a prime p. Thus the constructibility of a regular polygon with p sides, for an odd prime p, was the first problem to be considered. The reason why the special case $p = 17$ is of interest will become clear later.

Much of Part VII is devoted to a thorough investigation, extending well beyond questions of constructibility of polygons, of the subfields of what has become known as the *cyclotomic field* $\mathbb{Q}(\omega)$, with ω a primitive p-th root of unity, and p an odd prime. (The literal meaning of the word *cyclotomy* is *division of the circle*). The field $\mathbb{Q}(\omega)$ is generated by the roots of the polynomial equation

$$(1) \qquad x^p - 1 = (x - 1)(x^{p-1} + x^{p-2} + \cdots + x + 1) = 0.$$

Gauss introduced the notation X for the second factor, Ω for the set of roots of the equation $X = 0$, and pointed out some basic properties of the set of roots of unity Ω, namely that Ω consists of the complex numbers

$$\omega, \omega^2, \dots, \omega^{p-1},$$

and also the numbers

$$\omega^a, \omega^{2a}, \dots, \omega^{(p-1)a},$$

for any integer a not divisible by p. He continued with the observation that

$$1 + \omega^a + \omega^{2a} + \cdots + \omega^{(p-1)a} = 0.$$

From a modern point of view, the first of these remarks amounts to the statement that the set of pth roots of unity is a cyclic group, with generator ω, or ω^a, for any integer a not divisible by p, while the second is a consequence of the orthogonality relation, Theorem 1.1, for two characters of the cyclic group $\mathbb{Z}/p\mathbb{Z}$.

His plan for tackling the subfields of the cyclotomic field was stated as follows (in [**155**], §342).

> ... We intend to resolve X gradually into more and more factors, and in such a way that their coefficients are determined by equations of as low an order as possible. In so doing we will finally come to simple factors or to the roots Ω. We will show that if the number $p-1$ is resolved in any way into integral factors α, β, γ, etc. (we can assume that each of them is prime), X can be resolved into α factors of $(p-1)/\alpha$ dimensions with coefficients determined by an equation of degree α; each of these will be resolved into β others of $(p-1)/\alpha\beta$ dimensions with the aid of an equation of degree β etc. Thus if we denote by ν the number of factors α, β, γ, etc. The determination of the roots Ω is reduced to the solution of ν equations of degree α, β, γ, etc. For example, for $p = 17$ where $p - 1 = 2 \cdot 2 \cdot 2 \cdot 2$, there will be four quadratic equations to solve; for $p = 73$ three quadratic and two cubic equations.[7]

It is a nice exercise to interpret the preceding paragraph in terms of Galois theory (see §1). The first part states that if α is a factor of the order of the Galois group of $\mathbb{Q}(\omega)$ over $\mathbb{Q}$ (which in this case is cyclic of order $p-1$), then there exists a subfield K of $\mathbb{Q}(\omega)$ whose degree over $\mathbb{Q}$ is α. Moreover $\mathbb{Q}(\omega)$ is an extension of K of degree $(p-1)/\alpha$, and is the splitting field of an irreducible polynomial of degree $(p-1)/\alpha$ with coefficients in K. This polynomial and its conjugates are the irreducible factors of X in $K[x]$. As Gauss explained, the process can be repeated for the polynomial of degree $(p-1)/\alpha$, etc.

Gauss's own pre-Galois-theory method for carrying out his plan is equally interesting and began with the consideration of certain sums of roots of unity, called *periods*. Their definition is based on the choice of a primitive root mod p, which we shall denote by g. The set of roots of the equation $X = 0$, denoted by Ω, consists of the powers $\omega, \omega^2, \ldots, \omega^{p-1}$ of a single primitive p-th root of unity ω. It is clear that Ω also consists of the powers $\omega^g, \ldots, \omega^{g^{p-1}}$. The periods associated with a given factorization of $p - 1$ into positive integers, $p - 1 = m \cdot e$, are the complex numbers $\eta_0, \ldots, \eta_{m-1}$ defined by the formulas

$$(2) \qquad \eta_i = \sum_{j=0}^{e-1} \omega^{g^{i+mj}}, i = 0, \ldots, m - 1.$$

It is easily verified that the roots of unity appearing in the different periods form a partition of Ω into nonoverlapping subsets.

As an illustration, he considered the case $p = 19$, noting that 2 is a primitive root mod 19. The periods associated with the factorization $18 = 3 \cdot 6$, with $m = 3$ and $e = 6$, are

$$\eta_0 = \omega^1 + \omega^{2^3} + \cdots + \omega^{2^{15}} = \omega + \omega^7 + \omega^8 + \omega^{11} + \omega^{12} + \omega^{18}$$
$$\eta_1 = \omega^2 + \omega^3 + \omega^5 + \omega^{14} + \omega^{16} + \omega^{17}$$
$$\eta_2 = \omega^4 + \omega^6 + \omega^9 + \omega^{10} + \omega^{13} + \omega^{15}.$$

---

[7]Gauss, C. F., *Disquisitiones Arithmeticae*, English translation by A. A. Clarke, Yale University Press, New Haven, 1966. Used with permission.

He began his analysis of the factors of X and the subfields of the cyclotomic field corresponding to them with the following result, stated here in the language of field theory.

THEOREM 2.1. *The cyclotomic field $\mathbb{Q}(\omega)$ has a unique subfield E of degree m over $\mathbb{Q}$ for each factor m of $p-1$. The periods $\eta_0, \dots, \eta_{m-1}$ associated with the factorization $p-1 = m \cdot e$ are a basis of E over $\mathbb{Q}$.*

We shall give a sketch of the proof, using Galois theory. Gauss's proof, of course, was based entirely on results he had established earlier in the *Disquisitiones Arithmeticae*. The first step is to prove that the polynomial $X = x^{p-1} + \cdots + x + 1$ is irreducible over the rational field. Gauss proved this using one of several results now called Gauss's Lemma: *if a monic polynomial with integer coefficients has a nontrivial factorization over the rationals, then it has a nontrivial factorization over the integers.* The application of this result to prove the irreducibility of X is not difficult, and is omitted.

From the irreducibility of X it follows that the set of roots $\omega, \dots, \omega^{p-1}$ of the equation $X = 0$ form a basis over the rational field of the cyclotomic field $\mathbb{Q}(\omega)$. The field $\mathbb{Q}(\omega)$ is the splitting field of the irreducible polynomial X, and hence is a Galois extension of $\mathbb{Q}$ of degree $p-1$ over $\mathbb{Q}$. Its Galois group $\mathcal{G}$ is cyclic of order $p-1$, and consists of the powers of the automorphism σ, which takes ω to ω^g, for a fixed primitive root g mod p.

Let m be a factor of $p-1$, with $p-1 = m \cdot e$. The cyclic Galois group $\mathcal{G}$ has a unique cyclic subgroup of order e and index m, generated by the element σ^m. The subfield of $\mathbb{Q}(\omega)$ consisting of the elements left fixed by σ^m is the unique subfield E of $\mathbb{Q}(\omega)$ of degree m over $\mathbb{Q}$, by the fundamental theorem of Galois theory. It remains to show that this subfield has a basis over $\mathbb{Q}$ consisting of the periods $\eta_0, \dots, \eta_{m-1}$. This is proved as follows. We first observe that the elements $\omega^{g^0}, \dots, \omega^{g^{m-2}}$ also form a basis of the field $\mathbb{Q}(\omega)$ because g is a primitive root mod p. A typical element α of the field is a linear combination

$$\alpha = a_0 \omega^{g^0} + \cdots + a_{m-2} \omega^{g^{m-2}}$$

with rational coefficients $a_0, a_1, \dots$. As $\sigma \omega^g = \omega^{g^2}$ etc. we have

$$\sigma^m(\alpha) = a_0 \omega^{g^m} + \cdots + a_{m-2} \omega^{g^{m-2}}.$$

Therefore $\sigma^m(\alpha) = \alpha$ if and only if

$$a_i = a_{i+m} = \dots,$$

where we may identify a_i with a_j whenever $i \equiv j \pmod{p}$. It follows that an element α is left fixed by σ^m if and only if α is a linear combination of elements of the form

$$\omega^{g^i} + \omega^{g^{i+m}} + \cdots,$$

and these are the periods $\eta_0, \dots, \eta_{m-1}$. We noted earlier that the periods are sums of powers of ω belonging to nonoverlapping subsets of the basis elements $\omega, \dots, \omega^{p-1}$. Therefore the periods are linearly independent, and form a basis of E over $\mathbb{Q}$, completing the proof of the theorem.

3. Equations Defining Subfields of the Cyclotomic Field

Gauss's investigations led him to discover that what we would call subfields of the cyclotomic field were described by equations which he was able to calculate explicitly.

Let $p - 1 = m \cdot e$, and let E be the unique subfield of the cyclotomic field $\mathbb{Q}(\omega)$ of degree m, as in Theorem 2.1. The basis elements of E are the periods $\eta_0, \ldots, \eta_{m-1}$, and these are permuted by the generator σ of the Galois group of $\mathbb{Q}(\omega)$ over $\mathbb{Q}$ defined above:

$$\sigma(\eta_0) = \eta_1, \ \sigma(\eta_1) = \eta_2, \ldots, \sigma(\eta_{m-1}) = \eta_0.$$

Therefore the coefficients of the polynomial

$$\prod_{i=0}^{m-1} (x - \eta_i) = (x - \eta_0)(x - \eta_1) \cdots (x - \eta_{m-1})$$

belong to the field $\mathbb{Q}$. This follows from the fact that they are symmetric functions of the roots, and are consequently fixed by the elements of the Galois group of E over $\mathbb{Q}$. From this, it is readily shown that the polynomial is irreducible over the field of rational numbers, and is the minimal polynomial over the rational field of each of the periods η_i.

After making what amounted to these observations, Gauss continued his analysis of the factors of the cyclotomic polynomial X and subfields of $\mathbb{Q}(\omega)$ corresponding to them by setting out to determine as explicitly as possible the polynomial equations satisfied by the periods, and for this, it was clearly important to understand how the periods multiply. If one replaces g^i by j, the period η_i can be expressed in the form

$$\eta_i = \omega^j + \omega^{jg^m} + \cdots + \omega^{jg^{(e-1)m}} = \sum_k \omega^{jg^{km}},$$

where the sum is taken over an arbitrary set of representatives k of the residue classes mod e. Upon denoting the preceding sum by $\eta^{(j)}$, the multiplication of the periods $\eta^{(i)}$ and $\eta^{(j)}$ is given by the formula

$$\eta^{(i)} \cdot \eta^{(j)} = \sum_k \omega^{ig^{km}} \cdot \sum_\ell \omega^{jg^{\ell m}} = \sum_{k,\ell} \omega^{ig^{km}+jg^{\ell m}}$$

where the sums are taken over representatives of the residue classes mod e. After the substitution $\ell = \ell' + k$, the formula becomes

$$\sum_k \sum_{\ell'} \omega^{ig^{km}+jg^{km}g^{\ell' m}} = \sum_{\ell'} (\sum_k \omega^{ig^{km}+jg^{km}g^{\ell' m}}).$$

This proves:

THEOREM 3.1 (Gauss's Multiplication Formula). *The multiplication of the periods associated with the subfield E of degree m is given as follows:*

$$\eta^{(i)} \cdot \eta^{(j)} = \sum_\ell \eta^{(i+jg^{\ell m})}$$

where the sum is taken over a set of representatives ℓ of the residue classes mod e.

After taking account of the coincidences occurring on the right hand side of the multiplication formula resulting from the identities $\eta^{(i)} = \eta^{(ig^m)} = \cdots$ and $\eta^{(i)} = \eta^{(j)}$ whenever i and j are congruent mod p, it follows that the basis elements

$\{\eta_0, \ldots, \eta_{m-1}\}$ of the field E satisfy structure equations with integer structure constants c_{ijk}:

$$(3) \qquad\qquad \eta_i \cdot \eta_j = \sum_{k=0}^{m-1} c_{ijk}\eta_k.$$

From the multiplication formula and a closer analysis, it follows that the structure constants c_{ijk} are nonnegative integers, which count the number of solutions of congruences of the form

$$aX^m + bY^m \equiv c \pmod{p}$$

for certain integers a, b, c.

In modern terminology, the equations (3) define the structure of a certain commutative, associative algebra, by giving the multiplication of the elements of a basis. In this case, the algebra is the field E, with the basis consisting of the periods.

Gauss very naturally returned to the problem of the constructibility of the regular 17-gon in the new setting he had created, especially since it provided a nice illustration of his elegant multiplication theorem for the periods. The first step was to consider the cyclotomic field $\mathbb{Q}(\omega)$, with ω a primitive 17th root of unity, and the factorization $p - 1 = m \cdot e$, with $m = 2$. In this situation, 3 is a primitive root mod 17. By Theorem 2.1 there is a unique subfield E of degree 2 (that is, a quadratic extension of $\mathbb{Q}$), with a basis consisting of the periods

$$\eta_0 = \omega^{3^0} + \omega^{3^{0+2}} + \cdots = \omega + \omega^9 + \cdots = \eta^{(1)}$$

and

$$\eta_1 = \omega^{3^1} + \omega^{3^{1+2}} + \cdots = \omega^3 + \omega^{11} + \cdots = \eta^{(3)}.$$

By the multiplication formula, one has

$$\eta_0 \cdot \eta_1 = \eta^{(1)} \cdot \eta^{(3)} = 4\eta_0 + 4\eta_1.$$

The periods η_0 and η_1 also satisfy the relation

$$\eta_0 + \eta_1 = -1,$$

because the sum of all the pth roots of unity is zero, and $\eta_0 + \eta_1$ is the sum of the nonidentity ones. It follows that η_0 and η_1 are the roots of the quadratic equation

$$x^2 + x - 4 = 0,$$

and hence that $E = \mathbb{Q}(\sqrt{17})$.

It is not difficult to prove that the periods defined by the factorization $p - 1 = m \cdot e$, with $m = 4$, satisfy a quadratic equation with coefficients in the field $E = \mathbb{Q}(\eta_0, \eta_1)$; and that those for $m = 8$ satisfy a quadratic equation with coefficients in the field generated over $\mathbb{Q}$ by the periods for $m = 4$. It follows that all the 17th roots of unity are constructible numbers, in the sense that each of them belongs to a field E_t for which there is a tower of fields

$$E_0 \subset E_1 \subset \cdots \subset E_t,$$

with $E_0 = \mathbb{Q}$ and each field E_i a quadratic extension of the preceding one. This shows, as he explained, that the regular 17-gon is constructible with a ruler and compass. It also shows, in a special case, how he planned to use the periods and the results concerning them he had proved, to carry out his program for analyzing the factorization of the cyclotomic polynomial X.

4. The Quadratic Subfield of the Cyclotomic Field

Gauss solved the problem of finding the minimal polynomials of the periods over the field of rational numbers in two general cases. In the first, he found the quadratic equation satisfied by the two periods associated with the factorization $p - 1 = m \cdot e$, for an odd prime p and $m = 2$. In the second he determined the cubic equation satisfied by the three periods associated with the factorization $p - 1 = m \cdot e$, with $m = 3$, for primes p congruent to 1 modulo 3. We shall describe two approaches to the first problem, and include a few remarks on his solution of the second.

The solution of the first problem can be stated as follows.

THEOREM 4.1. *Let E be the unique quadratic subfield of the cyclotomic field $\mathbb{Q}(\omega)$, with ω a primitive p-th root of unity, for an odd prime p. Then E has a basis over the rational field consisting of the periods η_0 and η_1 defined above. They satisfy the quadratic equations*

$$x^2 + x - (p-1)/4 = 0 \quad \text{in case } p \equiv 1 \pmod 4$$

and

$$x^2 + x - (p+1)/4 = 0 \quad \text{in case } p \equiv 3 \pmod 4.$$

Thus we have, in all cases, $E = \mathbb{Q}(\sqrt{\pm p})$.

Gauss proved the theorem as follows, in [**155**], §356. From the previous discussion, the periods η_0 and η_1 are the roots of a quadratic equation

$$x^2 - Ax + B = 0,$$

with $A = \eta_0 + \eta_1 = -1$ and $B = \eta_0 \cdot \eta_1$. The product of the periods η_0 and η_1 was found using Theorem 3.1. He calculated the structure constants c_{ijk} in (3) using the theory of quadratic residues, and obtained the values of A and B given in the statement of the theorem.

We shall give a different proof of the theorem, based on formulas for the periods η_0 and η_1 which involve the characters of the additive and multiplicative groups of the finite field $\mathbb{Z}_p$. Gauss stated essentially the same results as an afterthought to his own proof of the theorem, but did not pursue them further, remarking that

> ... these matters are on a higher level of investigation, and we will reserve their consideration for another occasion.[8]

Here he was apparently referring to the use of certain sums of roots of unity, today called Gauss sums, not only in the proof of the theorem stated above, but also for their contributions to the proofs of reciprocity laws. These contributions began with his own proof using Gauss sums of the quadratic reciprocity law. A proof of the quadratic reciprocity law along these lines will be given later in Section 6.

A *Gauss sum* associated with the finite field $F = \mathbb{Z}_p$ is a complex number defined in terms of a character χ of the additive group of F, and a character ψ of the multiplicative group $F^* = F - \{0\}$ (see §1), by the formula:

$$(4) \qquad G = G(\chi, \psi) = \sum_{t \in F^*} \chi(t)\psi(t).$$

---

[8]Gauss, C. F., *Disquisitiones Arithmeticae*, English translation by A. A. Clarke, Yale University Press, New Haven, 1966. Used with permission.

The first result we require holds for an arbitrary Gauss sum $G = G(\chi, \psi)$, with nontrivial characters χ and ψ of F and F^* respectively (recall that nontrivial means that not all values of the character are equal to 1). It is the statement that

$$(5) \qquad\qquad G(\chi, \psi) \cdot \overline{G(\chi, \psi)} = p.$$

(Here, as usual, $\bar{z}$ denotes the complex conjugate of the complex number z.) The proof is as follows. The character χ is, by definition, a homomorphism from the additive group of the field F to the multiplicative group of the field of complex numbers. This means that $\chi(t + u) = \chi(t)\chi(u)$ for all t, u in F. Then $\chi(0) = 1$, $\chi(-t) = \chi(t)^{-1}$ for all $t \in F$, and the values $\chi(t)$ are roots of unity since $p \cdot t = 0$. It follows that $\overline{\chi(t)} = \chi(-t)$ for all t, because $z^{-1} = \bar{z}$ for a root of unity z. The same remarks apply to the multiplicative character ψ, and we have $\psi(tu) = \psi(t)\psi(u)$ and $\overline{\psi(t)} = \psi(t^{-1})$ for all elements t, u in F^*. Applying these remarks to the product of G with its conjugate, we obtain

$$G \cdot \overline{G} = \sum_{t \in F^*} \sum_{u \in F^*} \chi(t)\overline{\chi(u)}\psi(t)\overline{\psi(u)} = \sum_{t,u} \chi(t-u)\psi(tu^{-1}) = \sum_{x \neq 0} \psi(x) \sum_{u \neq 0} \chi(u(x-1)).$$

From the discussion in Section 1 we have $\sum_{x \neq 0} \chi(x) = -1$ and $\sum_{y \neq 0,1} \psi(y) = -1$ for the nontrivial characters χ and ψ of F and F^* respectively. It follows that the formula for $G \cdot \overline{G}$ becomes

$$p - 1 + \sum_{x \neq 0,1} \psi(x) \sum_{u \neq 0} \chi(u(x - 1)) = p - 1 - \sum_{x \neq 0,1} \psi(x) = p,$$

completing the proof.

We now examine the Gauss sum G in (4) in more detail, assuming as above that the characters χ and ψ are nontrivial. Fix a primitive root g modulo p, as in the definition of the periods (2), and consider the factorization $p - 1 = m \cdot e$ with $m = 2$. Letting $\psi(g) = \zeta$, we have $\psi(g^i) = \zeta^i$, and $\chi(g^i) = \omega^{g^i}$, for a primitive pth root of unity ω. Then

$$G = \sum_{i=0}^{p-2} \chi(g^i)\psi(g^i) = \sum_{i=0}^{p-2} \zeta^i \omega^{g^i}.$$

Now assume that ζ has order 2, so $\zeta = -1$. In this case, G becomes what is known as a *quadratic Gauss sum*, and is given by:

$$G = \sum_{i=0}^{1} \zeta^i \sum_{j=0}^{e-1} \chi(g^{i+2j}) = \sum_{i=0}^{1} \zeta^i \sum_{j=0}^{e-1} \omega^{g^{i+2j}} = \eta_0 - \eta_1,$$

where η_0 and η_1 are the periods associated with the factorization $p - 1 = m \cdot e$ with $m = 2$. Our aim is to prove the following formulas for the quadratic Gauss sum G:

$$(6) \qquad\qquad G = 1 + 2\eta_0 = \sum_s \omega^{s^2} = \sum_{t \neq 0} (t/p)\omega^t,$$

where (t/p) is the Legendre symbol (defined in §1), and s, t are representatives of the elements of $F = \mathbb{Z}/p\mathbb{Z}$ and F^*, respectively. The first part of the formula, and its application to the proof of Theorem 4.1, make up the afterthought of Gauss, mentioned above, to his own proof of the theorem. To prove (6), we begin with the formula $G = \eta_0 - \eta_1$ and use the fact that $\eta_0 + \eta_1 = -1$ to obtain $G = 1 + 2\eta_0$. By definition,

$$\eta_0 = \sum_{j=0}^{e-1} \omega^{g^{2j}} = \sum_{j=0}^{e-1} \omega^{(\pm g^j)^2}.$$

Since $\{\pm g^j\}$, for $0 \le j \le e - 1$, form a set of representatives of the nonzero residue classes mod p, the first part of the formula for G follows by plugging in the expression for η_0 in $G = 1 + 2\eta_0$.

The proof of the last part of the formula for G goes as follows. Let A denote a set of representatives of the quadratic residues mod p, and let B be a set of representatives of the quadratic nonresidues. By the previous formula for G, we have

$$G = 1 + 2 \sum_{a \in A} \omega^a.$$

Moreover,

$$1 + \sum_{a \in A} \omega^a + \sum_{b \in B} \omega^b = 0,$$

so

$$G = \sum_{a \in A} \omega^a - \sum_{b \in B} \omega^b = \sum_{t \ne 0} (t/p)\omega^t,$$

as required. (This argument appears in the account of the theory of Gauss sums in the first supplement to Dirichlet's lectures on number theory [**114**]).

It is now a simple matter to complete the proof of Theorem 4.1. By (6), we have

$$G = \sum_{t \ne 0} (t/p)\omega^t = 1 + 2\eta_0.$$

Then $G \in \mathbb{Q}(\omega)$. Moreover,

$$\overline{G} = \sum (t/p)\overline{\omega^t} = \sum (t/p)\omega^{-t} = (-1/p)G = (-1)^{(p-1)/2}G,$$

using the facts that the Legendre symbol defines a character of the multiplicative group F^*, and $(-1/p) = (-1)^{(p-1)/2}$, as is easily shown. This calculation and (5) imply that

$$G\overline{G} = (-1)^{(p-1)/2}G^2 = p.$$

From this result, and the formulas $\eta_0 = (G-1)/2$ and $\eta_0 + \eta_1 = -1$, the quadratic equations satisfied by η_0 and η_1, given in the statement of Theorem 4.1, are easily obtained. Finally,

$$G = \pm\sqrt{(-1)^{(p-1)/2}p},$$

and

$$\eta_0 = (G-1)/2 \in \mathbb{Q}(\sqrt{(-1)^{(p-1)/2}p}).$$

It follows that

$$E = \mathbb{Q}(\eta_0) = \mathbb{Q}(\sqrt{(-1)^{(p-1)/2}p}),$$

completing the proof of the theorem.

The problem Gauss solved with his proof of the preceding theorem was to determine the quadratic subfield of the cyclotomic field $\mathbb{Q}(\omega)$. The point of view can be reversed, with the conclusion that the quadratic fields $\mathbb{Q}(\sqrt{(-1)^{(p-1)/2}p})$ are all subfields of cyclotomic fields. A general result, proved later in the 19th century by Kronecker and Weber, is the deep theorem that all algebraic number fields which are Galois extensions of the rationals, with abelian Galois groups, can be embedded in cyclotomic fields (for some historical remarks about this theorem, with references, see Hasse [171]).

In §358 of [155], Gauss turned to the more difficult problem of finding the cubic equations with rational coefficients satisfied by the periods associated with the factorization $p - 1 = 3 \cdot e$, for primes p congruent to 1 modulo 3. His approach was based on a subtle argument involving the structure equations for the periods from Theorem 3.1. The structure constants c_{ijk} in (3) for the multiplication of the periods involved the number of solutions of congruences of the form

$$aX^3 + bY^3 \equiv c \pmod{p}.$$

He obtained the required information by an ingenious, but rather long, argument from first principles (which we shall not reproduce here), and finished with the comment:

> Although the problem we have solved in this article is rather intricate, we did not wish to omit it because of the elegance of the solution and because it gave occasion for using various devices that are fruitful also in other discussions.[9]

One of the steps in this elegant solution was stated as a theorem, and proved, along with some further comments, in Chapter 8 (§4 and the Notes at the end of the chapter) of Ireland and Rosen's book [188]. Their remarks draw on the historical analysis at the beginning of Weil's famous paper [281] on the numbers of solutions of equations over finite fields. With reference to Gauss's promise of fruitful use of the ideas in other discussions Weil commented:

> ... it is only much later, however, viz. in his first memoir on bi-quadratic residues [157] that he gave in print another application of the same method; there he treats the next higher case, finds the number of solutions of any congruence $ax^4 - by^4 \equiv 1 \pmod{p}$ for a prime p of the form $p = 4n + 1$, and derives from this the biquadratic character of 2 mod p, this being the ostensible purpose of the whole highly ingenious and intricate investigation.

5. Lagrange Resolvents and Gauss Sums

Towards the end of Part VII of [155], Gauss took up another problem concerning the equations satisfied by the periods, which had been considered previously, and in a more general setting, by Lagrange. In his fundamental work on the solutions of polynomial equations [201] published in 1770-71 (as Weil called it in [282], "la théorie de Galois avant la lettre"), Lagrange discussed the problem of finding conditions which would imply that the roots $x, x', \ldots$ of a polynomial equation

$$x^n - Ax^{n-1} + Bx^{n-2} - \cdots = 0$$

are in fact the roots of a so-called *pure equation* of the same degree,

$$x^n - R = 0.$$

This was a key step in the investigation of one of the main unsolved problems at that time, namely whether all polynomial equations are solvable by radicals, in the sense that their roots can be obtained by successively extracting kth roots

---

[9]Gauss, C. F., *Disquisitiones Arithmeticae*, English translation by A. A. Clarke, Yale University Press, New Haven, 1966. Used with permission.

of rational expressions involving the coefficients, ℓth roots of rational expressions involving the roots found at the first stage, etc. It was known that quadratic, cubic, and biquadratic equations were solvable by radicals, and there was great interest in the situation for equations of degree five and higher.

Lagrange explained his point of view in a later article [**202**] as follows:

> In the Mémoires de l'Academie de Berlin (années 1770 et 1771), I have examined and compared the principal known methods for the solution of algebraic equations, and I have found that these methods all reduce, in the last analysis, to the use of a secondary equation which one calls the resolvent equation, having a root of the form
>
> $$x' + \alpha x'' + \alpha^2 x''' + \cdots$$
>
> where $x', x'', \ldots$ are the roots of the proposed equation, and α is a root of unity, of the same order as the degree of the equation.

The root of the secondary equation given above is called the *Lagrange resolvent*, and a modern version of the main result concerning it can be stated and proved as follows.

THEOREM 5.1. *Let L be the splitting field over K of the irreducible polynomial equation*

$$P(x) = x^n - Ax^{n-1} + Bx^{n-2} - \cdots = 0.$$

Assume that the field K contains a primitive complex nth root of unity α, and let $x', x'', \ldots$ be the roots of the equation $P(x) = 0$. Assume further that the Galois group of L over K consists of the cyclic permutations of the roots $x', x'', \ldots$, (and hence is cyclic of order n). Then the Lagrange resolvent

$$t = x' + \alpha x'' + \alpha^2 x''' + \cdots$$

has the property that $t^n = u \in K$. Moreover, the field L is the splitting field of the pure equation $x^n - u = 0$.

The key step in the proof, as Lagrange observed in his memoir [**201**], is, in modern terminology, that the nth power of the resolvent, t^n, is fixed by the automorphisms of L over K defined by the cyclic permutations of the roots. To see this, let σ be the automorphism which takes x' to x'', x'' to x''' etc: It is easily checked that

$$\sigma(t) = x'' + \alpha x''' + \cdots = \alpha^{-1}t, \ \sigma(t^2) = \sigma(t)^2 = \alpha^{-2}t^2, \ldots.$$

Then $\sigma(t^n) = \sigma(t)^n = \alpha^{-n}t^n = t^n$. As σ is a generator of the Galois group of L over K, it follows that $t^n = u \in K$.

The roots of the polynomial equation $x^n - u = 0$ are $t, \alpha t, \ldots, \alpha^{n-1}t$, so $K(t)$ is the splitting field of the polynomial $x^n - u$, and we have $K \subseteq K(t) \subseteq L$. As $\sigma^k(t) = \alpha^{-k}t$ for $k = 1, 2, \ldots$, it follows that no nonidentity element of the Galois group of L over K fixes the elements of $K(t)$ pointwise. Therefore $K(t) = L$ by Theorem 1.3, completing the proof.

Towards the end of Part VII of [**155**], beginning with §359, Gauss presented his own account of the reduction of polynomial equations to pure equations, for the special case of equations satisfied by the periods. Referring to his derivation in §358 of the cubic equations satisfied by the periods associated with a factorization $p - 1 = 3 \cdot e$ for primes $p \equiv -1 \pmod 3$, he began §359 with the following remarks:

The preceding discussion had to do with the *discovery* of auxiliary equations. Now we will explain a very remarkable property concerning their *solution*. Everyone knows that the most eminent geometers have been ineffectual in the search for a general solution of equations higher than the fourth degree, or (to define the search more accurately) for the **reduction of mixed equations to pure equations**. And there is little doubt that this problem does not so much defy modern methods of analysis as that it proposes the impossible .... Nevertheless it is certain that there are innumerable mixed equations of every degree which admit a reduction to pure equations, and we trust that geometers will find it gratifying if we show that our equations are always of this kind.[10]

As Lagrange had done before him, he based his discussion on the resolvent for the periods $\eta_0, \ldots, \eta_{m-1}$ associated with a factorization $p - 1 = m \cdot e$, for an odd prime p as above. In this case, the resolvent has the form

$$\eta_0 + \alpha\eta_1 + \cdots + \alpha^{m-1}\eta_{m-1},$$

for a primitive mth root of unity α. It is worth noting that the Gauss sums (4) for the finite field $\mathbb{Z}_p$ are particular cases of the Lagrange resolvents. In fact, the preceding expression for the resolvent involving the periods coincides with the Gauss sum $G(\chi, \psi)$ for the finite field $\mathbb{Z}_p$ defined in (4), for a suitable choice of the additive and multiplicative characters χ and ψ respectively. Letting ω and g be defined as in (2), the right choice is to set $\psi(g^i) = \alpha^i$, and $\chi(g^i) = \omega^{g^i}$, for each i. Thus there is a legitimate question about the terminology: Lagrange resolvent vs. Gauss sums, but the weight of long established usage comes down on the side of Gauss sums.

Gauss derived the pure equations satisfied by the periods using, among other things, the fact from §1 concerning the characters of the cyclic group of order n, which had also been applied by Lagrange in his proof of Theorem 5.1, that

$$1 + \zeta + \zeta^2 + \cdots = n \text{ or } 0,$$

for an nth root of unity ζ, according as $\zeta = 1$ or $\zeta \neq 1$. At the end, in the last two sections of [**155**], he summarized the applications of the preceding theorems to the theory of geometric constructions as follows.

Thus by the preceding discussions we have reduced the division of the circle into n parts, if n is a prime number, to the solution of as many equations as there are factors in the number $n-1$. The degree of the equations is determined by the size of the factors. Whenever, therefore, $n - 1$ is a power of 2, which happens when the value of n is $3, 5, 17, 257, 65537$, etc. the sectioning of the circle is reduced to quadratic equations only, and the trigonometric functions of the angle $2\pi/n$, $4\pi/n$, etc. can be expressed as square roots which are more or less complicated (according to the size of n). Thus in these cases the division of the circle into n parts or the inscription of a regular polygon of n sides can be accomplished by geometric constructions. Thus, e. g., for $n = 17$, by §§354, 361 we get the

---

[10]Gauss, C. F., *Disquisitiones Arithmeticae*, English translation by A. A. Clarke, Yale University Press, New Haven, 1966. Used with permission.

following expression for the cosine of the angle $2\pi/17$:

$$-\frac{1}{16} + \frac{1}{16}\sqrt{17} + \frac{1}{16}\sqrt{[34 - 2\sqrt{17}]}$$
$$+ \frac{1}{8}\sqrt{[17 + 3\sqrt{17} - \sqrt{(34 - 2\sqrt{17})} - 2\sqrt{(34 + 2\sqrt{17})}]}.$$

The cosine of multiples of this angle will have a similar form, but the sine will have one more radical sign. It is certainly astonishing that although the geometric divisibility of the circle into three and five parts was already known in Euclid's time, nothing was added to this discovery for 2000 years. And all geometers had asserted that, except for those sections and the ones that derive directly from them (that is, division into $15, 3 \cdot 2^\mu, 5 \cdot 2^\mu$ and 2^μ parts), there are no others that can be effected by geometric constructions. But it is easy to show that if a prime number has the form $n = 2^m + 1$, the exponent m can have no other prime factors except 2, and so it is equal to 1 or 2 or a higher power of the number 2. For if m were divisible by an odd number ζ (greater than unity) so that $m = \zeta\eta$, then $2^m + 1$ would be divisible by $2^\eta + 1$ and so necessarily composite. All values of n, therefore, that can be reduced to quadratic equations, are contained in the form $2^{2^\nu} + 1$. Thus the five numbers $3, 5, 17, 257, 65537$ result from letting $\nu = 0, 1, 2, 3, 4$ or $m = 1, 2, 4, 8, 16$. But the geometric division of the circle cannot be accomplished for *all* numbers contained in the formula but only for those that are prime. Fermat was misled by his induction and affirmed that all numbers contained in this form are necessarily prime, but the distinguished Euler first noticed that this rule is erroneous for $\nu = 5$ or $m = 32$, since the number 4294967297 involves the factor 641.

Whenever $n - 1$ implies prime factors other than 2, we are always led to equations of higher degree, namely, to one or more cubic equations when 3 appears once or several times among the prime factors of $n - 1$, to equations of the fifth degree when $n - 1$ is divisible by 5, etc. **We can show with all rigor that these higher degree equations cannot be avoided in any way nor can they be reduced to lower degree equations.**[11]

The conclusion was that in order to divide the circle geometrically into N parts, it is necessary and sufficient that

N imply no odd prime factor that is not of the form $2^m + 1$ nor any prime factor of the form $2^m + 1$ more than once.[12]

Incidentally, a proof, to which he referred in the preceding quotations, of the impossibility of construction of a regular N-gon in case N is not of the required form, was not included in the *Disquisitiones Arithmeticae*.

---

[11]Gauss, C. F., *Disquisitiones Arithmeticae*, English translation by A. A. Clarke, Yale University Press, New Haven, 1966. Used with permission.
[12]Ibid.

6. Quadratic Reciprocity: a proof based on Gauss sums

The quadratic reciprocity law was first stated by Euler, and discussed by Legendre, who claimed to have given a proof. The result, which Gauss called the fundamental theorem in the theory of quadratic residues, was stated in §131 of [**155**] as follows:

> A prime number p of the form $4n + 1$, and, respectively, $-p$ in case p has the form $4n + 3$, is a quadratic residue of a given prime number, according as that prime number is a residue or a nonresidue of p.[13]

He added that,

> ... almost everything that can be said about quadratic residues rests on this theorem.[14]

His proof of the theorem in §§131-132 of [**155**] was entirely elementary, by a reduction to 8 cases, and their consideration one by one. He was fascinated by the result, and gave altogether 6 proofs during the course of his lifetime. The proof given below is a streamlined version of the sixth proof, the original version of which Gauss published in 1818 ([**156**]). It is based on the theory of quadratic Gauss sums (6), and, as he stated in the introduction to the paper, he hoped it would provide a basis for understanding the features quadratic reciprocity had in common with the theories of cubic and biquadratic reciprocity, which he had investigated for many years (since 1805). The ideas in the proof, along with other things such as Gauss sums (4), were indeed used to establish higher reciprocity laws, for cubic and biquadratic residues, by Eisenstein, Kummer, Jacobi, and Gauss himself, and the subject remained a focus of interest in number theory throughout the nineteenth century, and into the twentieth. A marvelous survey of these matters, with comments on how it all fitted together, was given in a Bourbaki seminar by Weil in 1974 [**282**].

In terms of Legendre symbols, the law of quadratic reciprocity can be stated as follows.

THEOREM 6.1. *Let p and q be distinct odd primes. Then*

$$(p/q)(q/p) = (-1)^{\frac{1}{2}(p-1)\frac{1}{2}(q-1)}.$$

For the proof, let G be the quadratic Gauss sum associated with the prime p, as in (6). It will be convenient to use the abbreviations

$$p' = \frac{1}{2}(p - 1), \; q' = \frac{1}{2}(q - 1).$$

Then

$$G^2 = (-1)^{p'}p,$$

by the proof of Theorem 4.1. The next step is to prove the congruence:

$$p^{q'} \equiv (p/q) \pmod{q}.$$

This can be shown as follows. Let a be a primitive root mod q, and write $p \equiv a^i$ (mod q), $x \equiv a^j$ (mod q), for x relatively prime to q. Then $x^2 \equiv p$ (mod q) if and

---

[13]Ibid.

[14]Ibid.

only if $2j \equiv i \pmod{q-1}$. As q is odd, the condition is equivalent to the statement that i is even, or that $p^{\frac{q-1}{2}} \equiv 1 \pmod{q}$, completing the proof. It follows that

$$G^{q-1} \equiv (-1)^{p'q'} p^{q'} \equiv (-1)^{p'q'} (p/q) \pmod{q},$$

where the latter congruence is taken modulo the principal ideal (q) in the ring of algebraic integers. On the other hand, equation (6) also states that

$$G = \sum_t (t/p)\omega^t,$$

and it follows that

$$G^q \equiv \sum (t/p)^q \omega^{qt} \equiv \sum (t/p)\omega^{qt} \pmod{q}.$$

The last expression becomes

$$\sum (t/p)\omega^{qt} = \sum (tq/p)(q/p)\omega^{qt} = (q/p)G,$$

and we obtain

$$G^q \equiv (q/p)G \pmod{q}.$$

The preceding formulas for G^{q-1} and G^q imply that

$$(-1)^{p'q'}(p/q)G \equiv (q/p)G \pmod{q},$$

and hence

$$(-1)^{p'q'}(p/q)G^2 \equiv (q/p)G^2 \pmod{q},$$

as a congruence in $\mathbb{Z}$. But $G^2 = \pm p$, so G^2 is not divisible by q, and G^2 can be cancelled from the preceding congruence. We are left with the equation

$$(-1)^{p'q'}(p/q) = (q/p),$$

which is the desired result. This completes the proof of the theorem.

Gauss's original version of the preceding argument was more algebraic, and involved congruences in the ring $\mathbb{Z}[x]$, instead of the ring of algebraic integers. The precise connection between the two approaches was explained by Ireland and Rosen in the notes following Chapter 6 of [**188**], and will not be repeated here.

7. Dirichlet's L-series and Characters mod k

Dedekind, in his remarks on characters of finite abelian groups made in his letter to Frobenius quoted at the beginning of the chapter, mentioned their use in P. G. Lejeune Dirichlet's great paper [**113**], on the existence of infinitely many primes in arithmetic progressions, presented to the Berlin Academy of Science in July, 1837. Although he contributed many, perhaps none of Dirichlet's ideas had a more lasting impact on mathematics than his introduction of what are now called Dirichlet series, and in particular, the version of them he called L-series, which appeared for the first time in his 1837 paper. The latter brought the theory of functions of a complex variable, or hard analysis, into the arena of number theory, and had a large role in creating the subject now known as analytic number theory. The aim of the discussion to follow is to understand the contribution of the theory of characters of finite abelian groups to this marriage of analysis and number theory.

The main result of Dirichlet's paper is the following:

THEOREM 7.1. *Each arithmetic progression of the form*

$$a + ik, \ i = 0, 1, \dots,$$

with a and k relatively prime, contains infinitely many prime numbers.

The theorem had been considered earlier, as Dirichlet noted in the introduction to his paper:

> There does not exist as yet a satisfactory proof of the theorem, although one is very much desired, because of its many number-theoretical applications. The only mathematician, as far as I know, who has attempted to prove it, is Legendre, for whom the investigation had the additional stimulus that he had already used the property of arithmetic progressions stated above, as a Lemma in a previous work.

His proof of the theorem, as he remarked, "is not purely arithmetic, but rests in part on the consideration of continuously varying quantities". The jumping-off point was an identity due to Euler, that for $s > 1$,

$$\sum \frac{1}{n^s} = \prod_p \frac{1}{1 - \frac{1}{p^s}},$$

where the product is taken over the set of primes p. The infinite series on the left hand side converges for $s > 1$, by the integral test. Each factor of the right hand side of the identity can be expanded as a convergent geometric series

$$\frac{1}{1 - \frac{1}{p^s}} = \sum_{i \geq 0} \frac{1}{p^{is}}.$$

Upon multiplying the series together, the identity follows from the fact that each term $\frac{1}{n^s}$ coincides with exactly one term on the right hand side, because n can be factored uniquely as a product of primes:

$$n = p_1^{a_1} \cdots p_r^{a_r}.$$

Dirichlet's proof began with a generalization of Euler's identity for certain infinite series which he called *L*-series. Their definition involved characters of finite abelian groups, and it is interesting to see exactly how he introduced them.

The starting point was Gauss's concept of the index of an integer n not divisible by p, with respect to a given odd prime p. The notion of the index was introduced in §57 of [**155**], and was thoroughly familiar, along with the rest of the *Disquisitiones Arithmeticae*, to Dirichlet. The *index* of n with respect to p, for $(n, p) = 1$, depends on the choice of a fixed primitive root c mod p, and is the exponent $\gamma \leq p - 1$ such that $c^\gamma \equiv n \pmod{p}$. The main property of the index, as Gauss remarked in §58 of [**155**], is that it behaves like a logarithm:

> The index of the product of any number of factors is congruent to the sum of the indices of the individual factors relative to the modulus $p - 1$.
>
> The index of the power of a number is congruent relative to the modulus $p - 1$ to the product of the index of the number by the exponent of the power.[15]

---

[15]Gauss, C. F., *Disquisitiones Arithmeticae*, English translation by A. A. Clarke, Yale University Press, New Haven, 1966. Used with permission.

Dirichlet denoted the index of n by γ_n. Taking a fixed root ω of the equation $\omega^{p-1} = 1$, a prime number q different from p, and a real number $s > 1$, he noted the identity given by the convergent geometric series

$$\frac{1}{1 - \omega^\gamma \frac{1}{q^s}} = 1 + \omega^\gamma \frac{1}{q^s} + \omega^{2\gamma} \frac{1}{q^{2s}} + \cdots,$$

where γ is the index of q. He then stated:

> If one considers the set of all prime numbers q different from p, and multiplies the resulting equations together, one obtains on the right hand side a series whose terms are easy to understand. Namely, if n is a number not divisible by p, and $n = q_1^{m_1} q_2^{m_2} \cdots$ with $q_1, q_2, \ldots$ different primes, then the general term has the form
>
> $$\frac{\omega^{m_1 \gamma_{q_1} + m_2 \gamma_{q_2} + \cdots}}{n^s}.$$

But now

$$m_1 \gamma_{q_1} + m_2 \gamma_{q_2} + \cdots \equiv \gamma_n \quad (\bmod\ p - 1),$$

and it follows [from the assumption that $\omega^{p-1} - 1 = 0$] that

$$\omega^{m_1 \gamma_{q_1} + m_2 \gamma_{q_2} + \cdots} = \omega^{\gamma_n}.$$

One obtains therefore the equation

$$\prod \frac{1}{1 - \omega^\gamma \frac{1}{q^s}} = \sum \omega^\gamma \frac{1}{n^s} = L,$$

> where the multiplication is taken over the set of all prime numbers different from p, and the sum over the integers from 1 to ∞, which are not divisible by p. The sign γ means γ_q on the left side, and γ_n on the right.

We now give a modern interpretation of the preceding paragraph. The last formula involves a character ψ of the multiplicative group $G(p)$ consisting of the residue classes mod p of the integers n not divisible by p. The character ψ is defined by setting $\psi(c) = \omega$, and for each integer n not divisible by p, one puts $\psi(n) = \omega^\gamma$, where γ is the index of n, and c and ω are as above. Upon extending the definition of ψ by putting $\psi(m) = 0$ if m is divisible by p, one obtains a complex valued function ψ defined on the set of all integers, and satisfying the conditions: $\psi(mn) = \psi(m)\psi(n)$ for all m, n and $\psi(a) = \psi(b)$ whenever $a \equiv b$ $(\bmod\ p)$. The first formula follows from the logarithm-like properties of the index mentioned earlier. The preceding remarks show that ψ can also be viewed as a function on the multiplicative group $G(p)$ consisting of residue classes $\overline{n} = n + p\mathbb{Z}$, with n not divisible by p. Then $\psi(\overline{mn}) = \psi(\overline{m})\psi(\overline{n})$ for all $\overline{m}, \overline{n} \in G(p)$, so that ψ is a character of this group. The series L in the preceding quotation can be expressed in the form

$$L = \sum \frac{\psi(n)}{n^s},$$

and is an example of what are now called L-series, or Dirichlet L-series. The multiplicative formula, or *Euler Product Formula*, for the series L follows from the fact that ψ is a character, as we shall prove in detail below in a more general situation.

After stating the multiplicative formula for L, Dirichlet discussed the convergence questions associated with it, which were not as routine as they are now. He emphasized the fact that for $s > 1$ the series L converges to a sum which is independent of the ordering of its terms, as we would verify today by noting that the series is absolutely convergent for $s > 1$. He then interrupted the discussion to give two examples of rearrangements of what we now call a conditionally convergent series,

$$1 - \frac{1}{2} + \frac{1}{3} - \frac{1}{4} + \cdots$$

and

$$1 + \frac{1}{3} - \frac{1}{2} + \frac{1}{5} + \frac{1}{7} - \frac{1}{4} + \cdots,$$

which both converge, but to different sums.

The proof of the theorem about arithmetic progressions requires a more general version of the preceding ideas, and some serious analysis. In what follows, we give a sketch, in modern terminology, of some of the steps. Let k be a positive integer. A *character mod* k is by definition a complex valued function ψ on the set of integers such that $\psi(mn) = \psi(m)\psi(n)$ for all integers m, n, $\psi(a) = \psi(b)$ whenever $a \equiv b \pmod{k}$, and $\psi(n) \neq 0$ if and only if n is relatively prime to k. As before, these functions define characters of the multiplicative group $G(k)$ of order $\varphi(k)$ consisting of the residue classes $\overline{n} = n + k\mathbb{Z}$ with n relatively prime to k. The trivial character 1 mod k is the character mod k which takes the value 1 at all integers n relatively prime to k. The next result gives an Euler Product Formula for an L-series associated with an arbitrary character mod k.

LEMMA 7.1. *Let ψ be a character mod k for some positive integer k. Then the L-series*

$$L(s, \psi) = \sum \frac{\psi(n)}{n^s},$$

viewed as a function of the complex variable s, is absolutely convergent if the real part of s is > 1. For all such values of s, one has

$$L(s, \psi) = \prod_q \frac{1}{1 - \frac{\psi(q)}{q^s}},$$

where the product is taken over the set of all prime numbers q.

We begin the proof by observing that the values of ψ are either zero or are the values of a character of the multiplicative group $G(k)$ of residue classes $n + k\mathbb{Z}$, with n relatively prime to k. As this group is finite, it follows that the nonzero values of ψ are roots of unity, and hence are complex numbers of absolute value one. It follows that the terms of the series $\sum |\frac{\psi(n)}{n^s}|$ are all less than or equal to the corresponding terms in the series $\sum |\frac{1}{n^s}|$, which is convergent whenever $Re(s) > 1$ by the integral test. This proves that the L-series $L(s, \psi)$ is absolutely convergent when $Re(s) > 1$.

The proof of the Euler product identity for $L(s, \psi)$ is an extension of the proof of Euler's identity, and of Dirichlet's argument, quoted above, for the L-series

$$L = \sum_{(n,p)=1} \frac{\omega^\gamma}{n^s},$$

associated with characters of the finite abelian group $G(p)$, for a prime p. As in the proof of Euler's identity, one begins with the absolutely convergent geometric series

$$\sum_{i=0}^{\infty} \frac{\psi(q^i)}{q^{is}} = \frac{1}{1 - \frac{\psi(q)}{q^s}}.$$

For a finite set of primes S, one can multiply together the series corresponding to the primes in S to obtain, for $s > 1$,

$$\prod_{q \in S} \frac{1}{1 - \frac{\psi(q)}{q^s}} = \prod_{q \in S} \sum \frac{\psi(q^i)}{q^{is}}.$$

As the series involved are absolutely convergent for $s > 1$, one can collect terms on the right hand side and use the multiplicative property of the character ψ to write it in the form $\sum \frac{\psi(n)}{n^s}$, where the sum is taken over all integers $n \geq 1$ whose prime factors are in S. As S increases, this series approaches the series for $L(s, \psi)$. It follows that the infinite product

$$\prod_q \frac{1}{1 - \frac{\psi(q)}{q^s}},$$

taken over all primes q, is convergent for $Re(s) > 1$, and is equal to $L(s, \psi)$.

The rest of the proof of Dirichlet's Theorem is based on an analysis of the behavior of the logarithm of an L-function $L(s, \psi)$, as $s \to 1$. Not all the details are included here, but the argument is carried far enough to introduce the important concept of *Dirichlet density* of a set of prime numbers. We shall also see a nice application of the orthogonality relations for characters mod k.

We begin with a discussion of the logarithm of the *zeta function*

$$\zeta(s) = \sum_{n=1}^{\infty} \frac{1}{n^s},$$

which may be viewed as an L-function associated with the trivial character mod k, for $k = 1$. We consider only the case of a real variable s. As noted in the proof of Lemma 7.1, the series for $\zeta(s)$ converges for $s > 1$, by the integral test.

THEOREM 7.2. *We have*

$$\lim_{s \to 1^+} (s - 1)\zeta(s) = 1.$$

Moreover,

$$\lim_{s \to 1^+} \frac{\log \zeta(s)}{\log \frac{1}{s-1}} = 1.$$

The proof uses only elementary calculus. The function t^{-s}, for fixed $s > 1$, is monotone decreasing for positive values of t, so

$$\frac{1}{(n+1)^s} < \int_n^{n+1} t^{-s} \, dt < \frac{1}{n^s}.$$

Upon summing these inequalities, and evaluating the improper integral $\int_1^\infty t^{-s} dt$, one obtains:

$$\zeta(s) - 1 < \frac{1}{s-1} < \zeta(s).$$

It follows that

$$1 < \zeta(s)(s-1) < s,$$

proving the first statement. For the second, let $F(s) = (s-1)\zeta(s)$; then

$$\log F(s) = \log(s-1) + \log \zeta(s),$$

so

$$\frac{\log \zeta(s)}{\log \frac{1}{s-1}} = 1 + \frac{\log F(s)}{\log \frac{1}{s-1}}.$$

The second statement is now clear, because as $s \to 1$, one has $F(s) \to 1$ and $\log(s-1)^{-1} \to +\infty$.

THEOREM 7.3. *For $s > 1$, one has*

$$\log \zeta(s) = \sum_p \frac{1}{p^s} + R(s),$$

where the sum is taken over all primes p, and the remainder term $R(s)$ is bounded as $s \to 1$.

The proof begins with a version of Euler's identity:

$$\zeta(s) = \left(\prod_{p \le N} \frac{1}{1 - \frac{1}{p^s}} \right) \lambda_N(s),$$

where $\lambda_N(s) \to 1$ as $N \to \infty$. Then

$$\log \zeta(s) = \sum_{p \le N} -\log(1 - \frac{1}{p^s}) + \log \lambda_N(s).$$

Using the series expansion, also from elementary calculus,

$$-\log(1-x) = x + \frac{x^2}{2} + \frac{x^3}{3} + \cdots, \quad \text{for } |x| < 1,$$

the preceding formula for $\log \zeta(s)$ becomes

$$\sum_{p \le N} \sum_{m=1}^{\infty} \frac{1}{m} \frac{1}{p^{ms}} + \log \lambda_N(s).$$

Taking the limit as $N \to \infty$, one has

$$\log \zeta(s) = \sum_p \frac{1}{p^s} + \sum_p \sum_{m=2}^{\infty} \frac{1}{m} \frac{1}{p^{ms}}.$$

The second sum is bounded by

$$\sum_p \frac{1}{p^{2s}} + \frac{1}{p^{3s}} + \cdots \le \sum_p \frac{1}{p^{2s}} \left(\frac{1}{1 - \frac{1}{p^s}} \right) \le 2\zeta(2),$$

as $s \to 1$, completing the proof.

DEFINITION 7.1. A set of primes $\mathcal{P}$ has *Dirichlet density* d provided that

$$\lim_{s \to 1^+} \frac{\sum_{p \in \mathcal{P}} \frac{1}{p^s}}{\log \frac{1}{s-1}}$$

exists, and is equal to d.

By the two preceding theorems, one has

$$\frac{\sum_p \frac{1}{p^s}}{\log \frac{1}{s-1}} = \frac{\log \zeta(s)}{\log \frac{1}{s-1}} - \frac{R(s)}{\log \frac{1}{s-1}} \to 1$$

as $s \to 1^+$, proving that the Dirichlet density of the set of all prime numbers is equal to 1. Notice also that in order to prove that a set of primes $\mathcal{P}$ is infinite, it is sufficient to prove that $\mathcal{P}$ has Dirichlet density greater than zero (so the calculation just described is another, somewhat roundabout, proof that the set of all prime numbers is infinite).

We are now ready to understand how some of the preceding ideas are used in the proof of Dirichlet's theorem. The arithmetic progression in Theorem 7.1 consists of all positive integers n such that $n \equiv a \pmod{k}$. The set of primes q which occur in the arithmetic progression is denoted by P_a; it is the set of primes q such that $q \equiv a \pmod{k}$. By the preceding discussion, the theorem will follow if one can prove that the Dirichlet density of the set of primes P_a is greater than zero. This involves comparing the function of a complex variable s (for $Re(s) > 1$) given by

$$g_a(s) = \sum_{q \in P_a} \frac{1}{q^s}$$

with $\log \frac{1}{s-1}$. The first step is to express $g_a(s)$ in terms of the functions

$$f_\psi(s) = \sum_q \frac{\psi(q)}{q^s},$$

where the sum is taken over the set of primes q such that $(q, k) = 1$, and ψ is a character mod k. The result is viewed today as an application of what is called "Fourier analysis on the finite abelian group" $G(k)$.

LEMMA 7.2. *For each character ψ mod k, the series*

$$f_\psi(s) = \sum_q \frac{\psi(q)}{q^s}$$

is absolutely convergent whenever the real part of s is greater than 1. Moreover,

$$g_a(s) = \frac{1}{\varphi(k)} \sum_\psi \psi(a)^{-1} f_\psi(s),$$

where the sum is taken over all characters ψ mod k.

The proof of convergence is the same as in the case of the L-series $L(s, \psi)$. By the definition of f_ψ, we have

$$\sum_\psi \psi(a)^{-1} f_\psi(s) = \sum_q \sum_\psi \frac{\psi(a)^{-1}\psi(q)}{q^s}.$$

The orthogonality relations (Corollary 1.1) for the finite abelian group $G(k)$ of order $\varphi(k)$, state that

$$\sum_\psi \psi(a)^{-1}\psi(q) = \varphi(k) \text{ if } a \equiv q \pmod{k}$$

and is 0 otherwise, where the sum is taken of the set of characters of $G(k)$. It follows that

$$\sum_{\psi} \psi(a)^{-1} f_{\psi}(s) = \varphi(k) g_a(s),$$

as required.

The most difficult part of the proof, which is not given here,[16] is based on a comparison (similar to the proof of Theorem 7.3) of each function $f_{\psi}(s)$ with the logarithm, suitably interpreted, of the corresponding L-function $L(s, \psi)$, expressed in its product form, from Lemma 7.1. From this analysis, it can be proved that the difference between $f_{\psi}(s)$ and $\log \frac{1}{s-1}$ remains bounded as $s \to 1$ in case ψ is the trivial character mod k, and that $f_{\psi}(s)$ remains bounded as $s \to 1$ for all other characters ψ mod k. By Lemma 7.2, it follows that the difference between $\varphi(k) g_a(s)$ and $\log \frac{1}{s-1}$ remains bounded as $s \to 1$. This means that the Dirichlet density of the set of primes P_a occuring in the given arithmetic progression is $\frac{1}{\varphi(k)}$, and finishes the proof of the theorem.

8. Characters associated with quadratic forms

In his introduction to the paper *Über Gruppencharaktere*, Frobenius discussed at some length a concept Gauss had introduced in §230 of the *Disquisitiones Arithmeticae*, which Gauss had called the *character* of a binary quadratic form. He noted that Dirichlet and Dedekind had reinterpreted Gauss's idea so that it can be viewed as an early, and particularly interesting, application of characters of finite abelian groups (see also the quotation from Dedekind's 1896 letter to Frobenius in the introduction to this chapter). This seems to be the origin of Frobenius's use of the mathematical term *Charaktere*. In this section, we shall first examine Frobenius's remarks, and then explain, in some detail, the ideas of Gauss, Dirichlet, and Dedekind on the composition of quadratic forms to which he referred, and how the notion of *character* came on the scene.[17]

Here are Frobenius's remarks, from [**131**].

> By a character of a quadratic form, Gauss, *Disqu. Arithm.* §230, meant a relation between the numbers represented by a form and the odd prime numbers p (or 4 or 8) dividing the determinant of the form. He indicated the relation by the signs Rp and Np. Dirichlet, *Recherches sur diverses applications de l'analyse infinitésimale à la théorie des nombres*, §3 (Crelle's Journal Bd. 19), replaced these symbols by the Legendre (and Jacobi) symbols $\left(\frac{m}{p}\right)$, which (next to the resolvent of Lagrange), may well be the oldest example of the application of characters of commutative groups. The superiority of this conversion is that the genus characters of Gauss describe only relations, while those of Dirichlet are numbers, with which one can calculate. Thus through the multiplication of these characteristic numbers, the composition of genera corresponds to composition of characters, §§246-248.

In order to understand these remarks, we need to review parts of Gauss's theory of binary quadratic forms. In particular, his definition of composition of quadratic

---

[16]See Ireland and Rosen [**188**] or Serre [**263**] for a full account.

[17]See Hawkins [**172**] for a more complete historical analysis.

forms implied that the set of classes of quadratic forms with a given determinant has the structure of a finite abelian group, and his notion of genus led to a problem requiring the use of characters of that group.

We shall be concerned with binary quadratic forms

$$F(x, y) = ax^2 + 2bxy + cy^2,$$

whose coefficients a, b, c are integers. Gauss's theory of binary quadratic forms is perhaps the centerpiece of the *Disquisitiones Arithmeticae*, and it is a pity that time and space force us to consider only the part of it directly relevant to our discussion of characters.

Gauss used the notation (a, b, c) to denote a quadratic form F as above. The expression

$$D = b^2 - ac$$

is called the *determinant* of $F = (a, b, c)$. In what follows, we shall consider only forms $F(x, y)$ whose determinant is not zero. An integer ℓ is said to be *represented* by F if

$$F(m, n) = \ell$$

for some integers m and n. A substitution

$$x = \alpha x' + \beta y', \ y = \gamma x' + \delta y',$$

with $\alpha, \beta, \gamma, \delta$ integers, carries $F(x, y)$ into a new quadratic form $F'(x', y')$. The transformation $(x, y) \rightarrow (x', y')$ from the set $\mathbb{Z} \times \mathbb{Z}$ to itself is invertible if and only if $\alpha\delta - \beta\gamma = \pm 1$. When this occurs, the forms $F(x, y)$ and $F'(x', y')$ are said to be *equivalent*. Equivalent quadratic forms have the same determinant, and represent the same set of integers. The relation of equivalence is reflexive, symmetric, and transitive (in other words, an *equivalence relation* in the modern sense), and partitions the set of quadratic forms $F(x, y)$ into disjoint subsets called *classes*.

A basic result, first proved by Lagrange, and reproved by Gauss[18] using different methods, is the following:

THEOREM 8.1. *The number of classes of binary quadratic forms with a given determinant is finite.*

The main results about characters of quadratic forms were first stated and proved in non-numerical form by Gauss in [**155**], §§228-230. The numerical version, to which Frobenius referred in the passage quoted above, was due to Dirichlet and appeared in the article cited by Frobenius, and later in Dirichlet's lectures on number theory [**114**].

Dirichlet was a keen student of the *Disquisitiones Arithmeticae*, and his own lectures contained fresh versions of large parts of it, along with important new results. The lectures [**114**] were published posthumously. When Gauss died in 1855, Dirichlet was called from Berlin to succeed him in Göttingen. He lectured on number theory there during the winter semester 1856-57. Fortunately, Dedekind, who was 26 years old at the time, attended the lectures, and made notes on them. In 1858, Dirichlet suffered a heart attack in Montreux, where he had gone to prepare a memorial address to honor Gauss. He died the following year. Dedekind wrote

---

[18]See Gauss [**155**] , §222, for a reference and discussion of Lagrange's proof of the theorem, and §223 for comments on his own proof. For the history of research on binary quadratic forms before Gauss, see [**283**], Chap. 4, §IV.

up his notes on Dirichlet's lectures for publication and published them in 1863. New editions appeared in 1871, 1879-80, and 1894. In addition to the lecture notes, Dedekind added supplements containing reports on some of Dirichlet's publications and new material of his own. The account of characters of quadratic forms given below is taken largely from Supplement X of the fourth edition ([114], 1894), with some streamlining of the discussion by L. E. Dickson ([112], Chap. V).

A quadratic form (a, b, c) is called *primitive* if the g.c.d. (greatest common divisor) of a, b, c is 1, and *properly primitive* if the g.c.d. of $a, 2b, c$ is 1. The first result related to characters can be stated as follows.

THEOREM 8.2. *Let*

$$F(x, y) = ax^2 + 2bxy + cy^2$$

be a properly primitive binary quadratic form of determinant $D = b^2 - ac$. *Let* p *be an arbitrary odd prime factor of* D. *Then the Legendre symbol* (n/p) *has the same value for all integers* n *relatively prime to* $2D$ *which are represented by* F. *The same is true of*

$$\delta = (-1)^{\frac{1}{2}(n-1)} \quad \text{if } D \equiv 0 \text{ or } 3 \pmod 4,$$
$$\varepsilon^{\frac{1}{8}(n^2-1)} \quad \text{if } D \equiv 0 \text{ or } 2 \pmod 8$$

and

$$\delta\varepsilon \quad \text{if } D \equiv 0 \text{ or } 6 \pmod 8.$$

The values of the Legendre symbols (n/p), for integers n relatively prime to $2D$ and represented by F, and p an odd prime divisor of D, together with δ, ε, and $\delta\varepsilon$ if they occur, are called *characters* of the quadratic form $F = (a, b, c)$. Clearly the value of a character depends only on the class to which F belongs. The set of all classes of forms F, with $F = (a, b, c)$ properly primitive of determinant D, on which the characters take a preassigned set of values, is called a *genus* (see Gauss [155], §231, and Dirichlet [114], §122). Thus a knowledge of the values of the characters of a properly primitive quadratic form F is equivalent to placing the class containing F in its genus.

Gauss's non-numerical descriptions of the characters in [155], §230, is as follows:

> Therefore all numbers that can be represented by a given primitive form F with determinant D will have a fixed relationship to the individual prime divisors of D (by which they are not divisible). And odd numbers that can be represented by F will also have a fixed relationship to the numbers 4 and 8 in certain cases .... We will call this type of relationship to each of these numbers the *character* or the *particular character* of the form F, and we will express it in the following manner. When only quadratic residues of a prime number p can be represented by the form F we will assign to it the character Rp, in the opposite case the character Np; similarly we will write $1, 4$ when no other numbers can be represented by the form F except those that are $\equiv 1 \pmod 4$.[19]

The statement, and the following proof, of Theorem 8.2 are taken from Dickson [112]; they are not essentially different, however, from the corresponding items in Gauss [155].

---

[19]Gauss, C. F., *Disquisitiones Arithmeticae*, English translation by A. A. Clarke, Yale University Press, New Haven, 1966. Used with permission.

The proof is as follows. Let

$$F(u, v) = au^2 + 2buv + cv^2 = n, \quad \text{and} \quad F(r, s) = m.$$

Then

$$nm = x^2 - Dy^2,$$

with

$$x = aur + bus + brv + cvs, \quad y = us - rv.$$

Now let n and m be relatively prime to $2D$, and hence to any odd prime factor p of D. Then

$$nm \equiv x^2 \pmod{D},$$

and hence

$$(nm/p) = 1.$$

It follows that $(n/p) = (m/p)$, because the Legendre symbol has the "character" property $(nm/p) = (n/p)(m/p)$. This proves the first statement of the theorem. Further analysis of the equation for nm in terms of x and y proves the remaining statements.

In [155], §234, Gauss stated:

> Now that we have explained the distribution of forms into classes, genera, and orders, and the general properties that result from these distinctions, we will go on to another very important subject, the *composition* of forms. Thus far no one has considered this point.[20]

Composition of binary quadratic forms is a process that expresses the product of two binary quadratic forms satisfying certain conditions as a third binary quadratic form. It introduces the structure of a finite abelian group on the finite set of classes of properly primitive binary quadratic forms with determinant D. The set of such classes is finite by Theorem 8.1. The abelian group structure is obtained from the properties of composition discussed below. It will then follow, as Dirichlet (and Dedekind) proved in [114], §152, and as Frobenius remarked in the quotation at the beginning of this section, that the characters defined in Theorem 8.2 are indeed characters on the finite abelian group of classes of forms in the sense of the definition in §1. The group structure organizes the properties of composition of forms, which, as we shall see, is a rather complicated and somewhat mysterious construction. The fact that the classes of quadratic forms belonging to this group are separated into genera by certain numerical-valued functions on the group illuminates another set of subtle ideas, and suggests new results, such as the Duplication Theorem stated below. It seems clear that the clarification of the theory of binary quadratic forms brought about by the introduction of group structure and characters of the resulting groups was what Frobenius wanted to emphasize in the passage quoted above from *Über Gruppencharaktere*.

Our presentation of the theory of composition of quadratic forms follows the account in Supplement X of Dirichlet's lectures [114]; it is simpler and more accessible than the original version of Gauss, which began with §234 of the *Disquisitiones Arithmeticae*. In view of the remarks made earlier about the publication of Dirichlet's lectures and the supplements to them, it follows that the discussion given below is based on the joint efforts of Gauss, Dirichlet, and Dedekind.

---

[20]This was not quite true. Legendre had anticipated him, but Gauss's approach is different (see Weil [283], pp. 333-335). Gauss, C. F., *Disquisitiones Arithmeticae*, English translation by A. A. Clarke, Yale University Press, New Haven, 1966. Used with permission.

LEMMA 8.1. *Let* $m, p_1, \ldots, p_n$ *be a set of integers with g.c.d. 1, and let* $q_1, \ldots, q_n$ *be another set of integers such that for all* i, j *one has*

$$p_i q_j - q_i p_j \equiv 0 \pmod{m}.$$

Then there is a unique residue class B *modulo* m *such that*

$$p_1 B \equiv q_1, \ldots, p_n B \equiv q_n,$$

where all the congruences are taken modulo m.

Here is the proof. By the hypothesis, it follows that there exist integers $\{x_i\}$ such that

$$\sum x_i p_i \equiv 1 \pmod{m}.$$

Then $B_0 = \sum q_i x_i$ satisfies the congruences

$$p_j B_0 = \sum_i p_j q_i x_i \equiv q_j \sum_i p_i x_i \equiv q_j \pmod{m},$$

using the second part of the hypothesis. Let B be another solution of these congruences; then upon multiplying the ith congruence by x_i and summing on i, we obtain $B \equiv \sum q_i x_i \equiv B_0 \pmod{m}$, completing the proof.

LEMMA 8.2. *Let*

$$b^2 \equiv D \pmod{a}, \quad (b')^2 \equiv D \pmod{a'},$$

and assume $a, a', b + b'$ *have g.c.d. 1. Then there exists a unique residue class* B *modulo* aa' *satisfying*

$$B \equiv b \pmod{a}, \quad B \equiv b' \pmod{a'}, \quad B^2 \equiv D \pmod{aa'},$$

and such that $a, a', 2B$ *have g.c.d. 1.*

It is readily checked that an integer B satisfies the congruences stated in the conclusion of the lemma if and only if the congruences

$$aB \equiv ab' \pmod{aa'}, a'B \equiv a'b \pmod{aa'}; (b+b')B \equiv bb' + D \pmod{aa'},$$

hold, using the identity

$$(B - b)(B - b') = B^2 - (b + b')B + bb',$$

in the case of the last one. The hypothesis of the lemma implies that $a, a', b+b', aa'$ have g.c.d. 1. We can now apply Lemma 8.1, with the p's given by $a, a', b + b'$, $m = aa'$, and the q's by $ab', a'b, bb' + D$. The required identities

$$p_i q_j - q_i p_j \equiv 0 \pmod{m}$$

hold because of the congruences $b^2 \equiv D \pmod{a}$ and $(b')^2 \equiv D \pmod{a'}$, from the hypothesis of Lemma 8.2; for example

$$a'(bb' + D) - a'b(b + b') = a(D - b^2) \equiv 0 \pmod{aa'}.$$

The existence of an integer B with the required congruence properties follows from Lemma 8.1 and the preceding remark. For the proof of the last statement, let d be an integer dividing $a, a', 2B$. Then

$$B \equiv b \equiv b' \pmod{d},$$

so

$$b + b' \equiv 2B \equiv 0 \pmod{d},$$

and hence d divides $a, a', b + b'$. Finally, $d = \pm 1$ by the hypothesis of the Lemma, and the proof is completed.

LEMMA 8.3. *Two quadratic forms (a, b, c) and (a, b', c') having the same first coefficient a and the same determinant D are equivalent whenever $b \equiv b'$ (mod a).*

This fact, from Dirichlet [**114**], §56, is proved as follows. Let $b' = b + ka$; then (a, b, c) is carried to (a, b', c') by the substitution $x = x' + ky'$, $y = y'$.

Composition of two binary quadratic forms (a, b, c) and (a', b', c'), both with determinant D, is defined whenever the forms are *concordant* [einig] in the sense that $a, a', b + b'$ have g.c.d. 1. In this situation, one has $b^2 \equiv D$ (mod a) and $(b')^2 \equiv D$ (mod a'). It follows from Lemmas 8.2 and 8.3 that there exist infinitely many equivalent quadratic forms (aa', B, C) of determinant D, such that $B \equiv b$ (mod a) and $B \equiv b'$ (mod a'). Indeed, let B be chosen as in Lemma 8.2; then $B^2 \equiv D$ (mod aa'), so $B^2 - D = aa'C$ for some integer C, and the quadratic form (aa', B, C) has determinant D. If B' is another solution of the congruences in Lemma 8.2, then $B' \equiv B$ (mod aa'), and the resulting quadratic form (aa', B', C') is equivalent to (aa', B, C), by Lemma 8.3. Each quadratic form (aa', B, C) of determinant D, obtained from the concordant forms (a, b, c) and (a', b', c') by the preceding construction, is called a *composite* of the forms (a, b, c) and (a', b', c'), and it said to be obtained from them by *composition*.

The meaning of the composition of two quadratic forms becomes clearer as a result of the following remarks from Dirichlet [**114**], §146. Let (a, b, c) and (a', b', c') be concordant forms, and let B and C be chosen as in the preceding paragraph. Then $B \equiv b$ (mod a) and $B \equiv b'$ (mod a'), so that, by Lemma 8.3, the quadratic forms (a, b, c) and (a', b', c') are equivalent to the forms $(a, B, a'C)$ and (a', B, aC), respectively. Moreover, the last two quadratic forms are concordant because $a, a', 2B$ have g.c.d. 1 by the last statement in Lemma 8.2, and (aa', B, C) is their composite. Now let x, y, x', y' be variables, and introduce the bilinear substitution:

$$X = xx' - Cyy', \quad Y = (ax + By)y' + (a'x' + By')y.$$

Letting $\pm\sqrt{D}$ denote the two square roots of D, it is easily verified that

$$(ax + (B + \sqrt{D})y)(a'x' + (B + \sqrt{D})y') = aa'X + (B + \sqrt{D})Y,$$

and a similar equation with $\sqrt{D}$ replaced by $-\sqrt{D}$. After multiplying the two equations, and deleting the factor aa', one obtains

$$(ax^2 + 2Bxy + a'Cy^2)(a'(x')^2 + 2Bx'y' + aC(y')^2)$$
$$= aa'X^2 + 2BXY + CY^2.$$

As Dirichlet noted, these observations prove

PROPOSITION 8.1. *The bilinear substitution defined above carries the quadratic form (aa', B, C) into the product of the quadratic forms $(a, B, a'C)$ and (a', B, aC).*

The next result, first proved by Gauss, was crucial for the later definition of a group structure on the set of classes of quadratic forms with a given determinant.

THEOREM 8.3. *Let the concordant quadratic forms (a, b, c) and (a', b', c') be equivalent to the concordant forms (m, n, ℓ) and (m', n', ℓ'), respectively. Then an arbitrary composite form (aa', B, C) of the first two is equivalent to any composite (mm', N, L) of the second two.*

For the proof, which is straightforward, we refer to Dirichlet [**114**], §146.

In order to apply the preceding theorem, it is important to have some natural condition for two quadratic forms (a, b, c) and (a', b', c') to be concordant. This is furnished by the concept of the *divisor* of a form (a, b, c); it is simply the g.c.d. of the coefficients $a, 2b, c$. The divisor of a form (a, b, c) is also the g.c.d. of the set of integers represented by the form, and is equal to the divisor of any other form in the equivalence class of (a, b, c). It is not difficult to prove (Dirichlet, §147) that if the divisors σ and σ' of two concordant forms (a, b, c) and (a', b', c') are relatively prime, then $\sigma\sigma'$ is the divisor of their composite (aa', B, C). As a kind of converse, if the divisors of two equivalence classes K and K' of determinant D are relatively prime, then it is always possible to choose a pair of concordant forms (a, b, c) and (a', b', c') from the classes K and K', respectively. All of these things apply, in particular, to classes K of forms with divisor 1, in other words, to classes of properly primitive forms introduced earlier, in Theorem 8.2.

The preceding discussion lays the groundwork for Dedekind's account of the abelian group of equivalence classes of binary quadratic forms, in [**114**], §§145 − 149. Let $\mathcal{H}$ denote the set of equivalence classes K of binary quadratic forms of determinant D, such that each class $K \in \mathcal{H}$ contains at least one properly primitive form. The set $\mathcal{H}$ is finite, by Theorem 8.1. By the remarks in the preceding paragraph, any two classes K and K' in $\mathcal{H}$ contain a pair of concordant forms, (a, b, c) in K, and (a', b', c') in K'. By Theorem 8.3, the definition of the composite of two concordant forms, and another statement from the preceding paragraph, their composite (aa', B, C) is defined, and belongs to a class L in $\mathcal{H}$, which is independent of the choice of the concordant forms from K and K', respectively. The class L is denoted by what Dedekind calls the symbolic equation [symbolische Gleichung]

$$L = KK'.$$

The operation of multiplication of classes in $\mathcal{H}$ defined in this way satisfies the commutative and associative laws:

$$KK' = K'K \quad \text{and} \quad (KK')K'' = K(K'K''),$$

for all classes K, K' and K'' in $\mathcal{H}$. The quadratic form $(1, 0, -D)$ represents a class in $\mathcal{H}$, called the *principal* class,[21] and denoted by 1, with the property that

$$1K = K,$$

for all classes K in $\mathcal{H}$. Finally, it is proved that, for each pair of classes K and K' in $\mathcal{H}$, there exists a class X in $\mathcal{H}$ such that

$$KX = K'.$$

---

[21]Integers represented by the form $(1, 0, -D)$ correspond to solutions of the diophantine equation $x^2 - Dy^2 = m$. The equation $x^2 - Dy^2 = 1$ is known as Pell's equation, after the 18th-century English mathematician John Pell. The latter equation has a long history. For example there is the famous Cattle Problem of Archimedes (see Davenport, [**101**], p. 107), involving seven linear equations in eight unknowns (representing different kinds of cattle). After simplification, one is led to the diophantine equation

$$x^2 - 4,729,494y^2 = 1,$$

the least solution of which (given by Amthor in 1880) involves an integer y with 41 digits. See also Weil [**283**] for the history of these equations.

These results[22] show that $\mathcal{H}$ is a finite abelian group.[23]

The connection between characters of quadratic forms, as in Theorem 8.2, and the finite abelian group $\mathcal{H}$, was stated by Dedekind ([114], p. 408) as follows.

> Let $\varepsilon, \varepsilon'$ be the values of a character C on the classes H, H' respectively; then $C = \varepsilon\varepsilon'$ for the class HH'.

In other words, if χ is one of the characters of the set of classes in $\mathcal{H}$ defined above, then

$$\chi(HH') = \chi(H)\chi(H')$$

for all elements H, H' in $\mathcal{H}$. A proof of this fact follows without too much difficulty from the preceding discussion.

Here is one example of how the characters of the group $\mathcal{H}$ give new information; it is also an illustration of Frobenius's remark, in the quotation at the beginning of this section, on the superiority of Dirichlet's idea of viewing the characters as complex valued functions on a group, with which one can calculate. Recall that a genus is the set of classes H in $\mathcal{H}$ on which the characters defined in Theorem 8.2 take a fixed set of values. The *principal genus* is the genus containing the principal class 1, and is the genus on which all characters take the value 1, since $\chi(1) = 1$ for all characters χ of the finite abelian group $\mathcal{H}$. The values of all the characters in Theorem 8.2 are ± 1; therefore $\chi(H^2) = \chi(H)^2 = 1$ for each character χ, and each class H in $\mathcal{H}$. This proves that, for each class H, H^2 belongs to the principal genus. According to Gauss (and Dirichlet) the class H^2 was obtained from the class H by *duplication*. The preceding result, proved using the characters of $\mathcal{H}$, states that any class obtained by duplication from another class belongs to the principal genus. A much deeper result is the Duplication Theorem of Gauss ([155], §286), which is a kind of converse to what has just been shown, and asserts that each class belonging to the principal genus can be obtained by duplication from some other class. Gauss's proof of this result used the theory of ternary quadratic forms.

---

[22]The same results were established by Gauss, [155], §§239 − 241, 243, 250. Gauss used the notation $K + K'$ instead of KK' ([155], §249).

[23]Dedekind uses the word *group* [Gruppe], and refers to Galois [154] as a source for this concept.

CHAPTER II

Frobenius and the Invention of Character Theory

The applications of characters of finite abelian groups to algebra and number theory surveyed in Chapter I give no hint of how to extend the theory to nonabelian groups. Frobenius solved this problem, and created much of what is known today as the character theory of finite groups, in his great series of papers published in 1896. In this chapter we shall examine their contents, along with some of his other work published before and after. As in the previous chapter, some of the mathematical parts of the discussion are presented from a modern point of view.

All four of the principal figures in the early development of representation theory had broad mathematical interests. The biographical sections at the beginnings of chapters are intended to give a sense of where the contributions to representation theory fitted in to their mathematical work, along with sketches from their correspondence and other writings, to show the motivation for some of their investigations, and their relationship with other mathematicians.

1. Frobenius in Berlin and Zürich: 1850-1900

Ferdinand Georg Frobenius (1849-1917) was a Berliner, the son of a clergyman, born in Berlin-Charlottenburg. He received his secondary education at the Joachim Gymnasium from 1860 – 1867 and began his university studies in Göttingen, where he attended lectures in analysis by Meyer and Stein, and in physics by Weber. He returned to Berlin after one semester, and completed his work for the doctorate there.

The University of Berlin had become a center for mathematical research in Europe during the first half of the 19th century, when Dirichlet, Eisenstein, and Jacobi were the main figures. They were succeeded by Kronecker, Kummer, and Weierstrass, who were the leaders in mathematical research in Berlin during Frobenius's student days there. He attended their lectures, and participated in the seminars of Kummer and Weierstrass. By any standards, he had a first-rate graduate education in the modern sense. He was awarded the doctoral degree *summa cum laude* in 1870 (at the age of 21) with a dissertation, written in Latin as was customary at the time, entitled *De functionum analyticarum unis variabilis per seres infinitas representatione*.

After teaching in a Gymnasium for a few years, he was appointed ausserordentlicher Professor at Berlin in 1874, thereby avoiding the period of apprenticeship as Privatdozent which was common at the time. This appointment, obtained without the Habilitation, was somewhat unusual, and would no doubt not have occurred without the strong support of Weierstrass.[1]

---

[1] The circumstances of this appointment were discussed in detail by Biermann [**14**], pp. 95, 96.

Ferdinand Georg Frobenius (1849-1917)

Publications began to appear regularly as soon as he received his degree, and showed from the start his versatility and power. These articles, and almost all of Frobenius's subsequent papers, up to the time he was elected to membership in the Prussian Academy of Science, were published in Crelle's *Journal für die reine und angewandte Mathematik*, a mathematical journal founded in 1826, and personally subsidized at the outset by the Berlin engineer August Leopold Crelle. This journal quickly became a respected publication, and a convenient outlet for the work of the Berlin mathematicians as well as other European contributors. For example, Volume 28 contains 25 articles by Eisenstein, published when he was 21 years old. In its 168th year, the journal is still going strong.

In 1875, after only one year in his new position in Berlin, Frobenius was appointed ordentlicher Professor at the ETH Zürich (Eidgenössische Technische Hochschule). As Biermann pointed out in ([**14**], p. 96), the ETH frequently recruited its professors from the Berlin school, and appointed Prym, Schwarz, Schottky, and Rudio in addition to Frobenius. In Zürich, Frobenius married Christiane Elis, and started a family. He enjoyed 17 productive years of mathematical research there.

In 1891, Kronecker died, and Frobenius was called back to Berlin to succeed him. Following his election in 1892 to membership in the Prussian Academy of Science, most of his publications from that time on appeared in the proceedings of the Academy. The following excerpts from the statements supporting his election by Weierstrass and Fuchs, and by Fuchs and Helmholtz, from the archives of the Academy, give a vivid impression of how Frobenius's work from 1870-1892 was viewed by his contemporaries.

> In the 22 years since his doctoral degree, Frobenius has published a large number of scientific articles, which have received wide recognition. It is pointless to introduce and discuss all of them here; in order to give an overview of them, we shall divide them into groups, according to their contents.
>
> 1. On the development of analytic functions in series ... [the subject of his dissertation].
>
> 2. On the algebraic solution of equations, whose coefficients are rational functions of one variable.
>
> 3. The theory of linear differential equations.
>
> 4. On Pfaff's problem.
>
> 5. Linear forms with integer coefficients.
>
> 6. On linear substitutions and bilinear forms, and related topics.
>
> 7. On adjoint linear differential operators, and related topics.
>
> 8. The theory of elliptic and Jacobi functions in arbitrarily many variables (partly joint work with Herrn Prof. Stickelberger [in Zürich]).
>
> 9. On the relations among the 28 double tangents to a plane curve of degree 4.
>
> 10. On Sylow's Theorem.
>
> 11. On the double congruence arising from two finite groups [double cosets].
>
> 12. On Jacobi's covariants ....

13. On Jacobi functions in three variables.
14. The theory of biquadratic forms.
15. On the theory of surfaces with a differential parameter.

The preceding will suffice to give the knowledgeable reader a picture of the extensive and many-sided scientific activity of Frobenius. All his works exhibit a sure mastery of the material; all either contain new results or known results in a new form. Moreover, Frobenius is a first-rate stylist, and writes clearly and understandably without ever attempting to delude the reader with empty phrases.

We believe, after all this, that the Academy will bring itself credit by his election to ordinary membership.

[signed] Weierstrass, Fuchs.

This document was accompanied by the following supplement, by Fuchs and Helmholtz:

Beginning with vol. 73 of our mathematical journal [Crelle's journal] up to the newest volume, which is number 110, we seldom find a volume in which not a single work of Frobenius appears, while several volumes contain more than one work.

The topics covered by these works are of a very diverse nature. At the outset there were function-theoretic tools which were useful to him in making his contributions to the theory of linear differential equations....

But soon Frobenius understood that his strength lay in the treatment of mathematical forms [matrices, bilinear forms, etc.], ... and he achieved a dominating mastery of forms, in which he is surpassed by no living mathematician. It would not be useful to list here the titles of his many works; it will suffice to indicate the different fields which are indebted to his creativity for new foundations or the enrichment of a particular discipline. These are the theory of differential equations, number theory, algebra, group theory, the theory of elliptic and abelian functions, and the theory of partial differential equations. In all the subjects he treated, he rebuilt the foundations, and in most of his works constructed anew from these foundations an entire discipline in an original way, and from a unified point of view. He succeeded in presenting the existing results of the discipline in a new light, filled gaps he encountered, and created a basis affording excellent opportunity for further investigation. Each of his larger works can rightly be regarded as a small compendium for the discipline concerned. ...

[signed] Fuchs, Helmholtz

Although Frobenius's mathematical work is readily available, his personal life and relations with friends and colleagues in the period under consideration are not well documented,[2] with the exception of his correspondence with Dedekind beginning at

---

[2]There is, however, a warm and appreciative recollection of Frobenius's often fast-paced lectures by Carl Ludwig Siegel (*Erinnerungen an Frobenius*, in *F. G. Frobenius, Gesammelte Abhandlungen*, Springer-Verlag, Berlin, 1968, I, iv-vi) along with an account of his personal

University of Berlin, 1992

about the time he began to work on character theory.[3] Biermann's book *Die Math-ematik und ihre Dozenten an der Berliner Universität* 1810 − 1920 [**14**] contains information based on faculty records about Frobenius's career as a professor at the University of Berlin. Biermann's report ([**14**], pp. 122, 123) portrayed Frobenius as an unbending, rather cantankerous, person, whose relations with his colleagues as well as the state ministry of education were sometimes strained. For example, Biermann remarked:

> He suspected at every opportunity a tendency of the ministry to lower the standards of the University of Berlin, in the exact words of Frobenius, "auf der Stufe des Technicum Mittweida" [to the rank of a technical school]. ... [4]

Biermann continued:

> ... Even so, Fuchs and Schwarz yielded to him, and later Schot-tky, who was indebted to him alone for his call to Berlin. Frobenius was the leading figure, on whom the fortunes of mathematics at Berlin University rested for 25 years. Of course, it did not escape him, that the number of Doctorates, Habilitations, and Dozents

interactions with Frobenius, when Siegel was a student at the University of Berlin in 1915, towards the end of Frobenius's career.

[3]Some of this correspondence is discussed later in this chapter. For a full account of it, and its connection with various stages of Frobenius's understanding of character theory, see Hawkins [**174**].

[4]Biermann, K. R., *Die Mathematik und ihre Dozenten an der Berliner Universität 1810–1920*, Wiley-VCH Verlag, Berlin, 1973. Used with permission.

slowly but surely fell off, although the number of students increased considerably. That he could not prevent this, that he could not reach his goal of maintaining unchanged the times of Weierstrass, Kummer, and Kronecker also in their external appearances, but to witness helplessly these developments, was doubly intolerable for him, with his choleric disposition.

Frobenius's lectures (on number theory, algebra, determinants, and analytic geometry) were attended by as many as 250 students, an indication confirming the recollections of one, that his lectures, in their variety and depth, were excellent. He taught number theory often, algebra almost always for 2 semesters, including in the first semester the elementary part, and in the second, ideal theory and the theory of discrete groups.

One of his greatest merits was that he recognized the strength and importance of I. Schur from the first day, and sought to advance him with every opportunity that presented itself. To have helped Schur reach a position from which he could freely exercise his power is always marked as an achievement of Frobenius.[5]

Frobenius served as "Gutachter" [member of the dissertation committee] of 19 doctoral students at Berlin. The complete list is given in Biermann [**14**]. The following excerpt from the list gives the names, titles, dates of the doctoral degree, and the names of the other expert, or Gutachter, besides Frobenius, of those students whose dissertations were in algebra or number theory. Schur's was the only dissertation in representation theory.

Radzig, Alexander, *Die Anwendung des Sylowschen Satzes auf die symmetrische und die alternierende Gruppe*, 1895, Fuchs.

Landau, Edmund, *Neuer Beweis der Gleichung*

$$\sum_{k=1}^{\infty} \frac{\mu(k)}{k} = 0,$$

1899, Fuchs.

Schur, Issai, *Über eine Klasse von Matrizen, die sich einer gegebenen Matrix zuordnen lassen*, 1901, Fuchs.

Steinbacher, Friedrich, *Abelsche Körper als Kreisteilungskörper*, 1910, Schwarz.

Remak, Robert, *Über die Zerlegung der endlichen Gruppen in indirekte unzerlegbare Faktoren*, 1911, Schwarz.

Stiemke, Erich, *Über unendliche algebraische Zahlkörper*, 1914, Schottky.

2. Research on Finite Groups and Number Theory: 1880-1896

Frobenius published five papers, totalling 111 pages, in the 1896 volume of the *Sitzungsberichte der Königlich Preussischen Akademie der Wissenschaften zu Berlin*, with the titles:

1. *Über die covarianten Transformationen der bilinearen Formen;*

2. *Über vertauschbare Matrizen;*

3. *Über Beziehungen zwischen den Primidealen eines algebraischen Körpers und den Substitutionen seiner Gruppe;*

---

[5]Biermann, K. R., *Die Mathematik und ihre Dozenten an der Berliner Universität 1810–1920*, Wiley-VCH Verlag, Berlin, 1973. Used with permission.

4. *Über Gruppencharaktere*;

5. *Über die Primfactoren der Gruppendeterminante.*

It is our aim to show what an extraordinary burst of creative energy these represent. We are mainly concerned with the last two, and will consider them in the next section. We begin here with a discussion of the third paper, and the finite group theory leading up to it.

Frobenius's investigations of finite groups began around 1880, when he became interested in a paper of Kronecker [**199**], and embarked, as we shall explain later, on a voyage of exploration in finite group theory and number theory culminating in the third 1896 paper. His first publication [**126**] on general finite groups[6] entitled *Neuer Beweis des Sylowschen Satzes*, was submitted to Crelle's Journal in 1884. In 1844-46, Cauchy (Exerc. d'analyse et de phys. math. tom. III, p. 250) had proved that a finite group whose order is divisible by a prime p contains an element of order p. Sylow's theorem (Math. Ann. **5** (1872)) is an extension of Cauchy's result, and has proved to be fundamental for research in finite group theory. Here is the statement.

THEOREM 2.1. *Let G be a finite group of order g, and assume that g is divisible by p^ν for a prime p. Then G contains a subgroup of order p^ν.*

Sylow's proof, and much of the literature on finite groups up to that time, was written in the language of permutation groups, viewed as subgroups of the symmetric group. Just as Dirichlet led his contemporaries by the hand into the realm of conditionally convergent series (see Chapter I, §7), Frobenius began his short paper on Sylow's theorem with a careful statement of the axioms for a finite group, and gently pushed his readers into the world of abstract groups with a remark that his proof depended only on the axioms, and not on the interpretation of group elements as permutations. For the axioms themselves he referred to Kronecker (Berl. Monatsber. 1870) and Weber (Math. Ann. **20**, 302). Frobenius's proof of the theorem is of particular interest to us because of its elegant application of conjugacy classes, which he began to study systematically in this paper and its sequel [**127**], and was to use later as the entering wedge in his first paper on character theory. It is a standard proof used today, and goes as follows.

He first assumed the theorem is true for groups of order less than the order of G, and all prime powers p^ν. Let Z denote the center of G, that is, the set of elements which commute with all the elements of G. Then Z is a subgroup of G whose order z divides the order g of G, by Lagrange's Theorem. The following two cases arise.

(1) *The order z of Z is divisible by p.* In this case, the abelian group Z contains an element x of order p. He proved the existence of x in the framework of finite abelian groups, without using Cauchy's result. The cyclic group $\langle x \rangle$ generated by x is a normal subgroup of G and the factor group $\overline{G} = G/\langle x \rangle$ has order g/p. The order of the factor group $\overline{G}$ is divisible by $p^{\nu-1}$, and consequently, by the

---

[6]There was one earlier joint paper with Stickelberger [**153**], published in 1879, on finite and finitely generated abelian groups, in which they gave their own proofs of the main structure theorems. They traced the origins of the main theorems to Gauss (*Démonstration de quelques théorèmes concernant les périodes des classes de formes binaires du second degré*, Werke Bd. II, 266) in the case of the primary decomposition theorem, and to Schering (*Die Fundamentalklassen der zusammensetzbaren arithmetischen Formen*, Göttinger Abh. Bd. 14), for the invariant factor theorem.

assumption made at the beginning of the proof, $\overline{G}$ contains a subgroup of order $p^{\nu-1}$. The union of the cosets belonging to this subgroup is a subgroup of G of order p^ν, so the theorem is proved in this case. Incidentally, the facts about normal subgroups and factor groups needed for this argument were known at that time, and for them he was able to refer to Kronecker [**198**] and Jordan [**192**].

(2) *The order z of Z is not divisible by p.* He defined two elements a and b of a finite group G to be *conjugate* (he used the word *ähnlich*) if $h^{-1}ah = b$ for some element $h \in G$, and observed that this relation divides the group into disjoint subsets called *conjugacy classes*. The conjugacy class containing a is the set of all elements $b \in G$ which are conjugate to a, and hence to each other. Note that each conjugacy class containing an element of the center Z consists of this element alone. The key step in this argument is the analysis of what we now call the *class equation*, which states that

$$g = z + h_1 + \cdots + h_m,$$

where g and z are as defined above, and $\{h_i\}$ are the numbers of elements in the different conjugacy classes containing more than one element. As p does not divide z, and does divide g (as we may assume), not all the h_j appearing in the class equation are divisible by p. The next step, and a crucial one, was to relate the numbers h_i to the subgroup structure of G. Each number h_j is the *index* (= number of left or right cosets in G) of the subgroup of G consisting of all elements of G commuting with some element x_j in the corresponding conjugacy class. This subgroup is called the *centralizer* of x_j, and is denoted by $C_G(x_j)$. By Lagrange's equation, one has

$$g = h_i |C_G(x_i)|.$$

Therefore, if p doesn't divide a particular h_i, and p^ν divides g, it follows that p^ν divides the order of $C_G(x_i)$. But $C_G(x_i)$ is a proper subgroup of G since h_i is greater than one, and the results follows, by the hypothesis that the Theorem holds for groups of order less than g.

His second paper on group theory [**127**], submitted two years after the first one, begins as follows.

> I was led to the investigations in the theory of groups which form the contents of this work through the study of a remarkable paper by Herrn Kronecker, *Über die Irreductibilität von Gleichungen* (Monatsber. der Berl. Akad. 1880, Seite 155), in particular through the attempt to find the indicated relations [die angedeuteten Relationen aufzufinden] at the end of page 157.

The motivation for this foray into group theory requires some explanation. Kronecker's paper begins with a statement of the following theorem.

THEOREM 2.2. *If $F(x)$ is a polynomial function of x with integer coefficients, and ν_p is the number of (equal or different) roots of the congruence*

$$F(x) \equiv 0 \pmod{p},$$

then the limiting value of the series

$$\sum_p \nu_p p^{-1-w},$$

for infinitely small positive values of w, is proportional to $\log \frac{1}{w}$, and indeed is equal to $\log \frac{1}{w}$ multiplied by the number of irreducible factors of $F(x)$.

He continued with some remarks on how this result is related to the densities of certain sets of prime numbers (see the discussion of Dirichlet density of sets of primes in Chapter I, §7).

> Since ν_p can have only the values $0, 1, \ldots, n$, if n denotes the degree of $F(x)$, each series [as in the theorem] can be expressed in terms of partial series
>
> $$\sum_{k=1}^{n} k \sum p_k^{-1-w},$$
>
> where p_k runs over the set of prime numbers for which k solutions of the congruence $F(x) \equiv 0$ exist. For the set of *all* prime numbers, it is known that the limit value of $\sum p^{-1-w}$ is $\log \frac{1}{w}$. Therefore, if one assumes the existence of a function which gives the density of [a set of] prime numbers, then one can formulate the above theorem simply as the statement that this density coincides with that which results, if each prime number is taken as many times as the number of roots of the congruence $F(x) \equiv 0 \pmod{p}$, under the hypothesis that $F(x)$ is irreducible.
>
> If one takes as measure the density of *all* prime numbers, and the density of the prime numbers p_k as D_k then the above theorem asserts that the equation
>
> $$\sum_{k=1}^{n} k D_k = 1$$
>
> is generally characteristic for irreducible equations $F(x) = 0$.

Miyake remarked ([**214**]) that the last equation is one of "die angedeuteten Relationen aufzufinden" which led Frobenius into group theory. Groups entered the picture, as Miyake said, through Frobenius's aim to express the densities D_k in terms of the Galois group $\mathcal{H}$ of the irreducible polynomial equation $F(x) = 0$. This required first of all some group theoretic preparation, namely some results on the distribution of elements of conjugacy classes of the Galois group $\mathcal{H}$ into subgroups, and is the subject of his second group theory paper [**127**], which we shall discuss presently.

Frobenius's contribution to the density problems [**130**], published in the same year as *Über Gruppencharaktere*, contains, as he stated in the introduction, work done well before 1896, and is independent of the theory of characters. Whether it was the incentive for Frobenius to create a theory of characters of general finite groups remains an unanswered question.[7] In any case, the paper [**130**] contained a number of things that proved to be fundamental, such as the definition of what has become known as the *Frobenius automorphism* in algebraic number theory, along with some further conjectures on densities involving conjugacy classes in Galois groups, which were proved after his death by Tchebotarev [**271**]. The theory of L-series with general group characters, towards which Frobenius's work pointed so clearly, was begun by Artin in 1923 [**7**] (see Chapter VII).

---

[7]Miyake [**214**] also raised this question, and added some comments on how the Dedekind ζ-functions, and their application to density questions involving sets of prime ideals in algebraic number fields, also pointed towards the need for character theory.

The paper [**127**], which laid the group theoretic foundation for his extension of Kronecker's density theorems, in fact contains much more. It is based on the theory of double cosets, which had been considered previously, in special cases, by Cauchy and Sylow. Frobenius developed the idea in full generality, and included applications of double cosets to the distribution of conjugacy classes in subgroups, and to an elegant new proof of Sylow's theorem on the conjugacy of what are now called Sylow subgroups of a finite group. The structure of finite groups had become a major new theme in his research, and he wrote a series of increasingly powerful articles on finite solvable groups before getting around to publishing his paper on density problems.

The paper [**127**] began with some comments about previous work, along with an admonition that these were abstract groups, whose elements were subject only to the axioms for a finite group. He put it this way ([**127**], Ges. Abh. II, p. 304):

> Concerning the *elements* to be considered, I make the same hypotheses that I summarized in my work "Neuer Beweis des Sylowschen Satzes" (in this journal, volume 100). For the case that any two elements commute, Herr Kronecker (Auseinandersetzung einiger Eigenschaften der Klassenanzahl idealer complexer Zahlen, Monatsber. 1870) called two elements a and b *equivalent* or *congruent* with respect to a group G [viewed as a subgroup of a group S],
>
> $$a \sim b \pmod{G},$$
>
> if ab^{-1} is contained in G. Herr Jordan (Sur la limite de transitivité des groupes non alternés, Bull. de la Soc. Math. de France, T. 1) applied this definition also to the cases where the elements a and b are permutable with the group G. I will use it here without making a hypothesis on the permutability of the groups and elements under consideration.

The number of elements of S which are different (mod G), or the number of *cosets* of S (mod G) as we would put it now, is given by Lagrange's Theorem, and is $|S|/|G|$, in case G is a subgroup of S, where $|S|$ and $|G|$ denote the orders of the finite groups S and G, respectively. He continued:

> For many investigations it appears advisable to call $a \sim b \pmod{G}$ if $a = bg$ and $g\ (= b^{-1}a)$ is an element of G. This remark indicates that the congruence relation defined above is only a special case of a very important concept for the study of groups of noncommuting elements. For two groups G and H, I call an element b and another element a congruent (modd G, H) if $gah = b$, where g belongs to the group G and h to the group H.

The relation Frobenius defined in the preceding paragraph is an equivalence relation, and he called the resulting equivalence classes *classes of congruent elements*. Today they are called *double cosets*, and the double coset containing the element a is denoted by GaH. The first problem he considered was an examination of the number of solutions of the equations

$$gsh = s, \ g \in G, h \in H, s \in S,$$

where G and H are subgroups of a given finite group S. The result is given in terms of the number m of double cosets, and can be viewed as a generalization of Lagrange's Theorem. Here is his solution of the problem.

Let $s_1, s_2, \ldots$ be the elements of S, and let d_i be the number of solutions of the equation $gsh = s$ with $s = s_i, i = 1, 2, \ldots$. Assume that $s_1, \ldots, s_c$ are the elements of the double coset Gs_1H. For each pair of elements s_i, s_j in the double coset, one has

$$g_i s_i h_i = s_1, \ g_j s_j h_j = s_1,$$

for elements g_i, g_j in G and h_i, h_j in H, so that for each solution $\{g, h\}$ of the equation $gs_1h = s_1$ one has $g's_ih' = s_j$ with $g' = g_j^{-1}gg_i$, $h' = h_ihh_j^{-1}$ in G and H respectively. Letting $s_i = s_j$, it follows that $d_1 = d_2 = \cdots = d_c$. By the same argument, it follows that the set of elements $\{gs_1h\}$ with $g \in G, h \in H$, contains each element $s_1, \ldots, s_c$ exactly d_1 times. Therefore

$$d_1 + \cdots + d_c = cd_1 = |G||H|,$$

and one has ([**127**], Ges. Abh. II, p. 308):

PROPOSITION 2.1. *Let G, H be subgroups of a finite group S. Then the number of solutions of the equation $gsh = s$ with $g \in G, h \in H$ and $s \in S$, is $|G||H|m$, where m is the number of double cosets GsH, and $|G|, |H|$ denote the orders of the finite groups G and H respectively.*

His ideas on an extension of Kronecker's Theorem 2.2 led Frobenius to the following general result on the distribution of conjugacy classes of S in the subgroups G and H. The key observation was that, if one replaces h in H by h^{-1}, the equation $gsh = s$ considered in the preceding proposition becomes $s^{-1}gs = h$, so the focus is switched from double cosets to conjugacy. He let ℓ denote the number of conjugacy classes in S, $a_1, a_2, \ldots$ the numbers of elements in the conjugacy classes, taken in some order, and $b_1, b_2, \ldots$ the numbers of elements in the intersections of the classes with G. Let $b_i > 0$ and let $g_1, \ldots, g_{b_i}$ be the elements belonging to the intersection of the ith class with G. He then examined the number of solutions of the equation

$$s^{-1}gs = h,$$

with $s \in S$, $h \in H$, and g one of the elements g_p belonging to the ith class. For one of these elements $g_p, 1 \leq p \leq b_i$, the set of all conjugates $s^{-1}g_ps$, with $s \in S$, contains each of the a_i elements of the ith class exactly $|S|/a_i$ times, as he had shown in his proof of Sylow's Theorem 2.1. This result is independent of the choice of the element g_p, so the set of all elements $s^{-1}gs$, with g one of the elements of the ith class, contains each element of the ith class exactly $(|S|b_i)/a_i$ times. Letting c_i denote the number of elements of the ith class belonging to the subgroup H, it follows that the number of solutions of the equation $s^{-1}gs = h$, with g in the ith class, is $(|S|b_ic_i)/a_i$. By the remark at the beginning of the discussion, the total number of solutions of the equation $s^{-1}gs = h$ is the same as the number of solutions of the equation considered in Proposition 2.1, and one obtains ([**127**], Ges. Abh. II, p. 310):

PROPOSITION 2.2. *Let G, H be subgroups of S, and m the number of double cosets GsH, as in the preceding result. Let $\{C_1, \ldots, C_\ell\}$ denote the conjugacy classes in S. Then one has*

$$\frac{|G||H|}{|S|}m = \sum_{i=1}^{\ell} \frac{|C_i \cap G||C_i \cap H|}{|C_i|}.$$

The preceding discussion follows Frobenius's line of reasoning exactly, and provides an illustration of his ability to extract information from a study of the number of solutions of equations in finite groups. The paper contains other applications of double cosets, and we shall take up one more of them before moving on to the number theory, namely a new proof of Sylow's second theorem. The first Sylow theorem (Theorem 2.1) implies that if p^a is the highest power of a prime p dividing the order of a finite group G, then there exists a subgroup P of order p^a; subgroups of this order are today called *Sylow p-subgroups*. We shall state Sylow's second theorem in terms of a finite group S, and two subgroups G, H, so the notation developed above can be used.

THEOREM 2.3. *Let p^a be the highest power of a prime p dividing the order of a finite group S. Then any two subgroups G, H of order p^a are conjugate: $xGx^{-1} = H$ for some element $x \in S$.*

Frobenius's proof goes like this ([**127**], Ges. Abh. II, p. 313). Let G, H be subgroups of S, as in the statement of the theorem. The preceding discussion shows that the set of elements gsh, with $g \in G$, $h \in H$, and s a fixed element of S, contains each element of the double coset GsH exactly d times, where d is the number of solutions of the equation $gsh = s$, or $s^{-1}gs = h$. Then $d = |H \cap s^{-1}Gs|$, and the number of elements c in the double coset GsH is given by the formula: $c = |G||H|/d$. By Lagrange's Theorem, d is a common divisor of both $|G|$ and $|H|$, and hence c is a common multiple of $|G|$ and $|H|$.

Now consider the double coset decomposition

$$S = Gs_1H \cup \cdots \cup Gs_mH.$$

From what has been said, we obtain $|Gs_iH| = f_i|G|$ for some integer f_i, $i = 1, \ldots, m$, and $d_if_i = |H|$, where $d_i = |H \cap s_i^{-1}Gs_i|$, for each i. To prove the theorem, we may assume that $|G|$ and $|H|$ both equal the highest power of p dividing the order of $|S|$. Then $|S|/|G|$ is not divisible by p. On the other hand, $|S|/|G| = f_1 + \cdots + f_m$ by the preceding formulas. It follows that at least one f_i is not divisible by p. For this choice of i, we have $|H| = d_if_i$, and hence $f_i = 1$, because $|H|$ is a power of p. Then $d_i = |H|$, or $|H| = |H \cap s_i^{-1}Gs_i|$. Therefore $s_i^{-1}Gs_i = H$, completing the proof.

Kronecker's paper containing Theorem 2.2 was published in 1880, and led to Frobenius's paper on double cosets, as we have explained. The sequel to it [**130**], in which the results on distribution of conjugacy classes in subgroups were applied to extensions of Kronecker's density theorem, was not published until 1896. As Frobenius was not a person to let grass grow under his feet, the question immediately arises as to the reason for the delay in publication. Frobenius himself addressed this point, in the introduction to the paper [**130**].

> I wrote the following article in November 1880, and communicated the results developed therein to my friends Stickelberger and Dedekind. Its foundation stands in close relation to the laws according to which the rational prime numbers are factored into prime ideals in an algebraic and especially a normal field. From some remarks in Dedekind's letter, I must assume that he has been engaged with the investigation of these laws for a long time, and in fact, on my inquiry, he sent me on 8 June 1882 the skeleton of this theory, that he has published on 10 September 1894 in the

Göttinger Nachrichten with the title *Zur Theorie der Ideale*. I had always wished that this sketch would appear before my own work, and this was the reason why I have decided on its publication only now.

If one assumes the relations explained in Dedekind's work as known, the following analysis can be significantly shortened. In this way Hurwitz found a Theorem developed in §5 of this work, as he communicated to me in a letter of 2 January 1896. This letter obliged me to reconsider my original intention to wholly rework the following investigation, and to publish it with some abridgment, exactly in the form I wrote it in 1880.

Adolf Hurwitz's letter[8] was an eye-opener for Frobenius, as he had associated Hurwitz with the mathematical school of Felix Klein at Göttingen. Hurwitz had been a student of Klein, and his early work was strongly influenced by Klein's program to continue the geometric approach to complex function theory inaugurated by Riemann.[9] Frobenius was skeptical of the big-picture geometric approach of Klein,[10] and his reply to Hurwitz, in a letter dated 3 February 1896, contains the passage:

> If you were emerging from a school, in which one amuses oneself more with rosy images than hard ideas, and if, to my joy, you are also gradually becoming emancipated from that, then ... old loves don't rust [alte Liebe rostet nicht]. Please take this joke facetiously.

H. Edwards remarked,

> Of course it is hard to judge what sensibilities were at the time, but to me it has a weird feel to *request* that someone take something as a joke. If nothing else, it certainly showed that Frobenius considered Hurwitz's mathematical origins as a negative thing from which he needed to be emancipated.[11]

The mathematics underlying these exchanges contributed important new ideas to algebraic number theory. Dedekind's paper [**102**] sketched, in outline form, some parts of what is called today the theory of ramification and Galois group action on prime ideals in algebraic number fields. Frobenius had a nice result to add, namely his construction of what is now called the *Frobenius automorphism*, which he needed for his density theorems. Our modern introduction to these ideas follows Janusz's book ([**191**], Chapter III), to which we refer for proofs of the statements made below.

Let Ω be a finite Galois extension of the rational field $\mathbb{Q}$, with Galois group $\mathcal{H}$, and let $\mathfrak{o}$ denote the ring of algebraic integers in Ω. Then $\mathfrak{o}$ is a *Dedekind domain*: each nonzero ideal in $\mathfrak{o}$ can be factored uniquely as a product of prime

---

[8]Hurwitz's letter was based on an untitled note on the subject which he had recorded in his diary in 1895. It began with an interpretation of some of Kronecker's work from the article [**199**], followed by some new results, which Frobenius also obtained and included in his paper [**130**]. Hurwitz's note [**187**] was published, essentially unaltered from his diary, with the same title as Frobenius's paper [**130**], after his death, in 1926.

[9]See Parshall and Rowe [**236**], Chapter 4, for an historical analysis of the work of Klein, and the Göttingen tradition in mathematical research, to which it belonged.

[10]In fact, Biermann remarked ([**14**], p. 123): "The aversion of Frobenius to Klein and S. Lie knew no limits ..." .

[11]I am indebted to H. Edwards for sharing his notes on this letter.

ideals. Moreover, each nonzero prime ideal $\mathfrak{p}$ in $\mathfrak{o}$ is a maximal ideal, so the quotient $\mathfrak{o}/\mathfrak{p}$ is a field. The *residue class field* $\mathfrak{o}/\mathfrak{p}$ is a finite extension of $\mathbb{Z}/(\mathfrak{p} \cap \mathbb{Z}) = \mathbb{Z}/p\mathbb{Z}$, where p is the rational prime number contained in $\mathfrak{p} \cap \mathbb{Z}$. It follows that $\mathfrak{o}/\mathfrak{p}$ is a finite field containing $q = p^f$ elements, and is a Galois extension with cyclic Galois group generated by the automorphism $\overline{\omega} \to \overline{\omega}^p$, for $\overline{\omega} = \omega + \mathfrak{p}$, $\omega \in \mathfrak{o}$.

Now let p be a rational prime, and $\mathfrak{p}$ a prime ideal in $\mathfrak{o}$ dividing p. Then, letting $(p) = p\mathfrak{o}$, one has the factorization

$$(p) = (\mathfrak{p}_1 \ldots \mathfrak{p}_t)^e,$$

with $\mathfrak{p} = \mathfrak{p}_1$. The rational prime p is called *unramified* if $e = 1$. The primes that ramify (i.e. $e > 1$) are those which divide the discriminant of the Dedekind ring $\mathfrak{o}$ (see Janusz, loc. cit. §7). In case the field $\Omega = \mathbb{Q}(\theta)$ is a simple extension generated by a root θ of an irreducible monic polynomial $f(X) \in \mathbb{Z}[X]$, Frobenius used the fact that p is unramified if and only if p does not divide the discriminant Δ of the polynomial $f(X)$, where

$$\Delta = \prod_{i > j} (\theta_i - \theta_j)^2,$$

and $\{\theta_i\}$ are the roots (which belong to $\mathfrak{o}$) of the polynomial equation $f(X) = 0$.

The *decomposition group* $G(\mathfrak{p})$ of a nonzero prime ideal $\mathfrak{p}$ is the subgroup of the Galois group $\mathcal{H}$ consisting of all automorphisms $\tau \in \mathcal{H}$ such that $\tau(\mathfrak{p}) = \mathfrak{p}$. Each element $\tau \in G(\mathfrak{p})$ defines an automorphism $\overline{\tau}$ in the Galois group of the residue class field $\mathfrak{o}/\mathfrak{p}$ over $\mathbb{Z}/p\mathbb{Z}$, where p is a rational prime contained in $\mathfrak{p}$. The automorphism $\overline{\tau}$ is defined by putting

$$\overline{\tau}(\overline{\omega}) = \overline{\tau(\omega)},$$

for $\overline{\omega} = \omega + \mathfrak{p} \in \mathfrak{o}/\mathfrak{p}$. It can be proved that the map $\tau \to \overline{\tau}$ is a surjective homomorphism from $G(\mathfrak{p})$ to the Galois group of $\mathfrak{o}/\mathfrak{p}$ over $\mathbb{Z}/p\mathbb{Z}$, and is an isomorphism in case p is unramified. As the latter group is cyclic, with generator $\overline{\omega} \to \overline{\omega}^p$, it follows, in particular, that one has:

PROPOSITION 2.3. *Let Ω be a Galois extension of $\mathbb{Q}$ with Galois group $\mathcal{H}$, and ring of integers $\mathfrak{o}$. Let p be an unramified rational prime, and $\mathfrak{p}$ a prime ideal in $\mathfrak{o}$ dividing p. Then there exists a unique element $\sigma \in \mathcal{H}$ such that*

$$\sigma(\omega) \equiv \omega^p \pmod{\mathfrak{p}}$$

for all elements $\omega \in \mathfrak{o}$.

Most of the preceding discussion appeared, in one form or another, in the papers of Dedekind and Frobenius cited above. The element $\sigma \in \mathcal{H}$ defined by the Proposition is called the *Frobenius automorphism* associated with the prime ideal $\mathfrak{p}$. As it is known (see Janusz, loc. cit. §6) that any two prime ideals $\mathfrak{p}$ containing p are conjugate by an element of $\mathcal{H}$, it follows that the Frobenius automorphisms associated with them are conjugate in $\mathcal{H}$. Thus each unramified rational prime is associated with a unique conjugacy class in $\mathcal{H}$. Frobenius's idea was to find the density of the set of rational primes associated in this way with a given conjugacy class in $\mathcal{H}$.

After giving his own version of the theory of ramification and construction of the automorphism σ in the first part of his paper [**130**], Frobenius put all the pieces together in §5, and proved his density theorem. We shall give an account, closely following Frobenius, of part of the story.

Keep the preceding notation, with $\Omega, \mathfrak{o}, \mathcal{H}$, etc. as above. Let ξ be an element of $\mathfrak{o}$, and let $\mathcal{G}$ be the stabilizer of ξ in $\mathcal{H}$, that is, the subgroup consisting of automorphisms $\tau \in \mathcal{H}$ which leave the element ξ fixed. Let $h = |\mathcal{H}|$, $g = |\mathcal{G}|$, and put $n = h/g$. Then there are exactly n distinct conjugates $\xi_1, \ldots, \xi_n$ of ξ by the elements of $\mathcal{H}$, and all of them belong to $\mathfrak{o}$. Now let

$$\psi(X) = \prod_{i=1}^{n} (X - \xi_i).$$

The coefficients of $\psi(X)$ are symmetric functions of the $\{\xi_i\}$, and are fixed by the elements of $\mathcal{H}$, so they belong to $\mathfrak{o} \cap \mathbb{Q} = \mathbb{Z}$ (see Chapter 1, §1). It follows that $\psi(X)$ is an irreducible monic polynomial with rational integral coefficients, that vanishes for $X = \xi$. Let

$$\varphi(X) = \prod_{\tau \in \mathcal{H}} (X - \tau(\xi));$$

then $\varphi(X)$ is also a monic polynomial with rational integral coefficients, and one has

$$\varphi(X) = (\psi(X))^g.$$

Frobenius's plan was to apply Kronecker's Theorem 2.2 to the polynomial $\varphi(X)$; this involved, for example, counting the number of solutions ν_p of the congruences $\varphi(X) \equiv 0 \pmod{p}$ for rational primes p.

Let p be an unramified prime, such that p does not divide the discriminant

$$\prod_{i > j} (\xi_i - \xi_j)^2$$

of the polynomial $\psi(X)$. Note that the discriminant is fixed by the automorphisms in $\mathcal{H}$ so it belongs to $\mathfrak{o} \cap \mathbb{Q} = \mathbb{Z}$.

We consider a prime ideal $\mathfrak{p}$ in $\mathfrak{o}$ containing p. Suppose a is a rational integer solution of the congruence $\psi(X) \equiv 0 \pmod{p}$; then one has

$$\prod (a - \xi_i) \equiv 0 \pmod{\mathfrak{p}}.$$

Then some factor $a - \xi_i \equiv 0 \pmod{\mathfrak{p}}$ because $\mathfrak{p}$ is a prime ideal, and the factor is unique, otherwise $\xi_i - \xi_j \equiv 0 \pmod{\mathfrak{p}}$ with $i \neq j$, and $\mathfrak{p} \cap \mathbb{Z} = p\mathbb{Z}$ divides the discriminant of $\psi(X)$, contrary to assumption. If $\xi_i \equiv a \pmod{\mathfrak{p}}$, then $\xi_i^p \equiv \xi_i \pmod{\mathfrak{p}}$. Conversely, from this congruence we obtain

$$\xi_i(\xi_i - 1) \ldots (\xi_i - p + 1) \equiv 0 \pmod{\mathfrak{p}},$$

and hence $\xi_i \equiv a \pmod{\mathfrak{p}}$ for some integer a such that $\psi(a) \equiv 0 \pmod{p}$. At this point, it has been shown that *the number of roots mod p of the congruence $\psi(X) \equiv 0 \pmod{p}$ is precisely the number of distinct conjugates $\{\xi_i\}$ of ξ such that $\xi_i^p \equiv \xi_i \pmod{\mathfrak{p}}$.*

From the remarks above, it follows that the number of roots mod p of the congruence $\varphi(X) \equiv 0 \pmod{p}$ is the number of conjugates $\tau(\xi)$ of ξ by elements $\tau \in \mathcal{H}$ which satisfy

$$\tau(\xi)^p \equiv \tau(\xi) \pmod{\mathfrak{p}}.$$

Now let σ be the Frobenius automorphism associated with the prime ideal $\mathfrak{p}$. By Proposition 2.3, the number of conjugates $\tau(\xi)$ satisfying the preceding congruence is equal to the number of conjugates satisfying the condition $\sigma\tau(\xi) = \tau(\xi)$. This statement amounts to saying that $\tau^{-1}\sigma\tau(\xi) = \xi$, and hence that $\tau^{-1}\sigma\tau \in \mathcal{G}$, remembering that $\mathcal{G}$ is the stabilizer of ξ. Now let C denote the conjugacy class in $\mathcal{H}$ containing σ; then C depends only on the rational prime p, and not on the choice of a prime ideal containing p, and will be referred to as the conjugacy class associated with p. From these remarks, we have:

PROPOSITION 2.4. *The number ν_p of solutions of the congruence $\varphi(X) \equiv 0$ (mod p), for an unramified prime p, is equal to hb/a, where $h = |\mathcal{H}|, a = |C|$, $b = |C \cap \mathcal{G}|$, and C is the conjugacy class in $\mathcal{H}$ associated with p.*

At this point, Frobenius brought Kronecker's Theorem 2.2 into the picture. For the polynomial $\varphi(X) = \psi(X)^g$, the number of irreducible factors is g, and the theorem can be stated in the form:

THEOREM 2.4. *Let $\{C_\lambda\}$ denote the conjugacy classes in $\mathcal{H}$, and let $\{p_\lambda\}$ denote the rational primes associated with the class C_λ. Let $a_\lambda = |C_\lambda|$, $b_\lambda = |C_\lambda \cap \mathcal{G}|$, and $h = |\mathcal{H}|$. Then*

$$\sum_\lambda \frac{h}{a_\lambda} b_\lambda \left(\sum p_\lambda^{-1-w} \right) = g \log\left(\frac{1}{w}\right) + P(w),$$

where $P(w)$ is a power series in positive integral powers of w which converges for small values of w.

Frobenius's density theorem, which he proved at the end of the paper, gives the Dirichlet density (see Chapter I, §7 for the definition) of the set of rational primes associated in the way described above to conjugacy classes belonging to what he called a *division* [Abteilung] of the Galois group $\mathcal{H}$. A division is the union of conjugacy classes of $\mathcal{H}$ containing a given element σ and its powers σ^r, for integers r relatively prime to the order of σ. His theorem can be stated as follows.

THEOREM 2.5. *Let σ be an element of $\mathcal{H}$, with t elements in its division T. Then the set of rational primes associated with conjugacy classes belonging to T has Dirichlet density t/h, where h is the order of $\mathcal{H}$.*

For a proof of a more general version of the theorem, we refer to Janusz (loc. cit. Chapter 5, §4). As an immediate application, the theorem implies that there exist infinitely many rational primes associated with a given division of the Galois group $\mathcal{H}$, because the Dirichlet density of a finite set of primes is zero. For other applications, and the role of the theorem in what Janusz calls the "direct approach" to class-field theory, see Janusz (loc. cit. Chapter V).

Frobenius stated as a conjecture a sharper version of his theorem, which would give the Dirichlet density of the set of rational primes associated with a single conjugacy class. This result was proved about 25 years later by Tchebotarev [271] (see also Janusz loc. cit. Theorem 10.4), and asserts that the Dirichlet density of the set of rational primes associated with a single conjugacy class C in the Galois group $\mathcal{H}$ is equal to $|C|/|\mathcal{H}|$.

3. Characters of Finite Groups

The fourth paper from 1896, *Über Gruppencharaktere* [131], lies at the heart of the modern theory of representations of finite groups. Never mind that Frobenius started with characters, and did not define representations until the following year [133]![12] The applications of characters of finite abelian groups to algebra and

---

[12]Frobenius's approach to the subject of representation theory, and in particular his definition of characters, given in §3 below, illustrate the fact that the origins of parts of representation theory bear little resemblance to the way we think of them today.

number theory mentioned in the introduction to his paper (see the opening remarks in Chapter I) provided ample incentive for him to create a theory of characters for general finite groups. But how should they be defined, and what properties should they have?

The following passage from a letter he received from Dedekind (dated 25 March 1896) was to lead, eventually, to Frobenius's definition of characters:

> ... Since I was once speaking about groups, I might yet mention another consideration to which I came in February 1886. To each group G of order n, I define a homogeneous polynomial H of degree n, in n variables, which I call the determinant of G: if $1, 2, \ldots, n$ are the elements of G, written in any order, then I let x_r be a variable corresponding to the element r of the group G, and form the determinant
>
> $$H = \begin{vmatrix} x_{11'} & x_{21'} & \ldots x_{n1'} \\ x_{12'} & x_{22'} & \ldots x_{n2'} \\ \ldots & & \\ x_{1n'} & x_{2n'} & \ldots x_{nn'} \end{vmatrix}$$
>
> where r' denotes the element inverse to r. If G is an abelian group, and $\psi', \psi'', \ldots$ are the characters corresponding to it (roots of unity), then the determinant H is decomposable, namely as the product of n linear factors
>
> $$\sum_r \psi^{(s)}(r)x_r = \psi^{(s)}(1)x_1 + \cdots + \psi^{(s)}(n)x_n,$$
>
> that correspond to the n values of s (a theorem, which in this generality, as I believe, has not been announced as yet). But if G is not an abelian group, then its determinant H possesses, as far as I have checked, besides linear factors (such as, for example, always $x_1 + \cdots + x_n$) also factors of higher degree, that are irreducible in the ordinary sense; but these will be further decomposable into linear factors, if one allows besides the ordinary numbers as coefficients also hypercomplex numbers [übercomplexe Zahlen] (with noncommutative multiplication), that correspond to the laws of the group G.[13]
>
> ...

There began a lively exchange of letters between Dedekind and Frobenius.[14] At the early stages, Frobenius pressed for more information, while Dedekind supplied some examples he had worked out. Dedekind also shared, in that letter dated 8 July 1896, the account of how he was led to the group determinant in the first place:

> I was led first to the concept of the general group determinant through the study of the discriminant of an arbitrary normal field Ω, in which I considered (very useful) bases of Ω, that consisted of conjugates of a single number ω (sometimes the system $\mathfrak{o}$ of all algebraic integers in Ω has such a basis, for example, if ω is an m-th root of unity, and m is divisible by no square, and the same holds for all subfields of Ω, for example all quadratic fields with odd fundamental number); this

---

[13]Dedekind, R., *Gesammelte Mathematische Werke*, Friedr. Vieweg & Sohn, Braunschweig, 1931. Used with permission.

[14]For a full account of this correspondence, see Hawkins [**172**], [**174**].

investigation probably took place around 1880 or still earlier, and at that time I may well have found Theorem A [stated in the previous quotation]; what induced me to return to the group-determinant in February 1886, I don't know.[15]

The kind of basis Dedekind referred to is today called a *normal basis* (see Roggenkamp [**241**] for some historical remarks about the form of the group determinant and the concept of a normal basis).

Dedekind had computed the group determinants in 1886 for two nonabelian groups, the symmetric group of order 6, and for the quaternion group of order 8, and he reproduced his calculations in a letter to Frobenius dated 6 April 1896. Here are the results he obtained for the symmetric group. Let the elements of the group of all permutations of $\{a, b, c\}$ be denoted by $1, 2, \ldots, 6$ as above, with 1 the identity element, 2 the 3-cycle (abc), $3 = 2^{-1}$, $4 = (bc)$, $5 = (ac)$, and $6 = (ab)$. Let $x_1, \ldots, x_6$ be indeterminates corresponding to the elements of the group. He let ρ be a solution of the equation $1 + \rho + \rho^2 = 0$, and put

$$\begin{aligned}
u &= x_1 + x_2 + x_3 & v &= x_4 + x_5 + x_6 \\
u_1 &= x_1 + \rho x_2 + \rho^2 x_3 & v_1 &= x_4 + \rho x_5 + \rho^2 x_6 \\
u_2 &= x_1 + \rho^2 x_2 + \rho x_3 & v_2 &= x_4 + \rho^2 x_5 + \rho x_6.
\end{aligned}$$

The group determinant factors in the following way:

$$\begin{vmatrix}
x_1 & x_3 & x_2 & x_4 & x_5 & x_6 \\
x_2 & x_1 & x_3 & x_5 & x_6 & x_4 \\
x_3 & x_2 & x_1 & x_6 & x_4 & x_5 \\
x_4 & x_5 & x_6 & x_1 & x_3 & x_2 \\
x_5 & x_6 & x_4 & x_2 & x_1 & x_3 \\
x_6 & x_4 & x_5 & x_3 & x_2 & x_1
\end{vmatrix} = (u+v)(u-v)(u_1 u_2 - v_1 v_2)^2.$$

The news about group determinants was the spark that kindled the flame. In an astonishingly short time, Frobenius found a way to define characters of general finite groups, proved the main theorems about them, applied the new theory to solve the problem of factoring a general group determinant into irreducible factors, and published all this in 3 papers, [**129**], [**131**] and [**132**], later in the same year.

Frobenius based his definition of characters on some results published separately in the first of these papers, *Über vertauschbare Matrizen*. These belonged to what we would call today the representation theory of commutative semisimple associative algebras. The starting point was a paper by Weierstrass [**280**], published in 1885, entitled *Zur Theorie der aus n Haupteinheiten gebildeten complexen Grössen*, and a sequel to it, with the same title, by Dedekind [**103**]. Weierstrass, beginning with his lectures in 1861, had begun to study an operation of multiplication defined on pairs of elements in a vector space E over $\mathbb{C}$, with a basis $\{e_1, \ldots, e_n\}$ (the *Haupteinheiten*). Multiplication of the basis elements was defined by the formulas:

$$e_j e_k = \sum_{i=1}^{n} \eta_{ijk} e_i,$$

---

[15]Dedekind, R., *Gesammelte Mathematische Werke*, Friedr. Vieweg & Sohn, Braunschweig, 1931. Used with permission.

for $1 \leq j, k \leq n$, and some set of constants (called *structure constants*) η_{ijk}, and was assumed to satisfy the conditions

$$e_j e_k = e_k e_j \quad \text{and} \quad (e_j e_k) e_\ell = e_j(e_k e_\ell),$$

for all j, k, ℓ. The multiplication of basis elements extends to a bilinear multiplication of *hypercomplex numbers* $\sum \xi_i e_i$ [überkomplexe Zahlen] belonging to the vector space E, satisfying the commutative and associative laws. The vector space E, with such a multiplication, was viewed as a generalization of the system of complex numbers with "Haupteinheiten" $\{1, i\}$, and would be called today a commutative, associative algebra.

According to Frobenius ([**129**], Ges. Abh. II, p. 705), the philosophical base of their work was the following result (see, for example, formula (90) in Dedekind's article [**103**]).

PROPOSITION 3.1. *Let* $\{a_{ijk}, i, j = 1, \dots, n; k = 1, \dots, m\}$ *be any set of* mn^2 *complex numbers, satisfying the conditions*

$$\sum_j a_{ijk} a_{jpq} = \sum_j a_{ijq} a_{jpk},$$

and set

$$a_{ij} = \sum_k a_{ijk} x_k.$$

Then the n-th order determinant $|a_{ij}|$ *is the product of* n *linear functions of the independent variables* $\{x_1, \dots, x_m\}$.

We shall first give a proof of the result, using some standard facts from linear algebra, and will then show how it applies to the commutative algebras E considered by Weierstrass and Dedekind. This is exactly the procedure Frobenius used in [**129**], but at that time the linear algebra was not as standard as it is now.

For each $k = 1, \dots, m$, let A_k denote the $n \times n$-matrix (a_{ijk}). The conditions in the statement of the theorem imply that the matrices A_k commute:

$$A_k A_\ell = A_\ell A_k,$$

for all k and ℓ. Then

$$(a_{ij}) = A_1 x_1 + \cdots + A_m x_m.$$

The commuting matrices $\{A_q\}$ can be simultaneously triangularized. This means that their eigenvalues can be ordered so that if $\rho_{q1}, \dots, \rho_{qn}$ are the eigenvalues of A_q for $q = 1, \dots, m$, then the eigenvalues of $(a_{ij}) = A_1 x_1 + \cdots + A_m x_m$ are $\rho_{1r} x_1 + \cdots + \rho_{mr} x_m$, for $r = 1, \dots, n$. One then obtains

$$|a_{ij}| = \prod_{r=1}^n (\rho_{1r} x_1 + \cdots + \rho_{mr} x_m),$$

which is simply the statement that the determinant of a matrix is the product of its eigenvalues. This completes the proof.

He then turned his attention to a commutative associative algebra E with basis $e_1, \dots, e_n$ and structure equations

$$e_j e_k = e_k e_j = \sum_{i=1}^n \eta_{ijk} e_i,$$

as above. As in the proof of the preceding theorem, let A_k be the matrix (η_{ijk}), for $k = 1, \ldots, n$. Then A_k is the matrix of the operation of left multiplication by e_k in the algebra E. By the associative law, it follows that the product of the basis elements e_k and e_ℓ is assigned to the product $A_k A_\ell$ of the corresponding matrices. In modern terminology, the map $e_k \to A_k$, extended by linearity, defines what is called today a homomorphism of algebras[16] from the algebra E to the algebra of $n \times n$ matrices, called the *regular representation* of the algebra E. From the structure equations for the algebra E, it follows that the matrices $\{A_k\}$ satisfy the identities:

$$A_j A_k = A_k A_j = \sum_{i=1}^n \eta_{ijk} A_i,$$

for all j, k. The commuting matrices $\{A_k\}$ can be simultaneously triangularized, and their eigenvalues ordered, as in the proof of the theorem given above. From the preceding identities, one has

$$\rho_{\ell j} \rho_{\ell k} = \sum_{i=l}^n \eta_{ijk} \rho_{\ell i},$$

where $\rho_{\ell 1}, \ldots, \rho_{\ell n}$ are the eigenvalues of A_ℓ for $\ell = 1, \ldots, n$.

The equations for the eigenvalues imply that the map

$$e_j \to \rho_{qj},$$

for fixed $q, 1 \leq q \leq n$, extends to a homomorphism of algebras from $E \to \mathbb{C}$, which we would call today a *one dimensional representation* of the commutative algebra E. An important question, whose answer was crucial to Frobenius's definition of characters, was under what circumstances are the representations $e_j \to \rho_{qj}$, for $q = 1, \ldots, n$, distinct, and linearly independent, in the sense that the n-tuples $(\rho_{q1}, \ldots, \rho_{qn})$, for $1 \leq q \leq n$, are linearly independent in $\mathbb{C}^n$? His answer was contained in the following theorem ([**129**], Ges. Abh. II, p. 717).

THEOREM 3.1. *Assume the n^3 complex numbers a_{ijk} satisfy the equations*

$$a_{ijk} = a_{ikj} \text{ and } \sum_j a_{ijk} a_{jpq} = \sum_j a_{ijq} a_{jpk},$$

and that the n-th order determinant formed from the elements

$$c_{k\ell} = \sum_{i,j} a_{ijk} a_{ji\ell}$$

is different from zero. Then the equations

$$r_j r_k = \sum_i a_{ijk} r_i$$

have exactly n different solutions $r_{q1}, \ldots, r_{qn}$, for $q = 1, \ldots, n$, and the n-th order determinant formed from these solutions is different from zero.

The theorem and Frobenius's proof of it are the first of several illustrations in this chapter, and in Chapter VII, §2, of Frobenius's grasp of subtle questions about the structure of algebras, and ways to approach them using linear algebra. In what follows, we present a modern version of the steps in Frobenius's proof of the theorem. The assumptions concerning the constants a_{ijk} imply that the matrices

---

[16]A homomorphism of algebras $f : A \to B$ is a linear map preserving the product operation.

$A_k = (a_{ijk}), k = 1, \ldots, n$ form a commutative set, as in the proof of Proposition 3.1, and that the vector spaces E with basis elements $e_1, \ldots, e_n$, and multiplication defined by

$$e_j e_k = \sum_{i=1}^{n} a_{ijk} e_i,$$

is a commutative, associative algebra.[17] We now consider the meaning of the assumption that the determinant $|c_{k\ell}| \neq 0$. In [**280**], Weierstrass proved that the condition implies that the algebra E contains no nonzero nilpotent elements, and with this as a starting point, Frobenius completed the proof.

Here is a modern proof of Weierstrass's result. A straightforward computation shows that $c_{k\ell} = \text{Trace}\ (A_k A_\ell)$ for all k and ℓ. By the remarks preceding the statement of the theorem, $e_k \to A_k$ is the regular representation of the algebra. The determinant $|\text{Trace}\ (A_k A_\ell)|$ is called today the *discriminant* of the algebra E, and its nonvanishing is a criterion for the algebra to be semisimple (containing no nonzero nilpotent ideals). To prove this, observe that the nonvanishing of the discriminant implies that the bilinear form (e, e') on E, whose matrix is $(e_k, e_\ell) = \text{Trace}\ (A_k A_\ell)$, is nondegenerate. If a nonzero element $e \in E$ generates a nilpotent ideal, then ee' is nilpotent for all elements $e' \in E$. Then the eigenvalues of ee' in the regular representation are all zero, and it follows that $(e, e') = 0$ for all $e' \in E$, contradicting the nondegeneracy of the form.

Frobenius finished the proof as follows. If the determinant $|r_{ij}|$ formed from the different solutions of the structure equations is zero, then there exists a nonzero element $e = x_1 e_1 + \cdots + x_n e_n \in E$ such that all the eigenvalues of e in the regular representation are zero, by the discussion preceding the statement of the theorem. But this means that the element e is nilpotent in the regular representation of E, and contradicts the result of Weierstrass proved above. This completes the proof of the theorem.

At this point, Frobenius was apparently unaware of the work of Theodor Molien (1861-1941) on hypercomplex numbers, which had been published earlier, and contained, among other things, the discriminant criterion for semisimplicity of noncommutative (as well as commutative) algebras.

Molien was born in Riga, Latvia, and educated at the University of Yurev (or Dorpat) in Estonia. After spending a short time in Leipzig, during the period when Klein was a professor there, he became a Dozent at Dorpat, remaining there until 1900 when he was appointed to a professorship at the technical institute in Tomsk, Siberia.[18] In his early research, published in 1893, he obtained the main structure theorems for semisimple associative algebras over the field of complex numbers, and in 1897 applied them to group algebras over $\mathbb{C}$. In this way he discovered much of the basic theory of representations of finite groups, independently of Frobenius.[19] After

---

[17]The commutative law follows from the assumption that $a_{ijk} = a_{ikj}$ for all i, j, k, while the associative law follows from the fact that the left multiplication by the basis element e_k commutes with the right multiplication by e_ℓ, for all k and ℓ. This, in turn, follows from the fact that $A_k A_\ell = A_\ell A_k$ and the commutative law.

[18]Frobenius learned about Molien's work in 1897, and shortly afterwards sent an apparently unsuccessful letter to Dedekind, quoted in §4, in which Frobenius expressed the hope that Dedekind might be able to help Molien find a better position.

[19]See Bourbaki ([**16**], Note historique) and Hawkins [**173**] for further discussion, with references to the literature, concerning Molien's work, and a parallel investigation of the structure of finite dimensional associative algebras by Élie Cartan ([**90**], 1898).

becoming familiar with Frobenius's work on character theory, Molien discovered an important application of character theory to polynomial invariants of finite groups, which was communicated by Frobenius to the *Berliner Sitzungsberichte* in 1898 (see Chapter III, §5).

The paper *Über Gruppencharaktere* began with an investigation of the multiplicative properties of the conjugacy classes $C_1, \ldots, C_n$ in an arbitrary finite group G of order g, leading to the construction of a commutative algebra whose structure constants satisfy the hypotheses of Theorem 3.1.

For each conjugacy class C_i, let $h_i = |C_i|$. As Frobenius had shown earlier in connection with the use of the class equation in the proof of Sylow's Theorem 2.1, h_i is a factor of g, and g/h_i is the order of the subgroup of G consisting of all elements which commute with x, for any fixed element $x \in C_i$. The set C_i^{-1} consisting of the inverses of the elements in C_i is one of the conjugacy classes, which he denoted by $C_{i'}$, for $i = 1, \ldots, n$. For each triple $\{i, j, k\}$ he let h_{ijk} be the number of solutions of the equation

$$abc = 1 \text{ for } a \in C_i, b \in C_j, c \in C_k,$$

and put

$$a_{ijk} = \frac{1}{h_i} h_{i'jk}.$$

From some elementary properties of the constants h_{ijk}, it follows that one has, as in the hypothesis of Theorem 3.1,

$$a_{ijk} = a_{ikj} \text{ and } \sum_j a_{ijk} a_{jpq} = \sum_j a_{ijq} a_{jpk}.$$

He proved that $|c_{k\ell}| \neq 0$, where

$$c_{k\ell} = \sum_{i,j} a_{ijk} a_{ji\ell},$$

by a further, and intricate, analysis of the constants h_{ijk}. Thus the discriminant of the commutative algebra E with structure constants a_{ijk} is different from zero, and Theorem 3.1 can be applied.[20]

From Theorem 3.1, and the discussion preceding it, the following result follows immediately.

PROPOSITION 3.2. *Let E be the algebra with basis $e_1, \ldots, e_n$, and multiplication defined by*

$$e_j e_k = \sum_i a_{ijk} e_i,$$

for the structure constants $\{a_{ijk}\}$ defined above. There exist n different solutions $(r_{q1}, \ldots, r_{qn})$ of the equations

$$r_{qj} r_{qk} = \sum_i a_{ijk} r_{qk},$$

such that the determinant $|r_{qi}| \neq 0$. Moreover, upon setting

$$a_{ij} = \sum_k a_{ijk} x_k \ i, j = 1, \ldots, n,$$

---

[20]A few years later, he observed that the algebra E is isomorphic to the center of the group algebra of the finite group G over the field of complex numbers.

for some set of indeterminates x_j, *and* $\delta_{ij} = 1$ *if* $i = j$ *and* 0 *otherwise (the Kronecker delta), one has*

$$|a_{ij} - r\delta_{ij}| = \prod_{q=1}^{n}(r_{q1}x_1 + \cdots + r_{qn}x_n - r),$$

for all complex numbers r.

The step from the preceding result to the factorization of the group determinant involved the definition of characters in the general case, and was not at all straightforward. For one thing, the solutions $(r_1, \ldots, r_n)$ of the structure equations, for nonabelian groups, may fail to satisfy the orthogonality relations (see Chapter I, Theorem 1.1). Frobenius described his struggle with the problem in letters he wrote to Dedekind in March and April, 1896 (see Hawkins [**174**]), and soon arrived at the definition of characters given below (from §§2-5 of [**131**]).

To each solution of the structure equations $(r_1, \ldots, r_n)$, he defined a set of complex numbers $\chi = (\chi_1, \ldots, \chi_n)$ by the formulas

$$r_i = \frac{h_i \chi_i}{f},$$

for $i = 1, \ldots, n$, with a factor f (independent of i) to be determined later. By Proposition 3.2, it follows that the equations

(7) $$h_j h_k \chi_j \chi_k = f \sum_i h_{i'jk} \chi_i$$

have exactly n different sets of solutions

$$\chi_i = \chi_i^{(q)}, \ f = f^{(q)}, \ q = 1, \ldots, n,$$

such that the determinant $|\chi_i^{(q)}| \neq 0$. For suitably chosen factors $f^{(q)}$, he called the resulting systems of n numbers $\chi^{(q)} = (\chi_1^{(q)}, \ldots, \chi_n^{(q)})$, for $q = 1, \ldots, n$, the *characters* of the finite group G. Thus a character χ of G can be viewed as a complex valued function on G, constant on the conjugacy classes $\{C_i\}$, in other words a *class function* on G, whose value on an element of the class C_i is χ_i. The number of different characters $\chi^{(q)}$ is equal to the number of conjugacy classes.

There was still some fine tuning to do. He first calculated the factor f associated with a character χ. Let C_1 be the conjugacy class containing the identity element of G, so that $h_1 = 1$ and χ_1 is the value of χ at the identity element. The factorization of the determinant in the proposition above can be restated in the following way:

$$\left| \sum_k h_{i'jk} x_k - h_{i'j1} r \right| = \prod_q h_q(\xi_q - r),$$

for all complex numbers r, where

$$f^{(q)} \xi_q = h_1 \chi_1^{(q)} x_1 + \cdots + h_n \chi_n^{(q)} x_n,$$

for $q = 1, \ldots, n$. Now let $x_1 = 1$ and $x_j = 0$ for $j \neq 1$. One obtains

$$(1 - r)^n = \prod_q \left(\frac{\chi_1^{(q)}}{f^{(q)}} - r \right)$$

for all complex numbers r, and hence

$$f^{(q)} = \chi_1^{(q)},$$

for $q = 1, \ldots, n$.

The next step was to find the generalizations of the orthogonality relations for finite abelian groups (Chapter I, Theorem 1.1). This was done as follows. By Proposition 3.2, the solutions r_{qj} of the structure equations satisfy $|r_{qj}| \neq 0$, so the matrix (r_{qj}) is invertible. This means that for each solution $(r_{q1}, \ldots, r_{qn})$ of the structure equations, there exists a *complementary system* $(s_{q1}, \ldots, s_{qn})$, with entries from the inverse matrix $(r_{qj})^{-1}$, satisfying the equations

$$(8) \qquad \sum_q r_{qk} s_{q\ell} = \delta_{k\ell} \text{ and } \sum_k r_{pk} s_{qk} = \delta_{pq},$$

where the δ's are Kronecker deltas. Using these formulas, one can solve the equations

$$r_{qj} r_{qk} = \sum_i a_{ijk} r_{qi}$$

for the structure constants a_{ijk}. The result is that

$$a_{ijk} = \sum_q s_{qi} r_{qj} r_{qk},$$

and hence the identities

$$s_{qi} r_{qk} = \sum_j a_{ijk} s_{qj}$$

hold. These identities determine the complementary systems uniquely up to scalar multiples. Letting $(s_1, \ldots, s_n)$ be the complementary system to $\left(\frac{h_1 \chi_1}{f}, \ldots, \frac{h_n \chi_n}{f} \right)$, for a character χ, the identities become

$$\sum_j \frac{h_{i'jk}}{h_i} s_j = \frac{h_k \chi_k}{f} s_i.$$

It follows that the systems

$$\left(\frac{h_i \chi_i^{(q)}}{f^{(q)}} \right), \text{ and } \left(\frac{e^{(q)} \chi_{i'}^{(q)}}{g} \right),$$

with $g = |G|$, and another proportionality factor $e^{(q)}$, are complementary. By the equations (8), and some simplifications, the following theorem is obtained ([**131**], Ges. Abh. III, p. 9).

THEOREM 3.2 (Orthogonality Relations). *Let χ be a character of a finite group G of order g, with proportionality factors e and f. Then*

$$\sum_i h_i \chi_i \chi_{i'} = \frac{gf}{e}.$$

If ψ is another character, different from χ, then

$$\sum_i h_i \chi_i \psi_{i'} = 0.$$

The orthogonality relations can be viewed as asserting that the product of two matrices is the identity matrix. Inverting the order of multiplication gave another useful set of formulas, the *Second Orthogonality Relations* ([**131**], Ges. Abh. III, p. 9):

$$(9) \qquad \sum_q \frac{e^{(q)}}{f^{(q)}} \chi_i^{(q)} \chi_j^{(q)} = \frac{g h_{ij1}}{h_i h_j}.$$

The $n \times n$ matrix $(\chi_j^{(i)})$, whose rows and columns are indexed by the conjugacy classes, and the characters, respectively, is today called the *character table* of the finite group G. The character table, whose rows and columns are related by the orthogonality relations, contains a wealth of information about the structure of the group G, as Frobenius demonstrated immediately following his proof of the orthogonality relations.

His first result in this direction ([**131**], Ges. Abh. III, p. 12) gave formulas for the integers h_{ijk}, which express the multiplicative relations satisfied by the conjugacy classes, in terms of the entries of the character table. Starting from equation (7),

$$h_j h_k \chi_j^{(q)} \chi_k^{(q)} = f^{(q)} \sum_\ell h_{\ell j k} \chi_{\ell'}^{(q)},$$

multiply both sides by $h_i \chi_i^{(q)}$, sum on q, and apply the second orthogonality relations (9). The result is:

$$(10) \qquad \frac{h_{ijk}}{h_i h_j h_k} = \sum_q \frac{e^{(q)}}{(f^{(q)})^2} \chi_i^{(q)} \chi_j^{(q)} \chi_k^{(q)}.$$

Thus the numbers of solutions of the equations $abc = 1$ with a, b, c belonging to given conjugacy classes can be computed in terms of the character values[21] $\{\chi_i^{(q)}\}$. Another remarkable formula ([**131**], §4), gave the number of ways a given element $s \in G$ can be expressed as a commutator:

$$s = aba^{-1}b^{-1},$$

in terms of the entries of the character table.

Frobenius solved the problem of factoring the group determinant in two stages. In the first, published in ([**131**], §5 and continued in §6), he factored the determinant $\Theta = |x_{ab^{-1}}|$ under the assumption that the indeterminates x_a corresponding to the elements $a \in G$ satisfy the condition $x_{ab} = x_{ba}$ for all a and b in G. In this case, he was able to prove that Θ is a product of n linear factors, each occurring with a certain multiplicity. The coefficients of each linear factor are given in terms of the values of one of the characters $\chi^{(q)}$, and that factor appears in Θ with multiplicity $e^{(q)} f^{(q)}$. We shall examine his reasoning in detail, as it throws more light on his definition of the characters of G as solutions of the equations (7), and was an important step along the way towards the factorization of the group determinant in the general case [**132**]. It is also without question one of the high points in the 1896 series of papers.

We begin by setting up some notation. As before, G denotes a finite group of order g, with conjugacy classes $C_1, \dots, C_n$, and C_1 the class containing the identity element 1. Introduce indeterminates x_a indexed by the elements $a \in G$, subject to the symmetry condition that $x_{ab} = x_{ba}$, for all elements $a, b \in G$. Because

---

[21]For computational purposes, it would have been a nuisance not to know the values of the constants $e^{(q)}$. This was a stumbling block (see Hawkins [**174**], §5), and it was only later, in the last 1896 paper [**132**], that he was able to prove that $e^{(q)}$ always coincides with $f^{(q)}$.

$ab = b^{-1}bab$, the symmetry condition on the indeterminates means that $x_c = x_d$ whenever the elements c and d are conjugate[22] in G. The object is to factor the determinant $\Theta = |x_{ab^{-1}}|$ of the matrix $(x_{ab^{-1}})$ whose rows and columns are indexed by the elements a and b respectively, for some fixed ordering of the elements of G. Evidently, Θ is a polynomial with integer coefficients in n independent indeterminates $x_1, \ldots, x_n$, and $\Theta = \Theta(x_1, \ldots, x_n)$, where x_i denotes any indeterminate x_a, with $a \in C_i$.

Frobenius's starting point was to consider two sets of indeterminates x_a and y_a, both indexed by the elements of G and satisfying the symmetry conditions, so that $x_{ab} = x_{ba}$ and $y_{ab} = y_{ba}$. He then defined a new set of polynomials z_a by setting

$$z_{ab^{-1}} = \sum_c x_{ac^{-1}} y_{cb^{-1}},$$

and observed (as it is a nice exercise to show) that the symmetry conditions imply:

$$z_{ab} = \sum_c x_{ac^{-1}} y_{cb} = \sum_d y_{ad^{-1}} x_{db} = \sum_d x_{bd} y_{d^{-1}a} = z_{ba},$$

for all a, b.

From these relations, it follows that for any two sets of complex numbers u_a and v_a indexed by the elements of G and satisfying the symmetry conditions, the matrices $(u_{ab^{-1}})$ and $(v_{ab^{-1}})$ commute. In particular, if one sets $x_a = 1$ for a belonging to a fixed conjugacy class C and $x_b = 0$ for $b \notin C$, one obtains n matrices $X_1, \ldots, X_n$ which commute, $X_i X_j = X_j X_i$, and satisfy

$$(x_{ab^{-1}}) = x_1 X_1 + \cdots + x_n X_n.$$

Thus he arrived at the situation considered in the proof of Proposition 3.1, and was able to conclude, by that result, that

$$\Theta(x) = \Theta(x_1, \ldots, x_n) = |x_{ab^{-1}}|$$

is a product of linear functions in the n variables $x_1, \ldots, x_n$.

In the determinant $|x_{ab^{-1}}|$, the n elements on the diagonal, and only these, are equal to x_1, so x_1^n appears with coefficient 1 in the polynomial $\Theta(x)$. He wrote the linear factors ξ of $\Theta(x)$ in the form

$$\xi = \frac{1}{f} \sum_i h_i \chi_i x_i,$$

with $h_i = |C_i|$ as usual, the χ_i viewed for the time being as coefficients to be determined, and $f = \chi_1$ so that the coefficient of x_1 is 1.

The polynomials z_a also satisfy the symmetry conditions, and

$$|z_{ab^{-1}}| = |x_{ab^{-1}}||y_{ab^{-1}}|,$$

by what has been shown. It follows that each linear factor of $\Theta(z)$ is a product of a linear function of $x_1, \ldots, x_n$ and a linear function of $y_1, \ldots, y_n$, so

$$\frac{1}{f} \sum_i h_i \chi_i z_i = \left(\sum_j a_j x_j \right) \left(\sum_k b_k y_k \right),$$

---

[22] At this point Frobenius had already observed that the characters χ of G, viewed as complex valued functions on G, satisfy the conditions $\chi(aba^{-1}) = \chi(b)$, and $\chi(ab) = \chi(ba)$, for all elements $a, b \in G$, so it was reasonable to experiment with the same conditions for the variables in the group determinant, assuming there was a connection between the factorization of the group determinant and the characters.

with $a_1 = b_1 = 1$. Now set $y_1 = 1$ and $y_k = 0$ if $k \neq 1$. Then $z_i = x_i$ for all i and

$$\frac{1}{f} \sum_i h_i \chi_i x_i = \sum_j a_j x_j,$$

and similarly for the factor involving $\sum b_k y_k$. Therefore

(11) $$f \sum_i h_i \chi_i z_i = \left(\sum_j h_j \chi_j x_j \right)\left(\sum_k h_k \chi_k y_k \right).$$

On the other hand, one has

$$z_c = \sum_{ab=c} x_a y_b$$

by the definition of the polynomials z_a. From this and the definition of the constants h_{ijk}, it is easily verified that

$$h_i z_i = \sum_{j,k} h_{jki'} x_j y_k.$$

Upon substituting this information in (11) and comparing coefficients of $x_j y_k$, it follows that the coefficients χ_i of a linear factor ξ of $\Theta(x)$ satisfy the equations

$$h_j h_k \chi_j \chi_k = f \sum_i h_{jki'} \chi_i.$$

But $h_{jki'} = h_{i'jk}$, so these equations are the same as the defining equations (7) of the characters, and it has been shown that the coefficients $\chi_1, \ldots, \chi_n$ of each linear factor ξ of $\Theta(x)$ are indeed the values of a character. It is not difficult to prove that for every character χ, the corresponding expression ξ occurs as a factor of $\Theta(x)$. This, and a further argument to determine the multiplicity with which a given linear factor ξ occurs in the factorization of $\Theta(x)$, complete the proof of the following theorem—certainly the main result of the paper [**131**]—and clearly a generalization of Dedekind's factorization of the group determinant for finite abelian groups given at the beginning of this section.

THEOREM 3.3. *Let G be a finite group, with conjugacy classes $C_1, \ldots, C_n$, and let $h_i = |C_i|$ for $i = 1, \ldots, n$. Let Θ be the determinant $|x_{ab^{-1}}|$, for a set of indeterminates x_a indexed by the elements of G, and satisfying the condition $x_{ab} = x_{ba}$ for all a, b. Then Θ is a polynomial with integer coefficients in the indeterminates $x_1, \ldots, x_n$, where $x_i = x_a$, for $a \in C_i$. The polynomial Θ is a product of linear factors; more precisely,*

$$\Theta(x_1, \ldots, x_n) = \prod_\chi \frac{1}{f_\chi}(h_1 \chi_1 x_1 + \cdots + h_n \chi_n x_n)^{e_\chi f_\chi},$$

where the product is taken over the set of characters χ of G, and e_χ, f_χ are the constants associated with χ, for each character χ.

Frobenius ended the paper *Über Gruppencharaktere* with the calculation of character tables of finite groups in a number of interesting cases, including the symmetry groups of the tetrahedron, the octahedron, and the icosahedron, and the infinite family of groups $\{PSL_2(p)\}$, for odd primes p, noting that the tetrahedral and icosahedral groups are members of the family in case $p = 3$ and $p = 5$, respectively. The group $PSL_2(p)$, of order $\frac{1}{2}p(p^2 - 1)$ consists of the linear fractional transformations

$$y \equiv \frac{\gamma + \delta x}{\alpha + \beta x} \quad (\text{mod } p),$$

whose determinant satisfies

$$\alpha\delta - \beta\gamma \equiv 1 \quad (\text{mod } p).$$

Its conjugacy classes had been determined by Gierster [**159**], and Frobenius was able to apply this information to solve the equations (7) for the characters in an elegant way. His work on the characters of the finite groups $PSL_2(p)$ remains of interest today, as it inaugurated the study of the character theory of finite groups of Lie type, a part of representation theory which became a focus of research in representation theory a century later (see Carter [**92**]).

The last paper [**132**] in the 1896 series was devoted to the factorization of the group determinant in the general case. The main result can be stated as follows ([**132**], Ges. Abh. III, p. 39).

THEOREM 3.4. *Let G be a finite group of order g, and let x_a be a set of independent indeterminates indexed by the elements $a \in G$. Then the group determinant $\Theta = |x_{ab^{-1}}|$ is a polynomial with integer coefficients in the indeterminates x_a, and factors over $\mathbb{C}$ as a product of irreducible polynomials Φ, each occurring with some multiplicity f:*

$$\Theta = \prod (\Phi)^f.$$

The number of irreducible factors Φ is the number of conjugacy classes in G. The degree of each factor Φ is equal to the multiplicity with which it occurs in the factorization of Θ:

$$deg\,(\Phi) = f.$$

If one sets $x_a = x_b$ whenever a and b are conjugate in G, then each factor Φ of the specialized polynomial Θ can be expressed in the form

$$\Phi = (\xi)^f,$$

where ξ is a linear function with coefficients given in terms of one of the characters χ of G, as in Theorem 3.3.

In the introduction to [**132**] Frobenius mentioned Dedekind's result on the factorization of the group determinant for abelian groups, and that he had himself proved an analogous theorem on the factorization of a determinant of order $r = 2^\rho$ as a product of polynomial factors in $2r$ variables, in connection with his investigation [**125**] of addition theorems satisfied by theta functions in several variables.

He singled out the result that the degree of each factor Φ is equal to its exponent f in the factorization of Θ as the *fundamental theorem in the theory of the group determinant*. It implies that for each character χ, one has

$$e_\chi = f_\chi = \chi(1),$$

answering a nagging question he had been unable to settle in the previous paper [**131**]. He called the positive integer $f_\chi = \chi(1)$ the *degree* [Grad] of the character χ. Another important consequence of the theorem is that the order of G is equal to the sum of the squares of the degrees of the characters:

(12) $$|G| = \sum_\chi f_\chi^2.$$

Frobenius's proof of the theorem above was, and still is, difficult to understand; moreover, he used the result and some parts of the proof as a basis for much of his subsequent work in representation theory. His successors Issai Schur, William Burnside, and Emmy Noether, found new ways to develop representation theory which were independent of the preceding theorem and incidentally gave new proofs of it (see Chapter IV, §2, for example), with the result that the group determinant is no longer an object of primary importance in representation theory. Nevertheless, interest in it has not faded completely. It was known from the beginnings of character theory that nonisomorphic finite groups may have the same character table, the first example occurring for the two nonisomorphic nonabelian groups of order 8. It was proved recently (by Formanek and Sibley [**123**]), however, that the group determinant is an invariant of finite groups: two groups having the same group determinant are isomorphic (for this and related developments, see also Roggenkamp [**241**]).

4. Group Representations and Characters

This was not a time to rest! Frobenius realized that the 1896 papers had opened the way to a new part of mathematics, and that his results on the factorization of the group determinant provided him with a powerful method for exploring the new territory. During the years $1897 - 1899$, he published 4 papers ([**133**], [**134**], [**135**], [**136**]) on the general theory, including two on representations of finite groups, and two more on what are now called induced characters, and tensor products of characters, respectively. These laid the foundations for his definitive papers ([**137**], [**138**]) on the characters of the symmetric and alternating groups (see §5), and his application of character theory to the structure of what are now called Frobenius groups [**139**], published in 1900 and 1901 (see §5 and §6).

Parts of the first four papers will be surveyed in this section, with emphasis on the new concepts and theorems they contain, and some indications of how the results were proved. While the theorems are of fundamental importance today, Frobenius's proofs often contained intricate arguments involving the irreducible factors of the group determinant, and were soon replaced by simpler ones (see, for example, Chapter IV, §2).

In [**133**], Frobenius defined representations as follows. He began by considering a finite set G' of *linear substitutions*

$$(A): \quad x_i = a_{i1}y_1 + \cdots + a_{in}y_n, \quad i = 1, \ldots, n,$$

with nonzero determinants, having the group property, that is, the product substitution [zusammengesetzte Substitution] $(C) = (A)(B)$ of two substitutions $(A), (B) \in G'$ belongs to G'. The coefficients of (C) are obtained by matrix multiplication:

$$c_{ij} = a_{i1}b_{1j} + \cdots + a_{in}b_{nj}$$

from the coefficients of (A) and (B). He attributed to Gauss the idea of omitting consideration of the variables, and simply viewing G' as a finite set of invertible matrices, closed under multiplication. He continued:

> Let G be an abstract group, $a, b, \ldots$ its elements. Assign to the element a the matrix (A), to the element b the matrix (B), etc. so that the group G' is isomorphic to the group G, that is, $(A)(B) = (AB)$ [where (AB) is the matrix assigned to ab]. Then I say that the matrices $(A), (B), \ldots$ *represent* the group G.

He noted that the "isomorphism," which assigns a matrix A to an element $a \in G$, need not be one-to-one, so that we would call it today a homomorphism. The *degree* of the representation $a \to (A)$ is, by definition, the size of the representing matrices (A), so that, for example, a representation by 2×2 matrices has degree 2. The element 1 in G is assigned to the identity matrix I because the identity matrix is the only matrix with nonvanishing determinant satisfying $I^2 = I$. The set of all elements of G assigned to the identity matrix is a normal subgroup H of G, and the group G' is isomorphic to the factor group G/H. If (P) denotes an arbitrary invertible $n \times n$ matrix, then the matrices $(P)^{-1}(A)(P), (P)^{-1}(B)(P), \ldots$ also represent the group G, and the corresponding substitutions are obtained from the original ones by a change of variables defined by the matrix (P). He called two such representations *equivalent*, and the set of all representations equivalent to a given one a *class* of equivalent representations.

The preceding remarks are taken without alteration from Frobenius's paper, and define matrix representations of finite groups, equivalence, etc. as we would do it today. At this point, however, the connection between representations and characters was not clear at all, because of his definition of characters as solutions of the equations (7). Frobenius found the link between representations and characters, namely that the characters were trace functions of certain representations, by investigating matrices with polynomial entries which are associated with representations of G.

He began by assigning $g = |G|$ independent variables $x_r, x_s, \ldots$ to the elements $r, s, \ldots$ of G. For each representation $r \to (R)$, he called the matrix with polynomial entries

$$(R)x_r + (S)x_s + \cdots = \sum (R)x_r,$$

the *matrix corresponding to the representation* $r \to (R)$ of the group G. As he had done in the theory of the group matrix $(x_{rs^{-1}})$ (see §3), he introduced another set of variables y_s for $s \in G$, and put

$$z_t = \sum_{rs=t} x_r y_s.$$

Then one has

$$\left(\sum (R)x_r\right)\left(\sum (S)x_s\right) = \sum (RS)x_r y_s = \sum (T)z_t,$$

and conversely, as he observed, this property characterizes matrices corresponding to representations. In other words, a function $r \to (R)$ from the elements $r \in G$ to matrices (R) is a representation whenever the matrix with polynomial entries $\sum (R)x_r$ satisfies the equation stated above. In particular, the group matrix $(x_{rs^{-1}})$ satisfies this condition, and is therefore associated with a representation whose degree is the order of the finite group.[23] Passing to the *determinant of the representation matrix* $F(x) = |\sum (R)x_r|$, one has $F(x)F(y) = F(z)$, and from this it follows, by further examination (in §1 of [**132**]) of the group determinant, that $F(x)$ divides a power of the group determinant

$$\Theta(x) = |x_{rs^{-1}}| = \prod \Phi^f$$

---

[23]As an exercise, the reader can show that the representation obtained from the group matrix is the *regular representation*, which assigns to each group element r the permutation matrix (R) describing the permutation of the group elements obtained by multiplying the elements of G (taken in some order) by the element r.

and hence is a product of the irreducible factors of Θ:

$$|\sum (R)x_r| = \prod \Phi^e,$$

for some exponents $\{e\}$.

The problem arises, whether there exists, for each irreducible factor of the group determinant, a representation of G whose determinant, as defined above, is the given irreducible factor. Using the full resources of his paper on the factorization of the group determinant [132], Frobenius proved the following result, which settles the problem ([133], Ges. Abh. III, p. 95).

THEOREM 4.1. *Let G be a finite group of order g, and let $X = (x_{rs^{-1}})$ be the group matrix of G. Then there exists an invertible $g \times g$ matrix P such that*

$$P^{-1}XP = U,$$

for a matrix U which is the direct sum of $\sum f$ submatrices U_i, and $\{f\}$ runs through the set of degrees of the characters of G. Each submatrix U_i is a matrix associated with a representation of G whose determinant is one of the irreducible factors Φ of the group determinant

$$\Theta = |x_{rs^{-1}}| = \prod \Phi^f.$$

Each factor of the group determinant is associated with a submatrix U_i, in this way. Moreover, let

$$U_i \leftrightarrow \Phi \leftrightarrow \chi,$$

with χ the character of degree f corresponding to Φ as in Theorem 3.4. Let $r \rightarrow (R)$ be the representation of G such that $U_i = \sum (R)x_r$. Then the coefficient of u^{f-1} in the characteristic polynomial $|U_i - uI|$ of the matrix U_i is $\sum \chi(r)x_r$ and one has

$$\chi(r) = \text{Trace } (R) = \sum r_{ii},$$

for the matrix $(R) = (r_{ij})$ corresponding to $r \in G$.

Frobenius defined a *primitive representation* of G to be a representation whose determinant is an irreducible factor of the group determinant, as in the preceding theorem. The first part of the theorem states that the regular representation of G is a direct sum of primitive representations. This result was a forerunner of Maschke's Theorem [213], published in 1899, which implies that every representation can be expressed as such a direct sum (See Chapter III, §4). Another consequence of the theorem is that the characters of the finite group G are the trace functions of the primitive representations. The proof of the theorem depends, as we noted, on the rather complicated machinery Frobenius had developed in order to factor the group determinant. We omit the details for now, and will return to them later when we consider simpler and more direct approaches to representation theory, by Burnside, Schur (Chapter IV, §2), and Noether (Chapter VI, §2). As an illustration of Theorem 4.1 Frobenius carried out the calculations for the symmetric group of order 6, whose group determinant was factored at the beginning of the preceding section. Keep the notation established there. Let X be the group matrix, and P,

U the matrices:

$$\begin{bmatrix} 1 & -1 & 1 & 0 & 0 & 1 \\ 1 & -1 & \rho^2 & 0 & 0 & \rho \\ 1 & -1 & \rho & 0 & 0 & \rho^2 \\ 1 & 1 & 0 & 1 & 1 & 0 \\ 1 & 1 & 0 & \rho & \rho^2 & 0 \\ 1 & 1 & 0 & \rho^2 & \rho & 0 \end{bmatrix}$$

and

$$\begin{bmatrix} u+v & 0 & 0 & 0 & 0 & 0 \\ 0 & u-v & 0 & 0 & 0 & 0 \\ 0 & 0 & u_1 & v_1 & 0 & 0 \\ 0 & 0 & v_2 & u_2 & 0 & 0 \\ 0 & 0 & 0 & 0 & u_1 & v_1 \\ 0 & 0 & 0 & 0 & v_2 & u_1 \end{bmatrix},$$

where $u, u_1, \dots$ are as defined in the previous section. Then

$$P^{-1}XP = U,$$

with submatrices $U_1 = (u+v)$, $U_2 = (u-v)$ etc. of U as in Theorem 4.1.

In §6 of [**133**], Frobenius defined an associative algebra [System hypercomplexer Zahlen] associated with a finite group G, today called the *group algebra* of G, defined, in modern terms, to be the vector space over $\mathbb{C}$ with a basis consisting of elements e_a, indexed by the elements $a \in G$. The basis elements are multiplied according to the group law in G, so that $e_a e_b = e_{ab}$, for all elements $a, b \in G$. While the group algebra is in general not commutative, its *center*, consisting of all elements which commute with all the basis elements e_a, is a commutative algebra with basis elements $\{c_i\}$ indexed by the conjugacy classes C_i of G, and given by

$$c_i = \sum_{a \in C_i} e_a.$$

He proved that the multiplication of the basis elements of the center is given by:

$$c_j c_k = \sum_i \frac{h_{i'jk}}{h_i} c_i.$$

It follows that the center of the group algebra is isomorphic to the commutative algebra E considered at the beginning of *Über Gruppencharaktere*, whose representations were described in Proposition 3.2. The end result of this discussion was the following theorem, which, with some changes in terminology, is a key point in any modern account of representations of finite groups ([**133**], Ges. Abh. III, p. 98).

THEOREM 4.2. *Let $r \to (R)$ be a primitive representation of G, associated with a prime factor of the group determinant and a character χ of degree f, as in Theorem 4.1. For each basis element c_i of the center of the group algebra of G, let (c_i) be the sum of the matrices (A) corresponding to the elements $a \in C_i$. Then*

$$(c_i) = \frac{h_i \chi_i}{f} I,$$

where I is the identity matrix, and $\chi_i = \chi(a)$ for some element $a \in C_i$.

The preceding theorem is important because it relates the characters of G, defined in terms of representations of the center of the group algebra, to representations of the group algebra itself. Frobenius's proof of it[24] was based on the important result that the f^2 entries of the matrix $(x_{ij}) = \sum(R)x_r$, associated with a primitive representation as in the statement of the theorem, are linearly independent. He obtained it, once again, as an application of results involved in the factorization of the group determinant. It implies that, by specializing the variables x_a, one obtains all $f \times f$ matrices over $\mathbb{C}$, so that the matrices (c_i) commute with all $f \times f$ matrices, and hence are scalar multiples of the identity matrix.[25] In connection with it, he acknowledged for the first time his awareness of the parallel investigations of Molien:

> This remarkable theorem, that there is a matrix associated with the group H whose f^2 elements are independent variables, has also been discovered by Molien in his first-class work *Über Systeme höherer complexer Zahlen* (Math. Ann. Bd. 41, S. 124), to which Study recently directed me. In a further work *Eine Bemerkung zur Theorie der homogenen Substitutionsgruppen*, Sitzungsberichte der Naturforscher-Gesellschaft zu Dorpat 1897, Jahrg. 18, S. 259, Molien has applied the general result above specifically to group determinants.

The following year, he wrote to Dedekind about Molien.[26]

> You will have noticed that a young mathematician, Theodor Molien in Dorpat, has considered the group determinant independently of me. In volume 41 of the *Mathematische Annalen* he published a very beautiful, but difficult, work *Über Systeme höherer complexer Zahlen*, in which he has investigated noncommutative multiplication and obtained important general results of which the properties of the group determinant are special cases. Since he was entirely unknown to me, I have made some inquiries regarding his personal circumstances. Details are still lacking. This much I have already learned: that he is still a Privatdozent in Dorpat; that his position there is uncertain and that he has not advanced as far as he would have deserved in view of his undoubtedly strong mathematical talent. I would very much like to interest you in this talented man; here and there you are virtually privy councillor; if an opportunity presents itself, please think of Herr Molien, and if you have time, look at his work.

It is not known whether Dedekind made any efforts on Molien's behalf.[27]

Frobenius's next two papers [**134**], [**135**] broke new ground altogether, and the first one especially has been fundamental for applications and computations with characters by Frobenius himself, and by his successors up to the present time. He introduced the first paper with the following remarks.

---

[24] A modern proof is included in the proof of Lemma 4.3 in Chapter IV, §4.

[25] These facts anticipate to some extent Burnside's Theorem and Schur's Lemma which, as we shall see later, are central to more direct approaches to representation theory.

[26] The following quotation is from a letter dated 24 February, 1898, and appears in Hawkins's article ([**174**], p. 240) on the Frobenius-Dedekind correspondence.

[27] Further information about Molien's career, up to the time of his appointment in 1900 to a professorship at the technical institute in Tomsk, Siberia, was obtained by Hawkins ([**174**], p. 241).

In my work *Über Gruppencharaktere* (Sitzungsberichte 1896) I have developed a general method for the calculation of the characters of a finite group of known structure, and illustrated its practical applicability by a series of simple examples. But since its application to complicated groups involves substantial difficulties, I have sought other ways to obtain the characters of a group and its primitive representations by linear substitutions, and I have found two wholly different methods, that in special cases accomplish these aims more simply than the general method.

In the first paper [**134**], Frobenius investigated the relations between the characters of a finite group G of order g and those of a subgroup H of order h. It contains a construction of what are called today *induced representations* and a proof of what is now known as the *Frobenius Reciprocity Theorem*. These ideas have lost none of their sparkle during the century since they were first published, and it is worthwhile to see exactly how they were initially developed. Frobenius approached the subject, as usual, starting from his work on irreducible factors of group determinants and their connection with representations, described earlier in this section.

Let

$$\Theta = \prod \Phi_j^{f_j}$$

be the factorization of the group determinant of G into irreducible factors, and

$$\Xi = \prod \Psi_k^{e_k}$$

the corresponding factorization of the group determinant of the subgroup H of G, as in Theorem 3.4. If in Θ one sets all variables $x_r, r \in G$, equal to zero except for those indexed by elements of H, then it is not hard to show that

$$\Theta = \Xi^n,$$

where $n = g/h$ is the index of H in G. From this it follows that each irreducible factor Φ_j of Θ, with the variables restricted as indicated above, can be expressed as a product of the irreducible factors of the group determinant Ξ of H:

$$\Phi_j = \prod_k \Psi_k^{r_{kj}}.$$

In order to understand the relation between the irreducible factors of Θ and Ξ, Frobenius began a close examination of the integers r_{kj}, and ways of expressing them in terms of the characters of G and H. This was to lead him to his Reciprocity Theorem, but that is getting ahead of the story. The first of several crucial points was his observation that if, in the preceding equation, one replaced the indeterminate x_1 indexed by the identity element 1 by $x_1 + u$, and compared coefficients of $u^{f_j - 1}$ one obtained the formula

(13) $$\sum_k r_{kj} \psi^{(k)}(p) = \chi^{(j)}(p), \; p \in H,$$

where $\chi^{(j)}$ and $\psi^{(k)}$ are the characters of G and H corresponding to the irreducible polynomials Φ_j and Ψ_k respectively. This can be shown from the information about the irreducible factors of group determinants given in Theorem 3.4. As we discussed in the previous section, each character of G is a class function on G, so its restriction to the subgroup H is a class function of H. Nevertheless, at this stage in the development of the theory, there was no clear connection between the characters of G restricted to H and the characters of H. The preceding equation, however, implies precisely such a relationship, and Frobenius was quick to capitalize on it. The key was to apply the first and second orthogonality relations for the characters of H and G, namely Theorem 3.2 and equation (9). Recalling that $e = f$ by Theorem 3.4, the first orthogonality relations for the characters of H assert that

$$\sum_{p \in H} \psi^{(k)}(p)\psi^{(k)}(p^{-1}) = h, \ \sum_{p \in H} \psi^{(k)}(p)\psi^{(\ell)}(p^{-1}) = 0, \ k \neq \ell.$$

Applying them to (13) one obtains

$$hr_{kj} = \sum_{p \in H} \psi^{(k)}(p)\chi^{(j)}(p^{-1}).$$

Upon multiplying the preceding equation by $\chi^{(j)}(r)$ with $r \in H$, and summing on j, one obtains

$$h\sum_j r_{kj}\chi^{(j)}(r) = \sum_p \psi^{(k)}(p)(\sum_j \chi^{(j)}(p^{-1})\chi^{(j)}(r)).$$

The second orthogonality relations for the characters of G assert that

$$\sum_j \chi^{(j)}(p^{-1})\chi^{(j)}(r) = 0$$

unless r and p are conjugate in G, and that the sum is equal to $g/|C_i|$ in case the elements p and r belong to the same conjugacy class C_i of G. It follows that

$$(14) \qquad \sum_j r_{kj}\chi^{(j)}(r) = \frac{g}{h|C_i|}\sum_{p \in C_i} \psi^{(k)}(p).$$

These equations contain the information obtained by *restriction* of the group determinant of G to the subgroup H.

The next step was to construct representations of G from representations of the subgroup H corresponding to irreducible factors of the group determinant of H. The idea was to use the characterization of representations given in [**133**], discussed earlier in this section. Let $p \to (P)$ be a representation of H of degree e, and let $X = \sum(P)x_p$ be the $e \times e$ matrix corresponding to the given representation; then the entries of the matrix X are linear combinations, with coefficients in $\mathbb{C}$, of the indeterminates x_p, for $p \in H$. Let Y be the matrix obtained by replacing x_p by y_p for each $p \in H$, and let Z be the matrix obtained by replacing x_q by $z_q = \sum x_{p^{-1}}y_{pq}$, where the sum is taken over $p \in H$. Then $Z = XY$, and, as he noted in [**133**], this property characterizes representations of H. A representation of G, today called an *induced representation*, was constructed using the matrix polynomial X, as follows.

Let

$$G = a_1 H \cup a_2 H \cup \cdots \cup a_n H$$

describe the partition of G into distinct left cosets $a_i H$, and write $a \in G/H$ to mean $a = a_i$ for any one of the given coset representatives. For $a, b \in G/H$, and each element $r \in G$ of the form $r = apb^{-1}$ for some element p in H, introduce a new indeterminate $x_r = x_{apb^{-1}}$. Put

$$X_{a,b} = \sum_{p \in H} (P)x_{apb^{-1}};$$

in other words, replace the matrix $X = \sum (P)x_p$ by $X_{a,b} = \sum (P)x_{apb^{-1}}$ with the new indeterminates $x_{apb^{-1}}$. The entries of $X_{a,b}$ are linear combinations of indeterminates x_r for those elements r in G such that $a^{-1}rb \in H$. Now form the $ne \times ne$ matrix $(X_{a,b})$, whose (i, j) entry is the matrix X_{a_i,a_j} for coset representatives $a = a_i$ and $b = a_j$, taken in some fixed order. Then the entries of the matrix $(X_{a,b})$ are linear combinations of the indeterminates x_r, for $r \in G$, and we shall prove, as Frobenius did, that $(X_{a,b})$ is a matrix corresponding to some representation of G. Let $(Y_{a,b})$ and $(Z_{a,b})$ be the matrices obtained from $X_{a,b}$ by replacing x_r by y_r and

$$z_r = \sum_s x_{rs} y_{s^{-1}},$$

respectively. We want to prove that

$$(Z_{a,b}) = (X_{a,b})(Y_{a,b}).$$

This follows directly from the following computation (with a, b fixed in G/H, and p, q denoting elements of H):

$$\sum_{n \in G/H} X_{a,n} Y_{n,b} = \sum_n \sum_{p,q} (P)x_{apn^{-1}}(Q)y_{nqb^{-1}}$$

$$= \sum_n \sum_{p,q} (PQ)x_{apn^{-1}} y_{nqb^{-1}}$$

$$= \sum_q (Q) \sum_{n \in G/H,\ p \in H} x_{ap^{-1}n^{-1}} y_{npqb^{-1}}$$

$$= \sum_q (Q) \sum_{r \in G} x_{ar^{-1}} y_{rqb^{-1}}$$

$$= \sum_q (Q) z_{aqb^{-1}}$$

$$= Z_{a,b}.$$

In the calculation, we have used the fact that the elements np, with $n \in G/H$, $p \in H$ run through the elements of G without repetition.

Starting from a subgroup H of index n in G, and a representation of H of degree e, the preceding analysis established the existence of an *induced representation* of G of degree ne, whose associated matrix is $(X_{a,b})$. Therefore its determinant is a product of the irreducible factors of Θ,

$$|(X_{a,b})| = \prod_j \Phi_j^{r_j}.$$

Assume now that the given representation $p \to (P)$ is associated with an irreducible factor Ψ of the group determinant $\Xi(x)$ and the character ψ of the subgroup H. The coefficient of u^{e-1} in the polynomial obtained from Ψ by replacing x_1 by $x_1 + u$ is $\sum \psi(p)x_p$. It follows that the sum of the diagonal elements of $X_{n,n}$, for $n \in G/H$, is

$$\sum_{p \in H} \psi(p)x_{npn^{-1}} = \sum_{r \in G} \psi(n^{-1}rn)x_r,$$

with the convention that $\psi(r)$ is set equal to zero for elements r not belonging to the subgroup H. By comparing coefficients of u^{ne-1} in the equation for $|(X_{a,b})|$, and setting $r_j = r'_{kj}$ if $\psi = \psi^{(k)}$, one obtains

(15)
$$\sum_{n \in G/H} \psi^{(k)}(n^{-1}rn) = \sum_j r'_{kj} \chi^{(j)}(r),$$

where the left hand side is equal to the value of the trace function of the induced representation at an element $r \in G$. Using the fact that ψ is a class function on H, Frobenius derived the formula

$$\sum_{n \in G/H} \psi(n^{-1}rn) = \frac{1}{h} \sum_{n,p} \psi(p^{-1}n^{-1}rnp) = \frac{1}{h} \sum_{s \in G} \psi(s^{-1}rs),$$

for the trace function of the induced representation. The trace function of such an induced representation is called today an *induced character*, and is denoted by ψ^G or $\operatorname{ind}_H^G \psi$. Substituting the preceding expression in the formula (15), he obtained

$$h \sum_j r'_{kj} \chi^{(j)}(r) = \sum_s \psi(s^{-1}rs),$$

for each element $r \in G$. Letting r denote an element of the conjugacy class C_i of G, the elements $s^{-1}rs$ run through the elements of C_i, with each element repeated $g/|C_i|$ times. In this case, the preceding equation becomes:

$$\sum_j r'_{kj} \chi^{(j)}(r) = \frac{g}{h|C_i|} \sum_{p \in C_i} \psi^{(k)}(p).$$

The integers r'_{kj} are uniquely determined by these equations. Therefore the integers r'_{kj} coincide with the integers r_{kj} appearing in the equations (14), and one obtains ([**134**], §3):

THEOREM 4.3 (Frobenius Reciprocity Theorem). *Let r_{kj} be the integers such that*

$$\Phi_j = \prod_k \Psi_k^{r_{kj}},$$

where the left hand side denotes the restriction to the subgroup H of an irreducible factor Φ_j of the group determinant of G, and the Ψ_k are the irreducible factors of the group determinant of H. For each k, let $(X_{a,b}^{(k)})$ be the matrix corresponding to the induced representation from a representation of H associated with the factor Ψ_k; then one has

$$|(X_{a,b}^{(k)})| = \prod_j \Phi_j^{r_{kj}},$$

with the same set of exponents r_{kj} as in the preceding formulas. For the characters $\chi^{(j)}$ and $\psi^{(k)}$ of G and H associated with the irreducible factors Φ_j and Ψ_k respectively, one has

$$\chi^{(j)}(p) = \sum_k r_{kj} \psi^{(k)}(p)$$

for all $p \in H$, while

$$\sum_j r_{kj} \chi^{(j)}(r) = \frac{1}{h} \sum_{s \in G} \psi^{(k)}(s^{-1}rs), \ r \in G.$$

In this equation, the right side is the induced character from $\psi^{(k)}$, evaluated at $r \in G$.

The last part of the reciprocity theorem states that the multiplicity r_{kj} with which a character $\psi^{(k)}$ of the subgroup H occurs in the restriction to H of the character $\chi^{(j)}$ of G is equal to the multiplicity with which the character $\chi^{(j)}$ occurs as a summand of the induced character $(\psi^{(k)})^G$.

For the special case of the principal character[28] $\psi = \psi^{(0)}$ of the subgroup H, the preceding theorem, combined with formula (14), asserts that one has:

COROLLARY 4.1. *Let ψ^G be the induced character from the principal character $\psi = \psi^{(0)}$ of the subgroup H of G. Then*

$$\psi^G(r) = \frac{1}{h} \sum_{s \in G} \psi(s^{-1}rs) = \frac{g|C_i \cap H|}{h|C_i|} = \sum_j r_{0j} \chi_i^{(j)},$$

where r is an element of a conjugacy class C_i of G.

Frobenius realized immediately the importance of the reciprocity law as it applied to the interaction of characters of a finite group G and induced characters from a subgroup H, entirely apart from any consideration of irreducible factors of the group determinant. In fact, he was able to announce, in the last section of [134], that with the help of the formula stated in the preceding Corollary, he had been able to determine all the characters of the symmetric group on n symbols. The resulting paper [137], published in 1900, and another paper [140] on what are now called Frobenius groups, published a year later, were two brilliant applications of the theory of induced characters to very different kinds of problems, and established that subject as one of central importance for representation theory. Both will be surveyed in the following two sections.

Frobenius concluded the paper [134] with the following simple application of the formula stated in the Corollary. It established a powerful link between character theory, in particular the theory of induced characters and the reciprocity law, and the theory of permutation representations of finite groups. In the formula

$$\sum_j r_{0j} \chi_i^{(j)} = \frac{g|H \cap C_i|}{h|C_i|},$$

one has $r_{00} = 1$, by the reciprocity theorem. The question arises as to when the rest of the expression

$$\chi_i = \frac{g|H \cap C_i|}{h|C_i|} - 1$$

represents the values of a single character. The answer is provided by the following result, whose proof brings together a number of other ideas Frobenius had considered in his previous work on finite group theory ([134], §5).

PROPOSITION 4.1. *A necessary and sufficient condition for the numbers χ_i, defined above, to be the values of a character of G is that the number of double cosets of the form HsH is equal to two. The condition is equivalent to the assertion that the permutation representation of G on the cosets of H is doubly transitive.*

We shall include only part of the proof here, to show how Frobenius made the connection between character theory and his investigation of double cosets leading up to his work on density problems, discussed in §2. Let us assume that the numbers

---

[28]The *principal character*, as Frobenius called it, is the character whose value is 1 at each element of the group. Induced characters from principal characters of subgroups will be called *induced permutation characters* in what follows.

χ_i defined above are values of a character. Then, by the orthogonality relations, one has

$$\sum_i |C_i| \left(\frac{g|C_i \cap H|}{h|C_i|} - 1 \right)^2 = g.$$

Upon simplification, this equation becomes

$$\sum_i \frac{|C_i \cap H|^2}{|C_i|} = 2 \frac{h^2}{g},$$

and comparison with Proposition 2.2 yields the conclusion that the number of double cosets of the form HsH is equal to 2. The proof that this statement is equivalent to the assertion that the permutation representation of G on the cosets of H is doubly transitive is an elementary exercise, and is omitted.

5. The Characters of the Symmetric Group

Frobenius's paper [137], published in 1900, introduced new combinatorial methods, involving symmetric functions and their connection with conjugacy classes and induced permutation characters, in order to solve the difficult problem of calculating the characters of the symmetric group S_n. In his dissertation published the following year ([247], 1901), Frobenius's doctoral student Issai Schur classified the polynomial representations of $GL_n(\mathbb{C})$ (see Chapter V, §2), by a method based in part on Frobenius's research on the characters of the symmetric group. Their work has been incorporated in the modern theory of symmetric functions ([210]), and in computational group theory.[29]

In the same year, the subject received a powerful lift when a remarkable application of Frobenius's work on the characters of S_n to the theory of Riemann surfaces was obtained by Adolf Hurwitz. Hurwitz had proved in 1891, when he was at Königsberg, that the number of n-sheeted Riemann surfaces having exactly w simple branch points is equal to the number of solutions of the equation

$$t_1 t_2 \ldots t_w = 1,$$

where $t_1, \ldots, t_w$ are transpositions in the symmetric group S_n ([184], §2). But the problem of how to calculate the number of solutions of these equations was left unsolved. The following year, in 1892, Hurwitz was appointed to ETH Zürich to fill the position left vacant when Frobenius was called to Berlin as Kronecker's successor. He corresponded from time to time with Frobenius, and kept in touch with his work. In 1900, he submitted a second paper on the subject to the *Mathematische Annalen* ([186]), this time from Zürich. In the introduction, he stated:

> When I spoke about this subject last summer with E. Lasker, the well-known world chess champion and mathematician, I was led, by a clever remark of Lasker, to a consideration which shows directly that the number [of solutions of the equation] is a coefficient in the expansion of a certain rational function, that is closely connected with the group determinant of the symmetric group.

---

[29]The latter is a relatively new part of group theory, in which computer programs are designed to investigate questions about generators, subgroups, conjugacy classes, and characters of finite groups.

Thanks to the investigations of Frobenius on the group determinant, he continued, he was now able to give an explicit formula for the number of solutions of the equation. He first observed that, since the transpositions in S_n are all conjugate to each other, the number of solutions of the equation is a number $h_{i...i}$, introduced in *Über Gruppencharaktere* ([**131**], §1), as the number of solutions of an equation $a_1 a_2 \cdots a_w = 1$, where the a_i all belong to a given conjugacy class in S_n. The numbers $h_{ij...}$ were computed in terms of the character values in [**131**], equation (2), §4 (see the formula for the numbers h_{ijk} in (10), §3). To obtain the solution of the problem about Riemann surfaces, it was only necessary to incorporate the new information on the values of the characters of S_n at a transposition, from Frobenius's paper [**137**]. This was all carried out by Hurwitz. A new proof was given, with some further discussion, by Frobenius in his second paper [**143**] on the characters of S_n.

We now turn to Frobenius's calculation of the characters of the symmetric group. The first step was to review some of the properties of conjugacy classes in the symmetric group $G = S_n$, of order $|G| = n!$, viewed as the group of permutations of the symbols $\{1, 2, \ldots, n\}$. Two permutations in S_n are conjugate if and only if their factorizations as products of disjoint cycles contain the same number of cycles of each order. From this it is not difficult to calculate the order of the centralizer of an element with α cycles of order one, β cycles of order two, etc., and hence to obtain a formula for the number of elements in the conjugacy class C containing such an element, using the fact that this number is the index of the centralizer of an element in the conjugacy class (see the proof of Theorem 2.1). The result, which, as Frobenius noted, goes back to Cauchy, is that

$$|C| = \frac{n!}{1^\alpha \alpha! 2^\beta \beta! \cdots}.$$

The number k of conjugacy classes in S_n is equal to the number of solutions of the equation

$$n = \alpha + 2\beta + \cdots,$$

with $\alpha, \beta, \ldots$ positive integers or zero. The number of characters of S_n is also equal to k, and the character table of S_n consists of the set of k^2 complex numbers giving the values of the characters at elements of the conjugacy classes.

In order to calculate the entries in the character table, Frobenius introduced a family of subgroups $\{G_{n_1, n_2, \ldots}\}$ of $G = S_n$, parametrized by sets of positive integers $n_1, n_2, \ldots$ with the property that

$$n = n_1 + n_2 + \cdots.$$

The subgroup $H = G_{n_1, n_2, \ldots}$ associated with $n_1, n_2, \ldots$ consists of the elements of S_n which permute the first n_1 symbols among themselves, and permute the second n_2 symbols among themselves, etc. Then $|H| = n_1! n_2! \cdots$, and the number of such subgroups H is equal to the number of conjugacy classes. Each permutation r in H can be factored as a product of permutations $r_1 r_2 \cdots$, with r_1 a permutation of the first n_1 symbols, r_2 a permutation of the next n_2 symbols, etc. Let r_1 have α_1 cycles of order 1, β_1 cycles of order 2, etc. Then

$$n_1 = \alpha_1 + 2\beta_1 + \cdots, n_2 = \alpha_2 + 2\beta_2 + \cdots, \ldots,$$

and if r belongs to the class C of S_n defined above, we also have

$$\alpha = \alpha_1 + \alpha_2 + \cdots, \beta = \beta_1 + \beta_2 + \cdots.$$

The number of permutations in C belonging to H is therefore

$$|C \cap H| = \sum \frac{n_1!}{1^{\alpha_1} \alpha_1! 2^{\beta_1} \beta_1! \cdots} \frac{n_2!}{1^{\alpha_2} \alpha_2! 2^{\beta_2} \beta_2! \cdots} \cdots,$$

summed over all solutions of the preceding two equations. Of course, if there are no solutions of these equations, then $|C \cap H| = 0$. From these equations, it follows that

$$\frac{g}{|C|} = 1^{\alpha} \alpha! 2^{\beta} \beta! \cdots,$$

and

$$\frac{g|C \cap H|}{h|C|} = \sum \frac{\alpha!}{\alpha_1! \alpha_2! \cdots} \frac{\beta!}{\beta_1! \beta_2! \cdots},$$

where $g = |G| = |S_n| = n!$ and $h = |H|$.

The equation stated above, as he had observed in [134], §3, (4) and (5) (see Corollary 4.1 in the preceding section), gives a formula for the value of the induced permutation character of $G = S_n$ from the principal character of the subgroup H at elements in the conjugacy class C. He continued with the important remark that the preceding sum, and hence the value of the induced permutation character from H at elements of the class C, is also equal to the coefficient of the monomial $x_1^{n_1} x_2^{n_2} \cdots x_m^{n_m}$ in the homogeneous polynomial of degree n given by

$$(x_1 + \cdots + x_m)^{\alpha} (x_1^2 + \cdots + x_m^2)^{\beta} \cdots$$

with $x_1, x_2, \ldots$ indeterminates and m greater than or equal to the number of summands $n_1, n_2, \ldots$ of n.

This brought symmetric functions, that is, polynomials invariant under permutations of the variables, into the picture. He then multiplied the symmetric polynomial, given above, by a skew-symmetric, or alternating, polynomial $\Delta(x_1, \ldots, x_m)$ of degree $\frac{1}{2} m(m-1)$, given by the formula:

$$\Delta(x_1, \ldots, x_m) = (x_2 - x_1)(x_3 - x_1) \cdots (x_m - x_{m-1}).$$

The result is a homogeneous polynomial of degree $n + \frac{1}{2} m(m-1)$ with the property that the coefficient of each monomial term is now the value of an integral linear combination of induced permutation characters from subgroups of the form $G_{n_1, n_2, \ldots}$, evaluated at an element of the conjugacy class C. A further property of these polynomials is that the monomials appearing with nonzero coefficients contain the variables x_i to different exponents; this is an easy consequence of the fact that alternating polynomials change sign if two of the variables are interchanged.

He next put $m = n$, and proved that the sets of exponents of monomials occurring in the above alternating polynomial (of degree $\frac{1}{2} n(n+1)$) give a parametrization of the irreducible characters of S_n. First he recalled that the exponents $(\lambda) = (\lambda_1, \ldots, \lambda_n)$ of such a monomial are sets of nonnegative integers λ_i such that $\lambda_i \neq \lambda_j$ for all $i \neq j$, and

$$\lambda_1 + \cdots + \lambda_n = \frac{1}{2} n(n+1).$$

Letting

$$\lambda_1 < \lambda_2 < \cdots < \lambda_n,$$

he observed that each such set (λ) defines a partition of n with summands

$$\lambda_1 \leq \lambda_2 - 1 \leq \cdots \leq \lambda_n - n + 1,$$

and conversely. The partitions of n label the conjugacy classes of S_n, and hence the characters, so that the characters of S_n can be labeled by the sets (λ) as above, and will be denoted by $\chi^{(\lambda)}$. A sign $[\lambda_1, \ldots, \lambda_n] = \pm 1$ was attached to each set (λ), whose value is ± 1 according as $\Delta(\lambda_1, \ldots, \lambda_n)$ is positive or negative. He stated his main result as follows.

THEOREM 5.1. *Keep the preceding notation, with $m = n$. Let $C = C_j$ be a conjugacy class of S_n, and let $\chi_j^{(\lambda)}$ denote the value of the character $\chi^{(\lambda)}$ at elements of the class C_j, for each set (λ). Then the character values are given by the formula:*

$$(x_1 + \cdots + x_n)^\alpha (x_1^2 + \cdots + x_n^2)^\beta \cdots \Delta(x_1, \ldots, x_n) = \sum_{(\lambda)} [\lambda_1, \ldots, \lambda_n] \chi_j^{(\lambda)} x_1^{\lambda_1} \cdots x_n^{\lambda_n}.$$

Moreover, each character $\chi^{(\lambda)}$ can be expressed as a linear combination, with integer coefficients, of induced permutation characters from the subgroups $G_{n_1, n_2, \ldots}$, with coefficients given by the preceding formulas.

The complex valued functions $\chi^{(\lambda)}$ on $G = S_n$ defined in the statement of the theorem are linear combinations of induced permutation characters from certain subgroups of G, by the discussion preceding the statement of the theorem. The induced permutation characters themselves are linear combinations with integer coefficients of the characters of G, by the Reciprocity Theorem, or Corollary 4.1. After making these remarks, Frobenius stated that, in order to prove that the functions $\chi^{(\lambda)}$ are the characters, it was sufficient to prove that they satisfy the orthogonality relations

$$\sum h_j \chi_j^{(\lambda)} \chi_{j'}^{(\lambda)} = |G| \text{ and } \sum h_j \chi_j^{(\lambda)} \chi_{j'}^{(\mu)} = 0 \text{ if } (\lambda) \neq (\mu),$$

and that $\chi_1^{(\lambda)} > 0$, where as usual, $h_j = |C_j|$ for a conjugacy class C_j, $C_{j'} = C_j^{-1}$, and C_1 is the class containing the identity element. This general principle, stated here by Frobenius for the first time, has been a standard method for computing characters of finite groups ever since. As he explained, the argument goes as follows. We are given a family of k class functions $\chi_1, \ldots, \chi_k$ on G satisfying the orthogonality relations, each of which is a linear combination of the characters with integer coefficients. The orthogonality relations imply that the class functions χ_j are distinct, and that the sum of the squares of the coefficients of χ_j, in the expression of χ_j in terms of the characters, is equal to one. As the coefficients are integers, this implies that $\pm \chi_j$ is a character, for each $j, 1 \leq j \leq k$. The assumption that $\chi_j(1) > 0$ shows that each class function χ_j is in fact a character, for each j, completing the proof.

In order to establish the orthogonality relations for the class functions $\chi^{(\lambda)}$ he had obtained, and thereby complete the proof that these class functions are indeed the irreducible characters of S_n, Frobenius made another interesting excursion into the theory of symmetric functions. He introduced a second set of indeterminates $y_1, \ldots, y_m$, and, keeping the notation introduced at the beginning of the discussion, considered the sum

$$\sum_j \frac{|C_j|}{n!} (x_1 + \cdots + x_m)^\alpha (x_1^2 + \cdots + x_m^2)^\beta \cdots (y_1 + \cdots + y_m)^\alpha (y_1^2 + \cdots + y_m^2)^\beta \cdots,$$

where the conjugacy classes C_j are associated with partitions of n given by

$$\alpha + 2\beta + \cdots = n,$$

as above. If one abandons temporarily the restriction on $\alpha, \beta, \ldots$, and sums $\alpha, \beta, \ldots$ from 0 to ∞ in the preceding formula, one obtains

$$\sum_{\alpha,\beta,\ldots} \frac{1}{1^\alpha \alpha!}(x_1 + \cdots + x_m)^\alpha (y_1 + \cdots + y_m)^\alpha \frac{1}{2^\beta \beta!}(x_1^2 + \cdots + x_m^2)^\beta (y_1^2 + \cdots + y_m^2)^\beta \cdots,$$

which is equal to

$$e^{(x_1 + \cdots + x_m)(y_1 + \cdots + y_m) + \frac{1}{2}(x_1^2 + \cdots + x_m^2)(y_1^2 + \cdots + y_m^2) + \cdots} = e^{\log \prod_{i,j}(1 - x_i y_j)^{-1}},$$

where formal series are used for the exponential and logarithm functions so that questions of convergence may be ignored.

The next step was to apply:

LEMMA 5.1 (Cauchy's Lemma). *Let* $\Delta(x) = \Delta(x_1, \ldots, x_m)$ *be the alternating function defined above. Then*

$$\frac{\Delta(x)\Delta(y)}{\prod_{i,j}(1 - x_i y_j)} = \left| \frac{1}{1 - x_i y_j} \right|,$$

where the expression on the right hand side is the determinant of the $m \times m$ matrix whose (i,j)-entry is $(1 - x_i y_j)^{-1}$.

It is possible to prove the Lemma in a straightforward way by induction on m. A more conceptual proof appears in [**210**].

By the Lemma, the expression $\prod(1 - x_i y_j)^{-1}$ preceding it, multiplied by $\Delta(x)\Delta(y)$, becomes

$$\left| \frac{1}{1 - x_i y_j} \right| = \sum [\mu_1, \ldots, \mu_m] \frac{1}{(1 - x_{\mu_1} y_1)(1 - x_{\mu_2} y_2) \cdots}.$$

The sum is taken over permutations $\mu_1, \ldots, \mu_m$ of $1, \ldots, m$, and we have used the fact that $[\mu_1, \ldots, \mu_m]$ is the sign of the permutation. Upon expanding the factors $(1 - x_{\mu_j} y_j)^{-1}$ of the preceding expression in geometric series, the formula becomes

$$\sum [\mu_1, \ldots, \mu_m] x_{\mu_1}^{\lambda_1} y_1^{\lambda_1} x_{\mu_2}^{\lambda_2} y_2^{\lambda_2} \cdots.$$

With the obvious abbreviations, and after rearranging terms, the preceding expression is equal to

$$\sum [\kappa][\lambda] x^\kappa y^\lambda.$$

In this expression, the terms appearing with nonzero coefficients are those for which the sets $\{\kappa_i\}$ and $\{\lambda_j\}$ consist of distinct integers, which are permutations of each other.

The stage was now set for the finale! Frobenius multiplied the original sum by $\Delta(x)\Delta(y)$, and separated out the finite sum consisting of terms for which

$$\alpha + 2\beta + \cdots = n.$$

By the definition of the class functions $\chi^{(\lambda)}$, this is equal to

$$\sum_{\kappa,\lambda} \left(\sum_j \frac{|C_j|}{n!} \chi_j^{(\kappa)} \chi_j^{(\lambda)} \right)[\kappa][\lambda] x^\kappa y^\lambda,$$

where the sets $\{\kappa_i\}$ and $\{\lambda_j\}$ consist of distinct integers such that

$$\sum \kappa_i = \sum \lambda_j = n + \frac{1}{2}m(m-1).$$

Comparing this expression with the formula obtained from Cauchy's Lemma, Frobenius was able to conclude that

$$\sum_j \frac{|C_j|}{n!} \chi_j^{(\kappa)} \chi_j^{(\lambda)} = 0$$

unless the sets (κ) and (λ) are permutations of each other, and in that case, the sum is equal to 1. These are the orthogonality relations for the class functions $\chi^{(\lambda)}$. A streamlined version of the preceding, somewhat amazing, calculation can be found in [210].

Frobenius returned to the character theory of the symmetric group in a second paper [143], entitled *Über die charakteristischen Einheiten der symmetrischen Gruppe*, published in 1903. In the introduction, he acknowledged that the formula for the character values stated in Theorem 5.1 involved complicated considerations ["umständliche Betrachtungen"], and announced that he had found another approach to the problem of calculating the character values $\chi_j^{(\lambda)}$. We include here a brief survey of the new method, as it was closely related to research begun by Alfred Young in 1900; this led, in turn, to the standard construction of irreducible representations of S_n used today, based on "Young tableaux" (see James and Kerber [189]).

Frobenius began by recalling that the characters $\chi = \chi^{(j)}$ of a finite group G were the trace functions of the primitive representations $r \to (R)$ of G, namely those for which the determinant of the matrix $\sum(R)x_r$ was an irreducible factor of the group determinant (see Theorem 4.1). He called these characters "simple," and defined a composite character [zusammengesetzter Charakter] to be a linear combination

$$\varphi = \sum_j m_j \chi^{(j)}$$

of the simple characters with nonnegative integer coefficients m_j. He remarked that the trace function of an arbitrary representation, not necessarily primitive, was a composite character, as such a representation was equivalent to a direct sum of primitive representations.

A set of complex numbers a_r, not all zero, indexed by the elements $r \in G$, was called a *characteristic unit* [characteristische Einheit] if it satisfies the condition $\sum a_r a_s = a_t$ for all $t \in G$, where the sum is taken over all solutions of the equation $rs = t$. As he observed, the condition is equivalent to the statement that the matrix $A = (a_{rs^{-1}})$ is idempotent: $A^2 = A$. He then proved, by the following ingenious argument, attributed to I. Schur, that for such a matrix, the product $X\bar{A}$ is the matrix corresponding to a representation of G, where X is the group matrix, and $\bar{A}$ is the matrix $(a_{s^{-1}r})$. Introduce the matrix $Y = (y_{rs^{-1}})$ in a new set of variables y_r. Then the matrices X and $\bar{Y} = (y_{s^{-1}r})$ commute. If $Z = XY$, then $\bar{Z} = \bar{Y}\bar{X}$, and if $A^2 = A$, then $\bar{A}^2 = \bar{A}$. From the equation $XY = Z$ and the preceding remarks, it follows that $X\bar{A} \cdot Y\bar{A} = Z\bar{A}$. By the criterion[30] stated at the beginning of §4

---

[30] Frobenius had also used this way of checking whether an assignment of matrices to group elements was a representation in his construction of induced representations [134] (see §4).

(from [**133**]), he deduced that $X\bar{A}$ is a matrix corresponding to a representation of G. Its trace is $\sum_{r,s} a_{s^{-1}r^{-1}s}x_r$. From this he obtained:

PROPOSITION 5.1. *Let $\{a_r\}$ be a characteristic unit for the finite group G. Then the class function*

$$\varphi(r) = \sum_s a_{s^{-1}rs}$$

is a composite character of G.

If the character φ in the preceding result is a simple character, then Frobenius called $\{a_r\}$ a *primitive characteristic unit*; clearly the knowledge of a primitive characteristic unit gave the values of the character associated with it.

It is an interesting exercise for the modern reader to give a proof of Proposition 5.1 along the following lines. A set of complex numbers $\{a_r\}$ is a characteristic unit if and only if $a = \sum a_r r$ is an idempotent element in the group algebra $\mathbb{C}G$. The idempotent a is primitive if and only if $\mathbb{C}Ga$ is a simple left module for the group algebra, associated with an irreducible character χ. It is not difficult to prove, using the Wedderburn theorems applied to the semisimple algebra $\mathbb{C}G$ (see Chapter VI, §2) that $\chi(r)$ is given by the formula in the statement of the Proposition, for each $r \in G$. The extension to an arbitrary, not necessarily primitive, idempotent, is immediate.

With Proposition 5.1 in hand, Frobenius set out to determine the primitive characteristic units for the symmetric group S_n. Let

$$n = \alpha_1 + \cdots + \alpha_\mu, \text{ with } \alpha_1 \geq \alpha_2 \geq \cdots \geq \alpha_\mu > 0$$

be a partition of n, denoted by (α). Assume that β_1 of the integers α_i are ≥ 1, β_2 of them are ≥ 2, etc. Then it is evident that

$$n = \beta_1 + \beta_2 + \cdots \text{ and } \beta_1 \geq \beta_2 \geq \cdots,$$

so that (β) is another partition, which Frobenius called the partition associated with (α). For each pair of associated partitions, he defined a pair of subgroups $\mathfrak{P}$ and $\mathfrak{Q}$ of S_n as follows. Choose a distribution of the integers $1, 2, \ldots, n$ into disjoint subsets containing α_1 integers, α_2 integers, etc. The subgroup $\mathfrak{P} = \mathfrak{P}_{(\alpha)}$ was defined as the set of elements of S_n that permute the elements in each of the subsets among themselves. Now form an array of integers a_{ij} with $a_{11}, \ldots, a_{1\alpha_1}$ the first subset in the distribution, $a_{21}, \ldots, a_{2\alpha_2}$ the second subset in the distribution, etc. He called such an array a "Schema", or "graph", after Sylvester.[31] As $\alpha_1 \geq \alpha_2 \geq \cdots$, each row of the array contains more ($\geq$) entries than the next. It follows that the array obtained from the first one by interchanging rows and columns is an array whose rows contain β_1 entries, β_2 entries, etc. where $(\beta) = \beta_1, \beta_2, \ldots$ is the partition associated with (α). Frobenius defined the subgroup $\mathfrak{Q}$ to be the set of elements in S_n which permute the elements of the rows of the second array. He noted that $\mathfrak{P} \cap \mathfrak{Q} = 1$, so that the factors $p \in \mathfrak{P}$ and $q \in \mathfrak{Q}$ of elements $pq \in \mathfrak{P}\mathfrak{Q}$ are uniquely determined. His main result ([**143**], §8, III) can be stated as follows.

---

[31]Graphical methods in the theory of partitions were introduced by the British mathematician Norman M. Ferrers in unpublished work of 1853, and applied to several problems in the theory of partitions by James Joseph Sylvester and his students at the Johns Hopkins University around 1880. For further discussion and references to the literature, see [**236**], Chapter 3, *From invariant theory to the theory of partitions*.

THEOREM 5.2. *Let (α) be a partition of n, and let $\mathfrak{P} = \mathfrak{P}_{(\alpha)}$ and $\mathfrak{Q}$ be a pair of associated subgroups, as defined above. Let a_r be the set of complex numbers, indexed by the elements $r \in S_n$, defined by the conditions: $a_r = 0$ if $r \notin \mathfrak{P}\mathfrak{Q}$, and if $r = pq$, with $p \in \mathfrak{P}$ and $q \in \mathfrak{Q}$, then $a_r = 1$ or -1 according as q is an even or odd permutation. Then $\{(f/n!)a_r\}$ is a primitive characteristic unit of the symmetric group S_n, whose corresponding character $\chi^{(\alpha)}$, of degree f, is indexed by the partition (α), as in Theorem 5.1.*

By the remark following Proposition 5.1, the preceding theorem gave a new way to compute the characters of S_n. Frobenius's proof of Theorem 5.2, however, was not independent of his proof of Theorem 5.1, and used some of the same material.

After he obtained Theorem 5.2, Frobenius wrote that the properties of the function a_r, with reference to the subgroups $\mathfrak{P}$ and $\mathfrak{Q}$, had already been investigated in two "very remarkable" [sehr beachtenswerthen] papers by Alfred Young ([**291**], [**292**]) entitled *On quantitative substitutional analysis*, I and II, published in 1900 and 1901. Frobenius went on to say that Young had proved that the element $(f/n!)\sum_r a_r r$ in the group algebra of S_n was an idempotent, and had calculated the value of f in terms of the partition α, but that he (Frobenius) had shown the connection between the idempotent and the primitive representation and simple character corresponding to it.

Alfred Young (1873-1940) was born at Birchfield, Farnsworth, in Lancashire, received his early education in Bournemouth and at Monkton Combe School, near Bath, where his unusual talent for mathematics was recognized. He received a scholarship, and matriculated at Clare College, Cambridge, in 1892. He graduated in 1895, as tenth Wrangler (tenth place) in the Tripos examination.[32] He lectured at Selwyn College, Cambridge, from 1901-1905. Following his intention to take Holy Orders, he was ordained in 1908, and from 1910-1940, he served as the parish priest in Birdbrook, a village of Essex about twenty-five miles east of Cambridge. He maintained a keen interest in research and publication on symmetric functions, invariants, and combinatorial properties of the symmetric group throughout his life.

The two papers *On quantitative substitutional analysis* referred to above introduced the important concept of tableaux associated with the symmetric group S_n, and used them to investigate what we would call today idempotents in the group algebra of the symmetric group. The first paper was, as he said in the introduction, "rewritten and greatly enlarged at the request of the referees;" and continued, "my thanks are due to them—especially to Prof. Burnside—for many valuable criticisms and suggestions." He described his approach as follows (from the introduction to the second paper):

> The letters $a_1, a_2, \ldots, a_n$ are arranged in any manner in h horizontal rows, so that each row has its first letter in the same vertical column, its second letter in a second vertical column, and so on; there being α_1 letters in the first row, α_2 in the second, etc. and finally α_h in the last; the α's satisfying the relations
>
> $$\alpha_1 + \alpha_2 + \cdots + \alpha_h = n, \; \alpha_1 \geq \alpha_2 \geq \cdots \geq \alpha_h.$$
>
> From this table an expression
>
> $$S = \Gamma'_1 \Gamma'_2 \cdots \Gamma'_h G_1 G_2 \cdots G_k$$

---

[32]The biographical information is taken from the article on his life and mathematical work by H. W. Turnbull [**275**]. See Chapter III, §1, for further remarks about the Tripos examination.

is formed, such that Γ_1' is the negative symmetric group of the letters
of the first row, Γ_2' that of the letters of the second row, and so on;
G_1 is the positive symmetric group of the letters of the first column,
G_2 that of the second, and so on.

In the preceding definition of S, the positive symmetric group G in a set of letters
$a, b, \ldots$ denoted the sum of all elements in the subgroup of S_n consisting of the
permutations of the letters $a, b, \ldots$, viewed as an element of the group algebra of
S_n, while the negative symmetric group Γ' in a set of letters $u, v, \ldots$ denoted the
sum of the elements in the alternating group on the letters $u, v, \ldots$ minus the sum
of the elements in the symmetric group on the letters $u, v, \ldots$ which do not belong
to the alternating group on these letters. Young introduced the notation NP for
the element S defined above, where N is the product of the negative symmetric
groups in S, and P is the product of the positive symmetric groups. He proved, in
[292], that

$$(NP)^2 = \lambda NP,$$

and derived an interesting combinatorial formula for λ in terms of the partition
α ([292], p. 366). The idempotent element $\lambda^{-1}NP$ in the group algebra of the
symmetric group is the element $(f/n!) \sum a_r r$ obtained in Theorem 5.2, as Frobenius
observed.

A second main topic of discussion in the papers [291] and [292] was a study
of the elements

$$T_{\alpha_1, \ldots, \alpha_h} = \sum NP,$$

obtained as the sum of the expressions NP, over the different tableaux associated
with the partition (α), which are all possible ways of inserting the letters $a_1, \ldots, a_n$
in rows of lengths $\alpha_1, \ldots, \alpha_h$. Young proved that the elements $T_{\alpha_1, \ldots, \alpha_h}$ were multiples of idempotent elements belonging to the center of the group algebra. Frobenius
(in [143], §8) expressed the central idempotent obtained from $T_{\alpha_1, \ldots, \alpha_h}$, denoted
here by $\varepsilon_{(\alpha)}$, in terms of the character $\chi^{(\alpha)}$ of S_n of degree f_α indexed by the
partition (α):

$$\varepsilon_{(\alpha)} = \frac{f_\alpha}{n!} \sum_{r \in S_n} \chi^{(\alpha)}(r) r.$$

Today we recognize the elements $\lambda^{-1}NP$ as primitive idempotents in the group
algebra of S_n, and the central idempotents $\varepsilon_{(\alpha)}$ as the identity elements in the
Wedderburn components of the group algebra $\mathbb{C}S_n$ (see Chapter VI, §2). A simplified proof of Theorem 5.2 was obtained later by John von Neumann ([276], Second
Edition, vol. II, §129).

6. An Application of Character Theory: Frobenius Groups

While he was engaged in laying the foundations of character theory, Frobenius
never lost sight of his intention to apply it to the structure of finite groups. In the
introduction to *Über Gruppencharaktere* [131], he had remarked that the solution
of the problem concerning the group determinant

> has brought me to a generalization of the concept of character for
> arbitrary finite groups. I shall develop this concept here in the be-
> lief that through its introduction, group theory will be substantially
> enriched.

Frobenius's research in finite group theory, following his papers on Sylow's theorem and double cosets (see §3), was mainly published in a series of articles with the titles *Über auflösbare Gruppen* I-V. The papers III and IV of the series contained deep applications of character theory to finite groups. For example, paper IV in the series [140], to be surveyed in this section, contains a fundamental result on the structure of what are today called *Frobenius groups*. These papers followed by only a year a paper by Burnside [73], published in 1900, which, as Frobenius generously acknowledged,[33] "used the theory of group characters for the first time to investigate the properties of a group" (see Chapter III, §3).

The concept of a Frobenius group, defined below in the context of the subgroup structure of an abstract finite group, can also be approached from the standpoint of finite permutation groups. In that setting, Frobenius's result (Corollary 6.1 below) settled, in a definitive way, problems that had been studied earlier, with various additional hypotheses, first by E. Maillet, in his thesis *Recherches sur les Substitutions* (1892) and subsequent papers in the *Bulletin de la Société Mathématique de France*, and in 1900 by Burnside (*On transitive groups of degree n and class n - 1* [74]).

As we know, Frobenius was interested in all sorts of problems associated with the notion of conjugacy in a finite group, and he chose this frame of reference for the statement of the main theorem of [140]. At the end of the paper, he obtained the permutation group form of the theorem (Corollary 6.1). Both versions contributed new information to a subject he cared about very much, and he was surely pleased with them. He stated the main theorem as follows ([140]; Ges. Abh. III, p. 196):

THEOREM 6.1. *Let H be a subgroup of a finite group G with the following property: let $p \in H$ and $p \neq 1$; then, for all elements $r \in G$, $r^{-1}pr \in H$ only if $r \in H$. Then the subgroup H has a normal complement, that is, there exists a normal subgroup N of G such that $G = HN$, and $H \cap N = \{1\}$.*

The current terminology for the situation described in the theorem is as follows. Whenever one has a finite group G and a subgroup H which satisfy the hypothesis of the theorem, G is called a *Frobenius group*, the subgroup H is called a *Frobenius complement*, and the normal subgroup N whose existence is stated in the theorem, is called the *Frobenius kernel*. Examples of Frobenius groups are the dihedral groups of order $2q$ with q odd, with Frobenius complement any subgroup of order 2.

The hypothesis of the theorem implies that two nonidentity elements of H which are conjugate in G are already conjugate in H. Frobenius began the proof with the observation that the hypothesis of the theorem implies several other facts (labeled $(i) - (v)$ below) about the subgroup H. While straightforward to prove, they introduced important ideas which were essential for the proof of the theorem. Let us first introduce some notation: let $g = |G|$, $h = |H|$, and let n be the index of H in G, so that $g = hn$.

(i). The centralizer in G of each nonidentity element of H is contained in H.

(ii). Let $q \in G$ be an element such that $q^{-1}Hq \cap H \neq 1$; then $q \in H$, otherwise $q^{-1}hq = h' \neq 1$ for h, h' in H and q not in H, contrary to the hypothesis.

(iii). Let $G = H \cup Hq_1 \cup \cdots \cup Hq_{n-1}$. By (ii) it follows that the conjugates of H by elements of G are $H, q_1^{-1}Hq_1, \ldots, q_{n-1}^{-1}Hq_{n-1}$, and no two of them coincide.

---

[33]See [139], Ges. Abh. III, p. 180.

Moreover, the intersection of any two distinct conjugates of H is trivial[34] (i.e. equal to $\{1\}$).

(iv). There are exactly $n - 1$ distinct elements of G belonging to no conjugate of the subgroup H, by (iii).

(v). The order h of H divides $n - 1$, so that, in particular, the order and index of H are relatively prime. The information needed for the proof of this fact was at Frobenius's fingertips. He needed only to consider the distribution of one-sided cosets of H in double cosets. Each double coset HxH, with $x \notin H$, contains $|H : x^{-1}Hx \cap H|$ one-sided cosets (see the proof of Theorem 2.3), and this number is h by (iii). The number of one-sided cosets contained in double cosets different from H is $n - 1$, and the result follows.

The rest of the proof of the Theorem is based on the relation between the characters of G and those of the subgroup H, established in [**134**], and surveyed in §4. The normal subgroup N is exhibited as the intersection of the kernels of characters of G. While somewhat more conceptual proofs of the theorem have been found, all of them come down in the end to Frobenius's way of producing the normal subgroup N in terms of characters of G; no proof has been found which avoids the use of character theory. His argument, along with Burnside's proof of his $p^a q^b$ Theorem (Chapter III, §5), are models of how to prove theorems about the structure of finite groups using character theory.

Let $\psi^{(k)}$, $k = 0, 1, \ldots, s - 1$, be the characters of H, and $\chi^{(j)}$, $j = 0, 1, \ldots, t - 1$, the characters of G, with $\psi^{(0)}$ and $\chi^{(0)}$ the principal characters of H and G. Let e_k and f_j be the degrees of the characters $\psi^{(k)}$ and $\chi^{(j)}$, respectively, as in §4. For $p \in H$, with $p \neq 1$, let g_p and h_p denote the numbers of conjugates of p in G and H respectively. Then $g/g_p = h/h_p$, since the g/g_p is the order of the centralizer of p in G, while h/h_p is the order of the centralizer of p in H, and these coincide by (i). Moreover, the elements of H which are conjugate to p by elements of G belong to the conjugacy class of p in H, by the hypothesis of the theorem. With these preparations, equation (14) becomes

$$(16) \qquad \sum_j r_{kj} \chi^{(j)}(r) = \frac{gh_p}{hg_p} \psi^{(k)}(p) = \psi^{(k)}(p),\ ne_k\ \text{or } 0,$$

for each k, according as the element $r \in G$ is conjugate to an element $p \neq 1$ of H, or is 1, or is not conjugate to an element of H.

For the next step, he defined a new set of integers by the formulas:

$$s_{k\ell} = s_{\ell k} = \sum_j r_{kj} r_{\ell j}.$$

Upon multiplying equation (13) (in §4) by $r_{\ell j}$ and summing on j, one has

$$\sum_j \sum_k r_{\ell j} r_{kj} \psi^{(k)}(p) = \sum_j r_{\ell j} \chi^{(j)}(p);$$

and this becomes, by (16) and the definition of the integers $s_{\ell k}$,

$$\sum_k s_{\ell k} \psi^{(k)}(p) = \psi^{(\ell)}(p),\ p \neq 1,$$

---

[34]A subgroup of a finite group with this property is called a *TI-subgroup* (*TI* for trivial intersection). The concept, investigated here for the first time, has played an increasingly important role in late 20th-century work on finite group theory and character theory.

and for $p = 1$,

$$\sum_k s_{\ell k} e_k = n e_\ell.$$

As

$$\sum_k e_k \psi^{(k)}(p) = 0 \text{ or } h$$

by the second orthogonality relation (9), according as $p \neq 1$ or $p = 1$, the preceding equations can be combined, and yield

$$\sum_k s_{\ell k} \psi^{(k)}(p) = \psi^{(\ell)}(p) + \frac{n-1}{h} e_\ell \sum_k e_k \psi^{(k)}(p),$$

for all $p \in H$. Using these equations, it is easily verified, using the orthogonality relations, that

$$s_{k\ell} = \frac{n-1}{h} e_k e_\ell + e_{k\ell},$$

where $e_{k\ell} = 0$ or 1 according as $k \neq \ell$ or $k = \ell$. Note that this formula, with $k = l = 0$, gives another, somewhat more complicated, proof of (v).

Frobenius applied these calculations to obtain information about the relation between characters of H and characters of G in this special situation. He first observed that

$$\sum_j (r_{kj} - e_k r_{0j})^2 = s_{kk} - 2e_k s_{k0} + e_k^2 s_{00}.$$

If $k > 0$, the expression becomes

$$1 + \frac{n-1}{h} e_k^2 - 2e_k^2 \frac{n-1}{h} + e_k^2 (1 + \frac{n-1}{h}) = 1 + e_k^2.$$

Noting that $r_{k0} = 0$ if $k > 0$ and $r_{00} = 1$, by the Reciprocity Theorem, so that $r_{k0} - e_k r_{00} = -e_k$ for $k > 0$, it follows that, for $k > 0$, one has

$$\sum_{j>0} (r_{kj} - e_k r_{0j})^2 = 1.$$

This implies that, among the integers $r_{kj} - e_k r_{0j}, j > 0$, with a fixed choice of $k > 0$, one is ± 1 and the others are zero. A similar calculation shows that for $k, \ell \neq 0$, and $k \neq \ell$, one has

$$\sum_{j>0} (r_{kj} - e_k r_{0j})(r_{\ell j} - e_\ell r_{0j}) = 0.$$

This means that if $r_{kj} - e_k r_{0j} = \pm 1$ and $r_{\ell j} - e_\ell r_{0j} = \pm 1$ for a pair of integers k and ℓ both > 0, then $k \neq \ell$. The upshot is that, after changing notation, there is a bijection $\psi^{(k)} \to \chi^{(k)}$, $k = 1, \ldots, s-1$, from the set of all nonprincipal characters of H to a subset of the nonprincipal characters of G, with the properties that

$$r_{kk} - e_k r_{0k} = \pm 1,$$

and

$$r_{kj} = e_k r_{0j}, \; j \neq k,$$

for all values of j and k, with both > 0.

From equation (16), one obtains the following character formula:

$$\sum_j (r_{kj} - e_k r_{0j}) \chi^{(j)}(r) = \psi^{(k)}(p) - e_k \psi^{(0)}(p), \; k > 0,$$

for each element $r \in G$ which is conjugate to some element $p \in H$, including $r = 1$. The sum vanishes for the $n - 1$ elements belonging to no conjugate of H. It follows that, for $k = 1, \ldots, s - 1$, and r, p as above, one has

$$\pm \chi^{(k)}(r) - e_k \chi^{(0)}(r) = \psi^{(k)}(p) - e_k \psi^{(0)}(p).$$

Upon applying these formulas to $r = 1$, one obtains $\pm f_k = e_k$ for each $k > 0$, so the $+$ sign occurs in each case. Finally, for $k = 1, \ldots, s - 1$,

$$\chi^{(k)}(r) = \psi^{(k)}(p),$$

whenever r is conjugate to $p \in H$, and

$$\chi^{(k)}(r) = f_k,$$

in case r is one of the $n - 1$ elements belonging to no conjugate of H.

Let χ be a character of G of degree f. The set of elements $r \in G$ such that $\chi(r) = \chi(1)$ is a normal subgroup of G, and in fact is the kernel of the representation of G affording the character χ. This elementary fact, proved in [**134**], establishes the existence of the Frobenius kernel. Indeed, using it together with the character formulas proved above, one deduces that the set of elements $r \in G$ such that $\chi^{(k)}(r) = f_k$ for $k = 1, \ldots, s - 1$ is a normal subgroup N containing the $n - 1$ elements belonging to no conjugate of H and of course the identity element. But N contains no element of H different from 1. Indeed, an element $p \in H \cap N$ has the property that $\psi^{(k)}(p) = e_k$ for all characters $\psi^{(k)}$ of H, and is consequently equal to the identity element, by the second orthogonality relation (9). This completes the proof of the theorem.

While simpler, more conceptual, proofs of Theorem 6.1 have been found (see, for example, [**99**], §14A), all of them use, in one way or another, the ideas Frobenius introduced in his first proof, given above.

COROLLARY 6.1. *Let G be a transitive group of permutations acting on a set X of n elements, with the property that no element of G other than the identity element fixes two or more elements of X. Then there exist exactly $n - 1$ elements of G which have no fixed points in X, and these elements, together with the identity, form a normal subgroup of G.*

Let H be the subgroup consisting of the elements in G which leave some element of X fixed. Because the action of G on X is transitive, the G-action on X can be identified with the action of G by left translation on the set G/H of left cosets of H in G. As the set of elements fixing the left coset xH is xHx^{-1}, the assumption that no element of G fixes two or more elements implies that H is a TI-subgroup (in other words, H satisfies (*iii*) above). It is then easily verified that G is a Frobenius group, with Frobenius complement H, and the Corollary follows from the proof of the preceding theorem.

As Frobenius noted, special cases of the corollary had been investigated earlier by Maillet, and by Burnside, in the first edition of Burnside's book ([**69**], pp. 141-144), as well as in the references given above. Every finite group, in its regular representation, occurs as a permutation group in which no element of the group different from the identity has a fixed point, so the hypothesis of the Corollary can be viewed as a small perturbation of this situation. Burnside had also been aware of what the theorem meant in terms of abstract groups, and had proved Theorem 6.1 in [**74**], with the additional hypothesis that the Frobenius complements are solvable.

CHAPTER III

Burnside: Representations and Structure of Finite Groups

While several authors contributed to the theory of representations of finite groups immediately following Frobenius's series of articles from 1896-1898, Frobenius's most important rival at the early stages was Burnside. They began to follow each other's work closely, and their articles on the applications of characters to the structure of finite groups contained frequent references to the other. This chapter contains some biographical notes, followed by an account of Burnside's work on group representations and its applications to finite groups, together with related research done at about the same time by H. Maschke and E. H. Moore.

1. Burnside at Cambridge and Greenwich

William Burnside (1852-1927) was born in London, the son of William Burnside, a partner in a bookselling firm. He was left an orphan at the age of six, and received his education at Christ's Hospital, West Horsham, a school that admitted only those boys whose parents would not have been able to afford boarding school fees. He achieved the highest place in the mathematical school, and entered Cambridge University in 1871 with an entrance scholarship to St John's College, where he was regarded as the best man of his year. According to W. L. Edge,[1] he migrated to Pembroke College after a year, when he discovered that the standard of rowing at St. John's was so high that he could not get into the crew of eight who rowed in the first boat. Burnside graduated from Cambridge in the Mathematical Tripos of 1875 as Second Wrangler, and on that basis, was elected to a Fellowship at Pembroke College, which he held from 1875-1886. (The exit examination system in mathematics at Cambridge, called the Tripos, is described in an informative and entertaining way in the chapter about G. H. Hardy at Cambridge, in Kanigel's book *The Man Who Knew Infinity: A Life of the Genius Ramanujan* [**194**]).

As a fellow at Pembroke College, Burnside followed the tradition in applied mathematics at Cambridge, lecturing on hydrodynamics in an advanced course open to all at the University, and began to publish articles on elliptic functions and

---

[1]I am indebted to L. Solomon for sending me a copy of a letter he had received in 1979 from Professor W. L. Edge; it contained the information about Burnside's transfer from St. John's to Pembroke, and several other interesting remarks about Burnside's life and mathematical work. Among other things, Edge mentioned some correspondence between Burnside and H. F. Baker that was acquired by the library of St. John's College after Baker's death in 1956. The excerpts from that correspondence which appear later in this Chapter are quoted with the permission of St. John's College. I also wish to acknowledge assistance I received from Mrs. P. A. Judd, Assistant Librarian, Pembroke College, for arranging permission for me to quote from some material in the Pembroke College Library. The rest of the biographical information about Burnside is taken from the obituary article on William Burnside by A. R. Forsyth [**124**], and from a conversation I had with P. Neumann at Queen's College, Oxford, in November, 1992.

hydrodynamics in 1883, when he was 30. During that time, he also served as a coach, both for prospective Tripos examinees, and as a rowing coach.[2] Seminars, such as those at the University of Berlin in which Frobenius participated, and formal research training leading to a Ph.D. degree, were not available in Cambridge at that time.

In 1885, Burnside was appointed professor of mathematics at the Royal Naval College, Greenwich, a position he held throughout his teaching career. While there he married Alexandrina Urquhart (in 1886) and together they raised a family of two sons and three daughters. His official duties at Greenwich were entirely concerned with the training of naval officers. Burnside was a thorough and conscientious teacher, and made a strong impression on his students at the College. Near the time of his retirement, he was given a testimonial document full of praise for his teaching and personal qualities, signed by more than 100 "constructors" (naval architects) who had taken courses with him. The list of signers included, as noted by the Archivist at the College, several of the leading figures at the time in the field of naval architecture. Nevertheless, Burnside somehow found the time to follow the literature in mathematics, to carry out a vigorous program of research, and to publish several research articles each year until the last year of his life (for a bibliography of his published work, see Mosenthal and Wagner [**219**]). Forsyth remarked [**124**]: "There was a current belief now known to be justified by fact, that his old college had invited him to return to important office; but he remained at Greenwich." A handwritten note opposite this passage in the reprint of Forsyth's article in the Pembroke College Library states: "He was offered the mastership in 1903. G.B." (The initials G. B. were those of George Birtwistle, a fellow of the college at the time.)[3]

Burnside continued to work on hydrodynamics and elliptic functions until around 1891-1892, when he published two long papers on automorphic functions ([**63**], [**64**]). These led him to the study of discontinuous groups, and then to finite groups. In the years immediately following, he became deeply absorbed in finite group theory, culminating his work during this period with the publication of the first edition of his book [**69**], *Theory of Groups of Finite Order*, in 1897. It was the first treatise in English on finite group theory, and is still an important reference today. He remarked in the introduction:

> The present treatise is intended to introduce to the reader the main outlines of the theory of groups of finite order apart from any applications. The subject is one which has hitherto attracted but little attention in this country; it will afford me much satisfaction if, by

---

[2]As an undergraduate, he became well known as an oarsman. Forsyth [**124**] wrote: "While at St. John's College, even as a freshman, he had rowed in the Lady Margaret's First Boat which, with the famous Goldie as stroke, went head of the river in 1872." Forsyth, A. R., "William Burnside," *J. London Math. Soc.* **3** (1928), pp. 64–80. Used with permission. He maintained his prowess as an oarsman during the tenure of his fellowship, as shown by this excerpt from *Pembroke College Boat Club, 1831-1981* by M. B. Maltby: "1882. This was the first occasion on which a crew was entered for the Henley Royal Regatta. A four consisting of P. A. Ransom, J. B. Stack, W. Burnside and A. F. Simm met Reading Rowing Club and Caius [Caius College, Cambridge] in the first round of the Wyfold Cup. The race was won by Reading by four lengths from Caius, who themselves were two lengths ahead of Pembroke. ... "

[3]Forsyth, A. R., "William Burnside," *J. London Math. Soc.* **3** (1928), pp. 64–80. Used with permission.

means of this book, I shall succeed in arousing interest among English mathematicians in a branch of pure mathematics which becomes the more fascinating the more it is studied.

William Burnside (1852-1927)

As to the use of finite groups of linear transformations for proving theorems in finite group theory, he was pessimistic:

> Cayley's dictum that "a group is defined by means of the laws of combination of its symbols" would imply that, in dealing with the theory of groups, no more concrete mode of representation should be used than is absolutely necessary. It may then be asked why, in a book that professes to leave all applications to one side, a considerable space is devoted to substitution groups [permutation groups]; while other particular modes of representation, such as groups of linear transformations, are not even referred to. My answer to this question is that while, in the present state of our knowledge, many results in the pure theory are arrived at most readily by dealing with properties of substitution groups, it would be difficult to find a result that could most directly be obtained by the consideration of groups of linear transformations.

Within months after the publication of his treatise, he became aware of Frobenius's first papers on characters and the factorization of the group determinant, began experimenting with the subject, and published his own approach to the theory in 1898 ([**70**], [**71**]; see §3). While Frobenius was at least partly motivated by the potential applicability of characters to number theory, for example to density problems (see Chapter II, §2), Burnside focused on what information the new theory could provide about the structure of finite groups. He began his own investigation of representation theory, and found along the way important new results concerning representations and characters, together with the applications to finite group theory he was looking for.

His new interest had a strong influence on the second edition of his book [**87**], published in 1911; in fact, it was the first book to contain a comprehensive account of what was known at the time about representations of finite groups. Its preface began:

> Very considerable advances in the theory of groups of finite order have been made since the appearance of the first edition of this book. In particular the theory of groups of linear substitutions has been the subject of numerous and important investigations by several writers; and the reason given in the original preface for omitting any account of it no longer holds good. In fact it is now more true to say that for further advances in the abstract theory one must look largely to the representation of a group as a group of linear substitutions.

Burnside was elected a Fellow of the Royal Society in 1893, and served on the Council of that body from 1901-1903. He also served as a member of the Council of the London Mathematical Society from 1899-1917. These positions helped him keep up with mathematical developments in Europe and the United States, especially since his long-term position on the Council of the L.M.S. involved refereeing a large number of papers submitted to the journals of the Society. Forsyth remarked [**124**]:

> From 1906-1908 he served as President [of the London Mathematical Society]; while willingly allowing his name to be submitted for membership in the Council year after year, he accepted their highest office only with grave and characteristic reluctance. The honour in which he appeared to show most interest was conferred on him in 1900.

In that year he was elected an Honorary Fellow of his old college, Pembroke, and at the time of his death had become the senior on the small roll of Honorary Fellows. Yet, even in the few and far from fluent remarks of thanks which he made at the College dinner welcoming, by courteous custom, the newly elected honorary members of the foundation, he urged that the happy and successful pursuit of research was its own reward, ... [4]

Through his active participation in the affairs of the London Mathematical Society and the Royal Society, with regular attendance at meetings, and access to their libraries, his position at the Royal Naval College did not isolate him from the mathematical community as much as might be supposed. In addition to these resources, he made full use of an extensive personal library.[5] He also maintained an active correspondence with mathematicians abroad, notably Heinrich Weber,[6] and in the U.K., including a particularly interesting exchange, from around 1903 until his death in 1927, with Henry Frederick Baker (1866-1956), a few years junior to him at Cambridge. Baker was, at the time the correspondence began, a Fellow at St. John's College Cambridge, and was appointed Lowndean Professor of Astronomy and Geometry, at the University of Cambridge, in 1914. According to W. V. D. Hodge [**179**], Baker's geometry seminar, or "tea party," held at his home every Saturday during term, "was the prototype of the numerous seminars that are held nowadays, but was for a long time the only one of its kind." Burnside and Baker developed and maintained a keen interest in each other's research, and the two of them frequently exchanged ideas on a wide range of topics, including Galois theory, algebraic geometry, and the theory of invariants of finite groups of linear transformations, to which Burnside devoted a chapter in the second edition of his book.

For example, on January 2, 1907, Burnside wrote to Baker about his observation that the procedure Weber had laid out in his book [**278**] did not actually work. As Burnside put it:

... The paragraph in Weber is pp. 231-233 Vol. II (Second Edition)
"Aus diesem Satze können wir ... darstellbar sind." The first proof

---

[4]Forsyth, A. R., "William Burnside," *J. London Math. Soc.* **3** (1928), pp. 64–80. Used with permission. The attitude, expressed in the last sentence of the quotation, was to change. To give one illustration, André Weil, in *The Apprenticeship of a Mathematician*, Birkhäuser, 1992, p. 95, wrote:

> Fortunately for my finances, the Centre National de la Rercherche Scientifique (better known by its acronym, CNRS) was being inaugurated that year [1932], with some confusion as to the role it was to play. It occurred to someone to use it as a means of granting supplementary income to university faculty members who showed some inclination toward what was just beginning to be called "research": up until then, this had been called "personal work." I was informed that I had been selected for such a grant. As a result, every three months for several years I received a check drawn on the national treasury, which I would cash in Paris on the rue de Rivoli. When people asked my sister [Simone Weil] what her brother did, she would answer: "He does research." "In what?" "How to get money from the government." Weil, A., *The Apprenticeship of a Mathematician*, Birkhauser, Boston, 1992, p. 95. Used with permission of Springer-Verlag, New York.

[5]Burnside's mathematics collection was left to Pembroke College, Cambridge. An inventory of the collection can be found among Burnside's personal papers in the Pembroke College Library.

[6]See the first letter, dated Jan. 2, 1907, from Burnside to Baker, quoted below.

I take to be the general theorem of §57. In this second proof, and there only, does Weber put forward a definite process for forming a system of invariants of a finite group in terms of which all the invariants are rationally expressible. And his system won't work as the following simplest possible example shows. The group

$$x_1' = x_1 \quad x_1' = -x_1 \quad x_1' = x_2 \quad x_1' = -x_2$$
$$x_2' = x_2 \quad x_2' = -x_2 \quad x_2' = x_1 \quad x_2' = -x_1$$

when used to form the $\Theta(x)$ of bottom of p. 229 gives

$$\Theta = c_1 x_1 + c_2 x_2$$
$$\Theta_1 = -c_1 x_1 - c_2 x_2$$
$$\Theta_2 = c_1 x_2 + c_2 x_1$$
$$\Theta_3 = -c_1 x_2 - c_2 x_1.$$

The one condition on the c's is that no two of the Θ's shall be identical (middle of p. 230). This is satisfied if $c_1 = 1$, $c_2 = 0$, and then

$$\Theta(t) = t^4 - (x_1^2 + x_2^2)t^2 + x_1^2 x_2^2.$$

Now the invariants of the group are not rationally expressible in terms of $x_1^2 + x_2^2$ and $x_1^2 x_2^2$; for $x_1 x_2$ is an invariant.

Burnside also stated that he had pointed this out in a letter to Weber, "and he agrees and admits that so far no process has been given for forming a set of invariants of a finite group of linear substitutions in terms of which all others are rationally expressible."

Burnside and Baker were still actively discussing the problem of finding the invariants of finite groups in 1910 at the time Burnside was working on the second edition of his book. In his efforts to include new material on representations of finite groups and the theory of invariants, he solicited Baker's comments, gratefully incorporated them into his revision, and made further refinements following the mathematical give-and-take with his friend. On January 16, 1910, for instance, Burnside thanked Baker for his suggestions and forewarned him that, "When I have written my proofs in the light of your suggestions, I will ask you to glance at it again." He went on to offer a justification of the exposition that Baker had criticized:

The rather longwinded way of establishing the existence of n algebraically independent invariants was really chosen deliberately to cover all cases in one statement. If the group of substitutions is irreducible, the equation

$$\prod_g (t - x_1^g) = 0$$

certainly has n of its roots algebraically independent and therefore the same must hold for the coefficients. But if it is reducible this is not necessarily the case e.g. if the group is

identity and $(x_1 x_2)(x_3 x_4)$

the equation for x_1 is

$$t^2 - t(x_1 + x_2) + x_1 x_2 = 0$$

and has 2 algebraically indep. coeffs. This is a trivial case, but will probably make my meaning clear.

He closed the letter with a question, the answer to which he also hoped to include in his book:

> ..., but there is a case more nearly connected with what I am working at I would like to put forward. $x_0 x_1 \cdots x_{p-1}$ and $x_0', x_1' = \omega x_1, x_2' = \omega^2 x_2, \ldots, x_{p-1}' = \omega^{p-1} x_{p-1}$ generate a group of order p^3, p an odd prime. If $p > 3$, there are $p + 1$ linearly independent invariants of degree p, viz.
>
> $$\xi_m = \prod_{n=0}^{n=p-1} \left(\sum_{i=0}^{i=p-1} \omega^{mi(i-1)+ni} x_i \right), m = 0, 1, \ldots, p - 1$$
>
> and $x_\infty = x_0 x_1 \cdots x_{p-1}$. How to form and resolve if possible the equation connecting them (say for $p = 5$).

By January 20, Burnside had made some of his revisions, and had given some more thought to the problem of finding invariants:

> With this I enclose my statement rewritten. I don't think it will take you long to look at it again, and I should be most grateful if you would.
>
> ... I have come to the conclusion that for groups of finite order (linear substitutions) almost the simplest nontrivial test case (there must as you point out be not less than 4 variables to be sure of striking new ground) is that presented by the icosahedral group in 4 variables; which is really the same thing as alternating functions in 5 variables. I have satisfied myself that no process really analogous to that which succeeds with alternating functions of 4 variables is available. ...

Finally, four days later, Burnside acknowledged Baker's help once more, and was satisfied that his revision was sound: "Very many thanks. I feel now safe about the matter of three invariants. When the chapter is in type I will remember that you wish to see it." He had also worked out the calculations for the icosahedral group in four variables, and explained to Baker that:

> As regards the icos. group in 4 variables, it is merely what the alternating group in 5 symbols becomes if one assumes (as one always may with a transitive permutation group) that the sum of the symbols is zero. It can be generated by
>
> $$S \text{ or } \begin{array}{l} x' = y \\ y' = z \\ z' = u \\ u' = -x - y - z - u \end{array} \qquad \text{and } T \text{ or } \begin{array}{l} x' = y \\ y' = x \\ z' = u \\ u' = z \end{array}$$
>
> which satisfy $S^5 = E$, $T^2 = E$ and $(ST)^3 = E$. The invariants are the symmetric and alternating functions of $x, y, z, u, -(x+y+z+u)$; and the syzygy is got from the ordinary formula giving the discriminant of a quintic in terms of the coefficients by merely putting p_1 (the sum of the roots) $= 0$ and regarding p_2, p_3, p_4, p_5 and *discrim.* as the 5 connected invariants.

It was also clear that the end of his work on the second edition was in sight, for he asked Baker:

> Do you happen to know any young man in Cambridge who has taken any interest in groups of finite order? If such a one exists it would probably be an excellent thing for him to read my proof sheets carefully and I know it would be a great blessing for me. With a second edition one cannot come down on one's friends, but a keen young student would almost certainly benefit both himself and me.

Most of Burnside's research on representations of finite groups was done in the period from the publication of the first edition of his book, in 1897, through the preparation of the second, revised edition of 1911, and until the beginning of the 1914-18 World War. One might have expected that his Presidential Address, entitled *On the Theory of Groups of Finite Order* ([**85**], 1908), would have contained a summary of the new developments, but its tone suggests that Burnside felt that he was still swimming upstream, in terms of generating interest in group theory on his side of the English Channel:

> It has been suggested to me that I should take advantage of the present occasion to give an account of the recent progress of the theory of groups of finite order. That very considerable advance has been made in the last twenty, and especially in the last ten years, is undoubtedly the case. That advance, however, has been in a variety of directions, and it is probably too soon, as yet, to present its different parts in their proper proportion and perspective. It would not be possible to do at the present time for the theory of groups of finite order what has been done so ably for the theory of algebraic numbers by Prof. Hilbert in his report of 1897.
>
> But a more serious objection to any attempt on my part to give, on the present occasion, an account of the recent advance in the theory is that such an account would certainly be uninteresting to a considerable number of my audience. It is undoubtedly the fact that the theory of groups of finite order has failed, so far, to arouse the interest of any but a very small number of English mathematicians; and this want of interest in England, as compared with the amount of attention devoted to the subject both on the Continent and in America, appears to me very remarkable. I propose to devote my address to a consideration of the marked difference in the amount of attention devoted to the subject here and elsewhere, and to some attempt to account for this difference.

After carrying out some war-related research for the Navy, he gradually turned his attention to problems in applied mathematics and the theory of probability. Nevertheless, a few more papers on representation theory did appear between 1920 and 1925, and he continued his correspondence about algebraic geometry with Baker. In 1925, Baker had completed Volume IV of his treatise on algebraic geometry, *Principles of Geometry*, and on September 5, he sent a copy to Burnside accompanied by a letter, noting that he had reworked an earlier paper of Burnside in a chapter of the new volume:

I have great pleasure in being able to send you Vol IV of my book; I hope you will accept it. I suppose I am more concerned to know how it strikes you than for anyone else's opinion.

I have laboured for many years, very hard, at it; rearranging it, altering it, and muddling it about–and now it is done. It seems so slight and incomplete, and imperfect in so many ways. In particular you will truly say, I think, that the fact that Segre's variety has no absolute invariant is not sufficiently brought out.

I hope you will not mind the muddling about I have given, in Chap. IV, to your very remarkable paper. Or, worse, will not be able to convict me of a logical error. Often I have tried to bring the proof into briefer compass–because your own account still leaves one with the impression that the symmetry of the relations is not completely proven–and this is the best I could do till I gave it up. There still remains, I am sure, to find a "transformation" from the figure of the cubic surface.

But with all my complaints, there is no other book written from this point of view–I only hope it may do something to bring the hyperspace stuff into common use–which is my main purpose. My account, I know, needs spreading thinner; but I cannot help that, though I tried. And others will do it.

In a postscript, he added, "I wonder if you have seen Klein's collected papers? Three vols, 2000 odd pages. Full of an old man's reminiscences in small print—*as interesting as a fairy tale*—and very important for the historian of all that Klein was keen about."

Obviously pleased with Baker's remarks, Burnside replied on September 13, 1925. A comment at the end indicated that he was no longer closely following the German literature.

I have been away for a few days, and I found your Vol IV waiting me yesterday. I want to thank you very much for sending it to me; and I am sure I shall be even more interested in it than in the preceding three volumes.

May I say too that your letter that came with it is most gratifying to me. It must be years since I have re-read my little paper on 27 hyperplanes. I am going to do so before tackling your Chap IV; and I am more than prepared to find that what you call "muddling it about" is in fact putting the proof of the existence of the configuration on a proper logical basis and replacing assumptions by proofs. When I have done this much I will write you again at length. After that I will go on to study the part about Segre's variety.

So far as one can judge by examining the table of contents, I should say that you have given us all a practical introduction to the use of 4- and 5- dimensional geometry, and I am quite sure it will not seem "slight and incomplete" (your phrase in your letter) to your readers.

I seem to remember having put before you difficulties I had had with cubic invariants of linear groups in 5 homogeneous variables:

> but just now I cannot recall the point. Perhaps it may occur to
> me later.
>
> No, I have not seen Klein's collected papers. The real truth
> is that German is not so easy for me as it was 25 years ago and I
> should hesitate to buy 2000 pages for fear of making very little use
> of them.

Burnside continued to pursue the connections between finite group theory, algebraic geometry, and the theory of invariants through the following year. On
January 19, 1926, he wrote:

> I had a slight stroke on Dec. 22 and though I have gone on very
> well I am by no means out of the doctor's hands yet. Among other
> things he forbids any interest in mathematics. So with a view to a
> speedy recovery I have put them on one side. But later I hope to
> be able to send you a group of 2×25920 birational transformations
> of A, B, C, D, E for which E^2 is a rational function of A, B, C, D
> and E is not.

A few months later, he was able to fulfill this promise [**88**]. The group of order
25920 to which he referred is the rotation group in the group of the twenty-seven
lines on a cubic hypersurface, which he had investigated for the first time in 1909
([**86**]). This group has appeared and reappeared in various disguises throughout
the 20-th century. Perhaps its most important manifestation is its occurrence as
the rotation subgroup of the Weyl group of the exceptional simple Lie algebra of
type E_6 ([**182**], p. 45).

Burnside died in 1927. It is surprising that one published obituary contained
not a word about his mathematical research. The full text of it is as follows:[7]

> Rowing men will regret to hear of the death of W. Burnside, one
> of the best known Cambridge athletes of his day. He missed his
> Blue but captained the Pembroke boat.
>
> Appointed Professor of Mathematics at the Royal Naval Col-
> lege, Greenwich, Burnside became well known to many generations
> of Naval cadets. On one occasion he and a friend rowed from Cam-
> bridge to London.

Burnside's concern about the lack of interest in the theory of groups of finite
order among English mathematicians, expressed in his Presidential Address quoted
earlier, would almost certainly have been relieved if he had lived a little longer. In
1928 the first papers of Philip Hall (1904-1982) on group theory began to appear,
and were followed by a series of fundamental contributions to algebra and group
theory by Hall and his research students. In his obituary article [**243**], *Philip Hall*,
J. E. Roseblade wrote:

"By 1940 Hall's work commanded great respect and admiration; he was recog-
nised as knowing more about groups than anybody else and was thought of as the
very successor of Burnside. In 1942 he was elected to the Royal Society. Hall wrote
at the time:

> The aim of my researches has been to a very considerable extent that
> of extending and completing in certain directions the work of Burn-
> side. I asked Burnside's advice on topics of group-theory which would

---

[7]From a clipping in the Pembroke College Library.

be worth investigation and received a post-card in reply containing valuable suggestions as to worth-while problems. This was in 1927 and shortly afterwards Burnside died. I never met him, but he has been the greatest influence on my ways of thinking.

Burnside had more than group theory in common with Hall; he was also educated at Christ's Hospital and left as an Exhibitioner after winning the Thompson Gold Medal; his Obituary Notice appears in the same volume as Hall's short note of 1928."

2. Burnside's Early Research on Finite Groups: 1890-1900

Burnside's first publications on finite groups focused on two subjects: simple groups, in particular the order of a finite simple group; and permutation groups, especially doubly transitive groups. Today we can only marvel at his judgment in choosing for his first efforts in a new subject two areas of research which were to remain of central importance in finite group theory and computational algebra throughout most of the next century. While his predecessors in 19th-century group theory—Galois, Jordan, Netto, Hölder, and Frobenius—had all considered these subjects before him, Burnside advanced them to such an extent that he certainly deserves part of the credit for their place in 20th-century mathematics.

Simple groups are groups having no proper normal subgroups other than the subgroup containing only the identity element. The simple abelian groups of finite order are the cyclic groups of prime order. The finite nonabelian simple group of least order is the alternating group A_5 of order $60 = 2^2 \cdot 3 \cdot 5$, which consists of the even permutations of a set of 5 elements. C. Jordan and O. Hölder were the first to observe that every finite group $G \neq 1$ has a *composition series*, namely a chain of subgroups

$$G = G_0 \supset G_1 \supset \cdots \supset G_{s+1} = 1,$$

with the properties that each subgroup G_i is a normal subgroup of the preceding one, and each factor group G_i/G_{i+1}, called a *composition factor*, is a simple group. They proved the important facts that in two composition series, the numbers of composition factors are the same, and the composition factors in one series can be paired with the composition factors of the other series in such a way that corresponding composition factors are isomorphic. In this sense every finite group is built up in a certain way from a uniquely determined set of finite simple groups.

Solvable groups are finite groups whose composition factors are cyclic of prime order. These were defined by Galois, who used them to prove his great theorem that the polynomial equations of degree ≥ 5 are not in general solvable by radicals. To show this, he needed to exhibit some nonsolvable finite groups, and he remarked that symmetric groups S_n with $n \geq 5$, and the modular congruence groups $PSL_2(p)$ with $p \geq 5$ (defined in Chapter II, §3), were examples [**154**]. Following his new proofs of Sylow's theorems (see Chapter II, §2), Frobenius had begun a thorough investigation of the structural properties of finite solvable groups, and published a series of articles on the subject, with the titles *Über auflösbare Gruppen* I, II, ..., beginning in 1893.

Burnside's first paper [**65**] on finite simple groups, published in 1893, contained a proof of the result that, with the exception of A_5, there exists no finite simple group whose order is the product of 4 primes. He began by remarking that it was well known that there is no simple group whose order is the product of 2 primes, and that O. Hölder [**180**] had proved there were none whose order is a product of 3 primes. Burnside gave a new proof of Hölder's result based on Sylow's theorem, and extended the argument to handle groups whose orders are products of 4 primes. For good measure, he gave his own proof of Sylow's theorems, believing at the time that the elegant arguments he had found were new. Two years later, in [**67**], he was forced to admit:

> I take this opportunity of expressing my regret that in my former "Notes on the Theory of Groups of Finite Order" (Vol. XXV., pp. 9-18) I was led, by my ignorance of Herr Frobenius's investigations on the subject, to giving as new a proof of Sylow's theorem which was in fact six years old. His two papers [reference is made to [**126**] and [**127**]; see Chapter II, §2] contain a proof of the two parts of the theorem with which mine is, in essence, identical.

This is not the happiest way to become aware of a fellow contributor to one's area of research, but is all too familiar to active mathematicians.

The series of papers with the title *Notes on the Theory of Groups of Finite Order* all have a common theme, which he stated as follows:

> They are concerned chiefly with the proof of certain tests that may be applied in particular cases to determine whether it is possible for a simple group of a given finite order to exist.

For example, in [**67**], he proved, that if $p_1, p_2, \ldots$, are distinct primes, in increasing order, then:

(i) there are no simple groups whose orders are of the forms

$$p_1 \cdots p_{n-1} p_n^m, \; p_1 \cdots p_{n-2} p_{n-1}^2 p_n, \; p_1^m p_2, \; p_1 p_2^m p_3;$$

(ii) a group whose order is

$$p_1^{m_1} \cdots p_n^{m_n},$$

in which the subgroups of orders $p_1^{m_1}$, $p_2^{m_2}, \ldots$, are all cyclic, cannot be simple; and

(iii) the only simple groups whose orders consist of the product of five primes are the three known groups of orders $2^3 \cdot 3 \cdot 7$, $2^2 \cdot 3 \cdot 5 \cdot 11$, and $2^2 \cdot 3 \cdot 7 \cdot 13$.

The proofs were based mainly on the Sylow theorems, and of course, mathematical induction. Finite groups whose orders contain no repeated prime factors are special cases of (ii), and were proved not to be simple at about the same time by Hölder [**181**] and Frobenius [**128**].

In the next paper in the series [**68**], Burnside considered groups of even order, and proved the following result.

THEOREM 2.1. *Let G be a finite group of even order, whose Sylow 2-subgroups are cyclic. Then G cannot be a simple group.*

For the proof, he first assumed that G contains some element s, of odd order, that commutes with no elements of order 2. From this, it follows that the centralizer of s has odd order. The number of elements in the conjugacy class C of s is equal to the index of the centralizer of s, and is consequently $2^m r$, where 2^m is the order of a Sylow 2-subgroup of G, and r is an odd number. The group G acts by conjugation as a transitive permutation group on the conjugacy class C. Let A be a Sylow 2-subgroup of G; then A is cyclic of order 2^m, by the hypothesis of the theorem. Let a be a generator of A, and consider the cycle decomposition of the permutation of the set C given by conjugation by a. As a has order 2^m, the length of each cycle is 2^k, with $k \leq m$. But in fact, the length of each cycle is 2^m. Otherwise, the length of some cycle is 2^k, with $k < m$, and there is a subgroup of A of order 2^{m-k} which acts trivially on the elements of C making up the cycle. These elements, which are all conjugates of s, would commute with the elements of a nontrivial subgroup of A, contrary to the assumption that s commutes with no element of order 2. It follows that the permutation of C corresponding to a is a product of r cycles each of length 2^m, with r odd, and is consequently an odd permutation. From this it follows easily that G cannot be simple. In the other case that has to be considered, we may assume that there is no element of odd order which does not commute with a given element b of order 2. This involves the fact that all the elements of order 2 are conjugate in G, by the Sylow theorems, and the assumption that the Sylow 2-subgroups are cyclic. Then it is clear that b belongs to the center of G, and G cannot be simple in this situation either, completing the proof.

The proof given above, from [68], is included here as an illustration of how Burnside, and also Frobenius at about the same time in [128], applied the theory of permutation groups to prove theorems about the structure of abstract groups. Later, Burnside gave a much simpler proof, in the first edition of his book ([69], §250, p. 360), as follows. Let the order of G be $2^n r$, with $n \geq 1$ and r an odd number. Let A be a Sylow 2-subgroup of G, assumed cyclic as in the hypothesis of the theorem, and let a be a generator of A. In the regular representation of G, the element a is represented as the product of r cycles, each of length 2^a, and is consequently an odd permutation. Therefore G cannot be simple.

In §243 of the second edition of his book [87], Burnside proved a more general result, known today as:

> BURNSIDE'S TRANSFER THEOREM. *Assume that a Sylow p-subgroup P of a finite group G is contained in the center of its normalizer. Then G contains a normal p-complement, that is, a normal subgroup H of order prime to p, such that $G = PH$.*

Burnside proved the theorem using monomial representations of finite groups. For a proof using the transfer map, see M. Hall ([168], §14.3). That Burnside's Theorem 2.1 is indeed a special case of the Transfer Theorem is shown as follows. Let P be a Sylow 2-subgroup of a finite group G, such that P is cyclic of order 2^a. Then the automorphism group of P has order $\varphi(2^a) = 2^{a-1}$ (where φ is the Euler φ-function). Let N and C denote the normalizer and the centralizer of P, respectively; then the quotient group N/C has odd order, and is isomorphic to a subgroup of the automorphism group of P, which has order a power of 2. It follows that $N = C$, so that P is contained in the center of its normalizer, and the Transfer Theorem can be applied.

Another theme which Burnside developed in his early work on finite group theory was the proof of existence of some infinite families of finite simple groups. These were obtained from linear groups, and groups of linear fractional transformations, over finite fields.

In his papers on automorphic functions [63] and [64], he had already made a thorough study of groups of *linear fractional transformations*

$$z \rightarrow \frac{az+b}{cz+d}$$

in the complex plane, where a, b, c, d are complex numbers such that $ad - bc = 1$. He was interested in the study of infinite, but finitely generated, groups of linear fractional transformations in connection with the work of Henri Poincaré on automorphic functions published in the first, third, fourth, and fifth volumes of *Acta Mathematica*.[8] Poincaré had proved that, for a finitely generated group of suitably normalized transformations

$$z \to \frac{a_i z + b_i}{c_i z + d_i}, \ i = 1, 2, \ldots,$$

the infinite series

$$\sum |c_i z + d_i|^{-2m}$$

is always convergent, in case m is a positive integer > 1. It follows that the series

$$\sum_i H(\frac{a_i z + b_i}{c_i z + d_i})(c_i z + d_i)^{-2m}$$

converges uniformly, except for particular values of z, for an entire function $H(z)$, and defines what Poincaré called a theta-fuchsian or a theta-kleinian function $\Theta(z)$, depending on the type of the group. Such a function $\Theta(z)$ has the property that for each transformation

$$z \to \frac{az + b}{cz + d}$$

belonging to the group,

$$\Theta(\frac{az + b}{cz + d}) = (cz + d)^{-2m}\Theta(z).$$

The quotient of two of these functions, for a fixed value of m, is invariant under all the transformations in the group, and is a kind of *automorphic function*.[9] In his first paper [**63**], Burnside analyzed the convergence of the series defined above for the case $m = 1$, and used quite a lot of group-theoretic information for this purpose. It was the group theory, however, and not the analysis, that had caught his fancy, and within a year, he changed the direction of his research, and began his series of papers on finite groups.

Groups of linear transformations over the fields $\mathbb{Z}_p$, for a prime p, had been studied by Galois, Serret, and especially Jordan in his great work *Traité des substitutions et des équations algébriques* ([**192**], 1870).[10] In his first paper on linear groups over finite fields ([**66**], 1894), Burnside proved that the group of linear fractional transformations

$$z \to \frac{az + b}{cz + d},$$

with the coefficients a, b, c, d and the variable z taken from a finite field of p^n elements, was simple, of order $2^n(2^{2n} - 1)$, or $\frac{1}{2}p^n(p^{2n} - 1)$ for odd primes p, with the exceptions $p = 2, n = 1$ and $p = 3, n = 1$, extending a result proved by Jordan for the case $n = 1$. He believed that he had discovered an infinite family of new

---

[8]Finite groups of linear fractional transformations had been classified by Arthur Cayley [**93**] and Felix Klein [**196**], and were not involved in Burnside's work at this stage.

[9]In a footnote on p. 52 of [**63**], Burnside remarked: " I have used the phrase "automorphic function," as introduced by Professor Klein, to denote generally any function which is unchanged by the substitutions of a discontinuous group, whatever be the nature of the group."

[10]In a footnote on the second page of [**45**], Richard Brauer remarked, "It is interesting to note that Jordan in the introduction to his fundamental book [**192**] refers to the book as a "commentary" on the work of Galois. Historians of mathematics have remarked that this is one of the most modest statements ever made by a mathematician."

simple groups. A few months earlier, however, Eliakim Hastings Moore (1862-1932), who had been appointed Professor and Acting Head of the Department of Mathematics at the University of Chicago the previous year, published a paper [**217**] with the title *A doubly-infinite system of simple groups*, containing proofs of the same results. So, at best, Burnside had to settle for a share of the glory. He explained the situation in the first edition of his book [**69**] as follows: "For an independent proof of the existence of these simple groups and for an investigation of their properties, the reader is referred to the memoirs mentioned below [citing the papers by Moore and himself]." The addition of E. H. Moore to the list of persons whose work he needed to follow turned out to be useful to him when he began his research on representation theory a few years later.[11]

Further results on the theme of the possible existence or nonexistence of simple groups of a given finite order, due mainly to Burnside and Frobenius, were included, often in an improved form, and with new proofs, in the last chapter of the first edition [**69**] of Burnside's book *Theory of Groups of Finite Order*, published in 1897. By this time, he was beginning to think about some long range problems about simple groups. He mentioned them in a note at the end of the book:

> No simple group of odd order is at present known to exist. An investigation as to the existence or nonexistence of such groups would undoubtedly lead, whatever the conclusion might be, to results of importance; it may be recommended to the reader as well worth his attention. Also, there is no known simple group whose order contains fewer than three different primes. This suggests that Theorems III and IV, §§243, 244, may be capable of generalisation. Investigation in this direction is also likely to lead to results of interest and importance.

Burnside never lost interest in these problems. He solved the second one himself around 1904, using the theory of characters combined with some algebraic number theory. His result, called the $p^a q^b$-Theorem, states that all groups of order $p^a q^b$, with p and q primes, are solvable (see §5). Theorems III and IV, referred to in the quotation above, were special cases.

Progress on the first problem was a long time coming.[12] A breakthrough came with Michio Suzuki's proof [**268**] in 1957 that there are no simple groups of composite odd order having the property that the centralizers of all nonidentity elements are abelian. His proof was based, in part, on a subtle extension of Frobenius's work on induced characters, called the theory of exceptional characters (see Chapter VII, §3). The next step was taken by Walter Feit, Marshall Hall, and John Thompson [**120**], who proved that the same result held for groups with the property that centralizers of nonidentity elements are nilpotent.

The culmination of this line of research came in 1963, with the publication by Feit and Thompson of what has become known as the odd-order paper [**122**], containing one theorem: *All finite groups of odd order are solvable*. Although a purely group-theoretic proof (not using characters) of Burnside's $p^a q^b$-Theorem was found later (see §5), the proof of the odd-order theorem, and, to go farther back,

---

[11]For an account of the mathematical work of E. H. Moore, and his contribution to the emerging American school of mathematical research during the last part of the 19th century, see Parshall and Rowe [**236**], Chapters 6-7 and 9.

[12]The following two paragraphs are taken, more or less unaltered, from the author's article [**98**].

the proof of Frobenius's theorem on the structure of Frobenius groups (Chapter II, §6), both contain apparently unavoidable applications of character theory. Feit and Thompson's proof of the odd-order theorem takes about 250 pages of close reasoning, which to this day has resisted significant simplification, so perhaps the 65-year wait following Burnside's statement of the problem[13] is not surprising.

3. From Lie Groups to Representations of Finite Groups: 1898-1900

Burnside's interests in group theory were by no means restricted to finite groups. Among other things, he studied the publications of Sophus Lie, Eduard Study, Wilhelm Killing, and Élie Cartan on continuous groups of linear transformations and their Lie algebras. Owing to the lesson he had learned from publishing his first papers on finite group theory without sufficiently checking the literature (see §2), he was also keeping track of Frobenius's publications as they appeared in the *Berliner Sitzungsberichte*. When Frobenius's articles on character theory and the factorization of the group determinant appeared in 1896, Burnside saw in them a way to relate finite groups to continuous groups, and published two papers ([70], [71]) on the subject in 1898. An account of them is given here to show how Burnside found the group matrix of a finite group in the setting of Lie groups, and how he was led, at the end of the second paper, to the important concept of an irreducible representation of a finite group, which was to play a fundamental role in his approach to representation theory.

His idea was to use the group matrix of a finite group G to define a continuous group (or Lie group as we would say today) $\mathcal{G}$ associated with G, whose Lie algebra $\mathfrak{g}$ is the group algebra of G over the field of complex numbers, with the bracket operation $[X, Y] = XY - YX$, for $X, Y \in \mathfrak{g}$. The Lie group $\mathcal{G}$, or "transitive linear homogeneous group in n variables with n independent parameters," as it was called at that time, was defined by the transformations

$$x'_s = \sum_{i,k} a_{sik} x_i y_k, \ s, i, k = 1, 2, \ldots, n,$$

with $n = |G|$, and structure constants a_{sik} defined in terms of the elements $S_1, \ldots, S_n$ of G by setting $a_{sik} = 1$ or 0 according as $S_k S_i^{-1} S_s$ is equal to 1 or not. For the general form of the equations, he referred to Study [266]. In this case, the resulting Lie group $\mathcal{G}$ can be identified with the group of units in the set of transformations

$$x_1 S_1 + \cdots + x_n S_n \to x'_1 S_1 + \cdots + x'_n S_n,$$

where $x'_1 S_1 + \cdots + x'_n S_n$ is the result of multiplying $x_1 S_1 + \cdots + x_n S_n$ and $y_1 S_1^{-1} + \cdots + y_n S_n^{-1}$ in the group algebra. In other words, $\mathcal{G}$ is the group of units in the algebra consisting of the linear transformations R_y, where R_y is the operator on the group algebra defined by $x \to R_y(x) = xy$, and xy denotes the product of the elements x and y of the group algebra. The expression of $y = y_1 S_1^{-1} + \cdots + y_n S_n^{-1}$

---

[13]Burnside amplified his remarks on the problem in Note M, at the end of the second edition of his book [87], with the comment: "There is in some respects a marked difference between groups of even and those of odd order." He continued with a discussion of the possible existence of nonabelian simple groups of odd order, noting that he had shown, for example, that the number of possible prime factors of a simple group of composite odd order is at least 7. He concluded: "The contrast that these results shew between groups of odd and even order suggests inevitably that nonabelian simple groups of odd order do not exist."

in terms of the inverses of the group elements guarantees that the operators R_y satisfy the condition

$$R_{y_1 y_2} = R_{y_1} R_{y_2}$$

for all elements y_1 and y_2 of the group algebra. Burnside did not make this last point explicitly, and it is possible he had something else in mind in his use of the inverses. In any case, for y given as above, write the coefficients in terms of the group elements, such that $y_i = y_{S_i}$ for $i = 1, \ldots, n$. Then the matrix of the linear transformation R_y with respect to the basis $S_1, \ldots, S_n$ of the group algebra is precisely the group matrix

$$(y_{S_i S_j^{-1}}),$$

as Burnside observed at the beginning of the first paper.

After identifying the Lie algebra $\mathfrak{g}$ of $\mathcal{G}$ with the group algebra of G, as noted above, Burnside began a thorough investigation of the structure of the Lie algebra $\mathfrak{g}$. The first step was to prove (as it is a nice exercise to do) that the center of the Lie algebra $\mathfrak{g}$ is the subspace of the group algebra consisting of the linear combinations of group elements $\sum x_S S$ with the property that $x_S = x_T$ whenever the elements S and T are conjugate in G.

The methods used in his first paper on continuous groups [**70**] were taken from the dissertation [**89**], entitled *Sur la structure des groupes de transformations finis et continus*, of Élie Cartan, published in 1894. Burnside could not have found a better source, as the dissertation was (and still is) a masterful presentation of the structure of Lie algebras over the field of complex numbers, and especially semisimple Lie algebras. While root systems and simple Lie algebras were first classified by W. Killing (see Coleman [**94**]), there were enough mistakes and rough edges in Killing's account to justify a new version, which Cartan provided, together with new results which illuminated the general structure of Lie algebras in a beautiful way. So it was not surprising that Burnside included Cartan's dissertation in his self-study project on continuous groups, and applied the ideas in it to the Lie algebra $\mathfrak{g}$ defined above. The main results he obtained in [**70**] can be summarized in modern terminology as follows.

THEOREM 3.1. *Let $\mathfrak{g}$ be the Lie algebra whose underlying vector space is the group algebra of a finite group G, with the bracket operation $[X, Y] = XY - YX$. Then:*

(i) The center $\mathfrak{z}$ of $\mathfrak{g}$ has dimension r, where r is the number of conjugacy classes in the finite group G;

(ii) The center $\mathfrak{z}$ of $\mathfrak{g}$ coincides with the maximal solvable ideal[14] of $\mathfrak{g}$.

(iii) Assume G is not abelian. Then the derived algebra $[\mathfrak{g}, \mathfrak{g}]$ is semisimple, and $\mathfrak{g}$ can be expressed as the direct sum:

$$\mathfrak{g} = \mathfrak{z} \oplus [\mathfrak{g}, \mathfrak{g}].$$

In the second paper, the investigation of the Lie algebra $\mathfrak{g}$ was "brought to a definite conclusion" with the next result.

THEOREM 3.2. *Let $\mathfrak{g}$ and r be as in the previous theorem. There exist integers $\mu_1, \ldots, \mu_r$ such that*

$$\mu_1^2 + \cdots + \mu_r^2 = |G|;$$

---

[14]Today, the latter is called the *radical* of $\mathfrak{g}$.

moreover, $\mathfrak{g}$ can be expressed as a direct sum,

$$\mathfrak{g} = \mathfrak{g}_1 \oplus \cdots \oplus \mathfrak{g}_r,$$

of Lie algebras $\mathfrak{g}_i$ of dimensions $\mu_1^2, \ldots, \mu_r^2$, respectively. Each Lie algebra $\mathfrak{g}_i$ is isomorphic to the Lie algebra consisting of all linear transformations on a vector space of dimension μ_i, with the bracket operation $[X, Y] = XY - YX$.

By this time, Burnside had become familiar with Molien's research on the structure of associative algebras (see Chapter II, §4). His proof of the second theorem was based on the analysis of the characteristic polynomial of the general element[15] of the group algebra of the finite group G, acting by left multiplication on the group algebra. This was the method Molien had used (see Hawkins [**173**], §3), and Burnside acknowledged that some of his results are "equivalent to results obtained by Herr Molien in his memoir *Über Systeme höherer complexer Zahlen* (Math. Ann., XLI., pp. 83-156, 1892). The form in which the properties are here presented and the methods of proof are quite distinct from those of Herr Molien." We recognize the theorem as an early version (stated for Lie algebras instead of associative algebras) of the standard structure theorem for group algebras over the field of complex numbers, proved today using Wedderburn's theorem, as in Emmy Noether's paper [**233**] (see Chapter VI, §2).

The following quotation, from the introduction to the second paper [**71**], gives some insight as to the motivation for the preceding results.

> The simultaneous consideration of the discontinuous group G, and the continuous group $\mathcal{G}$ that is allied to it, appears to me likely to facilitate the discussion of certain of the properties of G. In particular, it obviously has a bearing on the question of the smallest number of variables in which G can be expressed as a group of linear substitutions, i.e. what Prof. Klein calls the degree of the normal-problem connected with G. Again, the properties of the group-determinant, when investigated in this connexion, appear in a particularly clear and simple light.

As he promised, he was able to prove his own version of Frobenius's theorem (Chapter II, Theorem 3.4) on the factorization of the group determinant of a finite group G of order n, with r conjugacy classes. He stated it as follows.

THEOREM 3.3. *The determinant formed from the multiplication table of G, when the symbols of the operations are regarded as n independent algebraical quantities, contains r distinct irreducible factors. Each of these factors enters into the determinant to a power equal to its degree; and any irreducible factor of degree m is expressible in the form*

$$\begin{vmatrix} z_{11} & z_{12} & \cdots & z_{1m} \\ \cdots & & & \\ z_{m1} & z_{m2} & \cdots & z_{mm} \end{vmatrix}$$

where the z's are independent linear functions of the n algebraical quantities.

---

[15]The *general element* of an algebra A with basis $a_1, \ldots, a_n$ is the element $a = x_1 a_1 + \cdots + x_n a_n$ of the algebra obtained by extending the base field $\mathbb{C}$ to $\mathbb{C}(x_1, \ldots, x_n)$, for a set of algebraically independent indeterminates x_i over $\mathbb{C}$.

Burnside's proof of the preceding result was not appreciably simpler than Frobenius's; it was, however, based on a different set of ideas,[16] closely related to the proof of Theorem 3.2. The irreducible polynomial factors of the group determinant were obtained from the structure equations of the continuous groups $\mathcal{G}_i$ corresponding to the Lie algebras $\mathfrak{g}_i$, in the same way that the group determinant of G itself was obtained from the structure equations for $\mathcal{G}$. This meant that each irreducible polynomial factor was associated with one of the groups $\mathcal{G}_i$, and a representation of $\mathcal{G}_i$ of degree μ_i, for one of the integers μ_i from Theorem 3.2. Burnside realized that the homomorphism which projects $\mathcal{G}$ onto $\mathcal{G}_i$ defined (by specializing the variables) a homomorphism of finite groups from G to some finite group G_i, and a representation of G_i of degree μ_i. He then raised the question:[17]

> Now G, and the groups of smaller order with which G is isomorphic, can, in general, be represented as groups of linear substitutions in a variety of ways, and the question arises as to which of these different modes of representation occur among the groups G_i.

His answer to the question was the last result in [71], and was stated as follows.

THEOREM 3.4. *If a discontinuous group G', with which G is simply or multiply isomorphic, can be represented as a group of linear substitutions performed on m symbols, and if it is impossible, by choosing new variables, to divide the variables into sets such that those of each set are transformed among themselves by every operation in G', then m must be one of the integers μ_i defined above.*

Representations with the property stated in Theorem 3.4 are today called *indecomposable representations*. In [136], Frobenius had also given a definition of an indecomposable representation, and asserted that the notion was equivalent to the concept of primitive representation which he had defined in terms of the irreducible factors of the group determinant (Chapter II, §4). The issue here was to find a more direct and accessible characterization of primitive representations. In his paper *On group-characteristics* [72], published in 1900, Burnside defined a group of linear substitutions in m variables to be *irreducible* if "it is impossible to choose $m' < m$ linear functions of the variables which are transformed among themselves by every operation of the group," and proved that the concept of irreducible representation and Frobenius's concept of primitive representation were equivalent.[18] The fact that irreducible and indecomposable representations of a finite group, in the field of complex numbers, are the same, follows at once from Maschke's Theorem [213], published in 1899 (see §4).

In the preface to the first edition of his book (see §1), Burnside had expressed doubts as to the usefulness of groups of linear transformations in finite group theory. Thanks to his reading of Frobenius's first papers on character theory, and his own work on the Lie structure of the group algebra of a finite group, these doubts evaporated like mist in the morning sun. He was now a believer, and began his

---

[16]These are explained in a thorough and informative way by Hawkins ([173], §4).

[17]Here, as Hawkins observed in [173], §4, Burnside was undoubtedly thinking of Klein's normal problem, mentioned earlier.

[18]In [173], Hawkins documented Molien's introduction (in 1893) of the concept of irreducible representation in the representation theory of associative algebras, and Molien's own approach to group representations using this idea, in his last papers on representation theory, published in 1897-98.

next paper on the subject [**72**], entitled *On group-characteristics*, published in 1900, with the words:

> In a series of memoirs published in the *Berliner Sitzungsberichte* [citations to [**131**], [**132**]], Herr Frobenius has developed a theory of group-characteristics which must have a far-reaching importance in connexion with groups of finite order.

He continued:

> The present paper has been written with the intention of introducing this new development in the theory of groups of finite order to English readers. It is not original, as the results arrived at are, with one or two slight exceptions, due to Herr Frobenius. The modes of proof, however, are, in general, quite distinct from those used by Herr Frobenius.

His definition of the *characteristic* of a representation $a \to A$ of a finite group G by matrices (or operations, as he called them) was as follows:

> Any two conjugate operations of the group when brought into canonical form have the same multipliers. The sum of these multipliers is called the *characteristic* of the set of conjugate operations, or of any one of them, for the mode of representing the group in question; and the characteristics of the different sets of conjugate operations for one and the same mode of representation are called a *set* of characteristics.

This is essentially the definition of the *character* of a representation $a \to A$ used today, namely a complex valued function on G whose value at $a \in G$ is the trace function (the sum of the eigenvalues, or multipliers, as Burnside called them) of the matrix A corresponding to a. In comparison, we recall that Frobenius defined characters in terms of representations of the center of the group algebra (Chapter II, §3), and proved later that they were trace functions of primitive representations (Chapter II, §4).

The main point of Burnside's paper *On group-characteristics* was to prove, in his own way, the orthogonality relations for characters of irreducible representations (called *irreducible characters*), and Frobenius's theorem (from [**132**], §12) that the degrees of the irreducible characters divide the order of the group.

Burnside had already proved that, up to equivalence, the irreducible representations of G were the representations $G \to G_i$ (see Theorem 3.4 and the discussion leading up to the theorem). This meant that each irreducible representation was associated with an irreducible factor of the group determinant, and it was a short step to prove that their characteristics, or trace functions, coincided with the characters as defined by Frobenius. A further consequence was that the irreducible representations and the primitive representations of Frobenius were the same. Burnside's first proof of the orthogonality relations for the irreducible characters, in *On group-characteristics*, was based on an analysis of the characteristic polynomial of the general element of the center of the group algebra of G, and was a continuation of his analysis of the Lie algebra $\mathfrak{g}$ and its center $\mathfrak{z}$ (see Theorem 3.1). The details are not given here, as it was soon superseded by a more direct proof (see Corollary 4.1).

The second part of *On group-characteristics* contained a new proof, using the orthogonality relations, of a theorem of Frobenius ([**132**], §12).

THEOREM 3.5. *The degree f of each irreducible representation, or irreducible character χ, of a finite group G divides the order of G.*

The proofs of the theorem by Frobenius and Burnside were important steps in the development of character theory, as they used, for the first time, ideas from algebraic number theory.

The starting point of Burnside's proof was to observe that each value $\chi(a), a \in G$, of an irreducible character is an algebraic integer. According to his definition, the value of the character at $a \in G$ of an arbitrary representation (not assumed to be irreducible) is the sum of the eigenvalues of the matrix A corresponding to a. From elementary group theory, one has $a^n = 1$, where $n = |G|$. The same relation holds for the matrix A corresponding to a in any representation, with 1 replaced by the identity matrix. It follows that the eigenvalues of A are nth roots of unity. Therefore their sum, which is $\chi(a)$, is a sum of roots of unity, and hence an algebraic integer (see Chapter I, §1). This step had been more difficult for Frobenius, as he had not yet established a connection between characters and trace functions of representations. His approach was a good example of how he found ways to use the factorization of the group determinant to obtain information about the characters. His starting point was the observation that the values of the characters of the cyclic subgroup H generated by a were clearly roots of unity. Then, letting Ψ denote the group determinant of the subgroup H, and Θ' the group determinant of G with the variables x_p corresponding to elements p not in H set equal to zero, he proved that

$$\Theta' = \Psi^m,$$

where m is the index of H in G. Using this relation, and the connection of the characters of G, as he had defined them, with the irreducible factors of the group determinant, he was able to deduce that the values of the characters of G were sums of roots of unity, and hence were algebraic integers.

The rest of Burnside's proof was based on Frobenius's theorem on the representations of the center of the group algebra (Chapter II, Theorem 4.2). As in Chapter II, §4, let $\{c_i\}$ be the basis of the center of the group algebra indexed, by the conjugacy classes $\{C_i\}$ of G, with c_i the sum of the elements in the class C_i. Frobenius had shown that the structure constants for this basis were nonnegative integers, so that

$$c_j c_k = \sum_i a_{ijk} c_i,$$

with $a_{ijk} \in \mathbb{Z}$ for all i, j, k. Now let $a \to A$ be an irreducible representation of G with character χ, let χ_i be the value of χ at an element belonging to the conjugacy class C_i, let $h_i = |C_i|$, and let $f = \chi(1)$ be the degree of the representation, as usual. By now it had been proved that the irreducible representations and the primitive representations were the same, so Frobenius's theorem (Chapter II, Theorem 4.2) could be applied. It states that, in the given irreducible representation, the basis elements c_i of the center of the group algebra are represented by the scalar matrices $\frac{h_i \chi_i}{f} I$, so the structure equations for the basis elements of the center hold for the corresponding matrices.[19] It follows that

$$h_j h_k \chi_j \chi_k = f \sum_i a_{ijk} h_i \chi_i.$$

---

[19]In his proof, Frobenius used the fact that the expressions $\frac{h_i \chi_i}{f}$ were algebraic integers, which he had obtained easily from his definition of characters (*Über Gruppencharaktere* ([**131**], p.

Now let j' index the conjugacy class C_j^{-1}, as usual. Note that $h_j = h_{j'}$, and that

$$a_{ijj'}h_i = a_{j'j'i}h_j,$$

by the definition of the structure constants. Substituting this information in the preceding equations, Burnside obtained the formulas

$$h_j^2 \chi_j \chi_{j'} = f \sum_i a_{ijj'} h_i \chi_i = f \sum_i a_{j'j'i} h_j \chi_i.$$

Cancelling h_j, and summing on j, it follows that

$$\sum_j h_j \chi_j \chi_{j'} = f \sum_{i,j} a_{j'j'i} \chi_i.$$

As the sum on the left hand side is equal to $n = |G|$ by the orthogonality relations, and the sum on the right hand side is an algebraic integer, it follows that the rational number $\frac{n}{f}$ is an algebraic integer, and hence f divides n in $\mathbb{Z}$, completing the proof.

Burnside's goal was to apply character theory to obtain new results about the structure of finite groups. He lost no time in making a start in this direction. In fact, the next paper to *On group-characteristics* in the 1900-volume of the *Proceedings of the London Mathematical Society* was another paper by Burnside [**73**], entitled *On some properties of groups of odd order*, which was hailed by Frobenius a year later (in [**139**]; see Chapter II, §5) as containing the first applications of character theory to finite groups.

It began with the following new theorem about characters.

THEOREM 3.6. *Let G be a finite group of odd order. Then no irreducible character, other than the principal character, assumes real values at all elements of G.*

Burnside gave two proofs, one based on a study of quadratic invariants, and a second, given below, which is essentially the one used today.

Let χ be an irreducible character of G, different from the principal character. Then by the orthogonality relations (Chapter II, Theorem 3.2) applied to χ and the principal character, one has

$$\sum_{x \in G} \chi(x) = 0.$$

Now assume that χ is real valued, so that $\chi(x) = \overline{\chi(x)}$ for all $x \in G$ (where $\overline{a}$ denotes the complex conjugate of a). The preceding formula becomes

$$\chi(1) = -\sum_{x \neq 1} \chi(x).$$

By Theorem 3.5, $\chi(1)$ divides the order of G, and is consequently an odd number, by the hypothesis of the theorem. As for the right hand side, no nonidentity element

---

9; see §5, Lemma 5.1 below). From the orthogonality relations, he derived the formula

$$\sum \frac{h_i \chi_i}{f} \cdot \chi_{i'} = \frac{|G|}{f}.$$

This proves that $\frac{|G|}{f}$ is an algebraic integer, and completes the proof of the theorem, exactly as it is done today, as soon as it is known that the character values are algebraic integers. But this last step was not easy to prove in Frobenius's scheme of things, as noted above, and gave Burnside reason to find another proof.

is equal to its inverse, again using the hypothesis that $|G|$ is odd. Therefore the sum on the right side can be written in the form $-\sum(\chi(x) + \chi(x^{-1}))$, where the sum is taken over a set of disjoint pairs $\{x, x^{-1}\}$. On the other hand, $\chi(x)$ is the sum of the eigenvalues of a matrix representing x, while $\chi(x^{-1})$ is the sum of the eigenvalues of the inverse of that matrix. The eigenvalues of these matrices are all roots of unity, and for a root of unity ω, one has $\omega^{-1} = \overline{\omega}$. Therefore $\chi(x) = \chi(x^{-1})$, because of the assumption that χ is real valued. It follows that

$$\frac{1}{2}\chi(1) = -\sum_x \chi(x),$$

which is impossible, as the left side is one half of an odd number, while the right side is an algebraic integer. This completes the proof.

The results that attracted Frobenius's attention a year later, were criteria for nonsimplicity of finite groups of odd order, and a theorem on permutation groups of prime degree. The first result was the following:

THEOREM 3.7. *Let G be a finite group of odd order, whose order is divisible by 3 but not by 9. Then G contains a normal subgroup of index 3.*

Burnside's proof of this theorem did not use the preceding theorem on nonreality of characters of groups of odd order, but a result equivalent to it. The result on nonreality of characters was used to prove the same theorem for the case $p = 5$. The equivalent result is that, in a finite group of odd order, no nonidentity element is conjugate to its inverse. Indeed, let $s \neq 1$ be an element in a group of odd order. Then clearly $s \neq s^{-1}$. If, however, $usu^{-1} = s^{-1}$, then $u^2su^{-2} = s$. But in a group of odd order, u is a power of u^2, and thus commutes with s if u^2 commutes with s, so we are led to the impossible conclusion that $s = s^{-1}$.

Here is a modern proof of the equivalence of the two results. The maps $\chi \to \overline{\chi}$ and $C \to C^{-1}$ define permutations of the rows, and columns, respectively, of the character table matrix $Z = (\chi_j^i)$, which are related by the formula

$$\overline{\chi}(h) = \chi(h^{-1}),$$

for each irreducible character χ and each element $h \in G$. This means that the permutation matrices P and Q corresponding to these permutations of the rows and columns of Z satisfy

$$PZ = ZQ.$$

The matrix Z is invertible, by the orthogonality relations, so $P = ZQZ^{-1}$, and the matrices P and Q have the same trace. The trace of a permutation matrix is the number of objects left fixed by the permutation. Thus the number of real-valued characters of a finite group G is equal to the number of self-inverse conjugacy classes, and this proves the equivalence of the two statements used by Burnside in the case of finite groups of odd order. The preceding argument about fixed rows and columns of the character table is a forerunner, in a special case, of a general result, proved in 1941 by Brauer ([**28**], Lemma 1), and known today as *Brauer's Permutation Lemma*.

Burnside's proof of the theorem stated above was as follows. By the preceding discussion, $C_k \neq C_{k'} = C_k^{-1}$ for each conjugacy class $C_k \neq \{1\}$ in a finite group G of odd order n, satisfying the hypothesis of the theorem. This means that, letting

χ_k^i denote, as usual, the value of the ith character at the class $C_k \neq \{1\}$, one has

$$\sum_i \chi_k^i \chi_k^i = 0,$$

and

$$\sum_i \chi_k^i \chi_{k'}^i = \frac{n}{h_k},$$

by the Second Orthogonality Relation. From these formulas, one has

$$-\sum_i (\chi_k^i - \chi_{k'}^i)^2 = \sum_i (\chi_k^i + \chi_{k'}^i)^2 = \frac{2n}{h_k}.$$

Now let C_k be a conjugacy class containing an element of order 3. Using the fact that character values are sums of eigenvalues of the representing matrices, Burnside obtained

$$\chi_k^i = a_0^i + a_1^i \omega + a_2^i \omega^2 \text{ and } \chi_{k'}^i = a_0^i + a_1^i \omega^2 + a_2^i \omega,$$

for some nonnegative integers a_0^i, a_1^i, a_2^i and a primitive cube root of unity ω. Then, as is easily checked,

$$\chi_k^i - \chi_{k'}^i = (a_1^i - a_2^i)\sqrt{-3}.$$

Summing the squares of these expressions, and using the identities proved earlier, he obtained

$$3\sum_i (a_1^i - a_2^i)^2 = \frac{2n}{h_k} = 2m_k,$$

where m_k is the order of the centralizer of an element of order 3 belonging to the class C_k (see Frobenius's proof of Sylow's Theorem in Chapter II, §3, for the result that $m_k = n/h_k$). As m_k is not a multiple of 27 by the hypothesis of the theorem, $a_1^i - a_2^i$ is not a multiple of 3, for some i. It follows that the product of the eigenvalues of a matrix A representing an element in the class C_k, in an irreducible representation with character χ^i for this choice of i, is ω or ω^2, and definitely not 1. Burnside then noted that, in any representation $a \to A$ of G, the map $a \to A \to \det(A)^t$, with t a fixed positive integer, is a homomorphism from G to the multiplicative group of non-zero complex numbers, by the multiplicative property of the determinant $\det(A)$. The final step was the observation that the homomorphism just defined, for the irreducible representation having the character χ^i, and $t = |G|/3$, has the properties that: (i) x^3 is in the kernel N of the homomorphism, for all elements $x \in G$; but that (ii), no element belonging to the class C_k lies in the kernel N. Therefore N is a normal subgroup of G of index 3, completing the proof.

In his paper containing the reference to Burnside's work [**139**], Frobenius proved, using character theory, the following result.

> THEOREM. *Let H be a finite group whose order h is divisible by the first power of a prime p, and not by p^2, and assume that $p - 1$ and h are relatively prime. Then H contains one and only one subgroup [which is necessarily a normal subgroup] of order h/p.*

This result, as Frobenius remarked, includes Burnside's theorems as special cases, for $p = 3$ and $p = 5$.

The second main result of Burnside's paper on groups of odd order [**73**] was a powerful theorem concerning permutation groups of prime degree. Interest in this subject can be traced back to Galois's letter to Chevalier mentioned in Chapter I, §1. The letter was known to Burnside, and was mentioned in the first edition of his book. In modern terminology, Galois proved that the linear fractional group $PSL_2(p)$, for a prime p, cannot act in a nontrivial way as a transitive permutation group on a set of fewer than $p + 1$ elements if $p > 7$. There does exist a nontrivial transitive permutation action of $PSL_2(p)$ on a set of p elements, in case $p = 3, 5, 7$, and each of these is responsible for an isomorphism between two apparently unrelated finite groups: $PSL_2(3) \cong A_4$, arising from the case $p = 3$, $PSL_2(5) \cong A_5$, for $p = 5$, and $PSL_2(7) \cong PSL_3(2)$ for $p = 7$ (see Burnside [87], §327, and Conway [95] for further discussion).

Burnside's theorem was a general result about permutation groups of prime degree, proved using character theory. Among other things, it throws a little more light on the examples arising from Galois's theorem.

THEOREM 3.8. *Let G be a transitive group of permutations on a set of p elements, for a prime p. Then either G is solvable or the permutation action is doubly transitive.*

For groups of odd order, the theorem implied:

COROLLARY 3.1. *Let G be a transitive group of permutations on a set of p elements, for a prime p. If G has odd order, then G is solvable.*

We can imagine that Frobenius's eyes sparkled as he read the proofs of these theorems; they were first steps towards fulfilling his dream of applying character theory to finite groups. Burnside liked them himself, and remarked at the end of the introduction to [73]:

> The results obtained in this paper, partial as they necessarily are, appear to me to indicate that an answer to the interesting question as to the existence or non-existence of simple groups of odd composite order may be arrived at by a further study of the theory of group characteristics.

Transitive permutation groups of prime degree occur as the Galois groups of the splitting fields of irreducible polynomials of prime degree, and, according to Brauer [31], "this is the reason that these groups have been the subject of a large number of investigations." New proofs of Burnside's theorem, not involving the use of character theory, were obtained by Burnside himself [84], and by Schur ([252], and Chapter IV, §6). Further results on them were obtained by Frobenius [141] in 1902, and more recently by Brauer, Ito, Neumann and others (for references, and further results, see Brauer [31] and Feit [118], Chapter VIII, §6).

4. Foundations of Representation Theory: 1900-1905

An important step towards the way representations of finite groups are studied today was taken by Heinrich Maschke (1858 − 1908), in a paper [213] published in 1899. Maschke had been a student of Klein, and, after teaching in a Gymnasium for some time, joined Klein's mathematical school in Göttingen around 1886. He had participated in the research on Klein's normal-problem (referred to in a quotation from Burnside in §3), and was interested in the theory of invariants for finite groups. In 1892 he was recruited, along with Oskar Bolza, another student of Klein, by the president, William Rainey Harper, of the fledgling University of Chicago, to join E. H. Moore as the nucleus of the mathematics department.[20] At Chicago,

---

[20]For the history of mathematics at the University of Chicago in its early years, see Parshall and Rowe [236].

he continued to work on Klein's program, but, in the words of Hawkins [**173**], in "a generality not characteristic of the Kleinian program, and that may have reflected Moore's influence." In a paper entitled *Über den arithmetischen Charakter der Coefficienten der Substitutionen endlicher linearer Substitutionsgruppen* [**212**], published in 1898, Maschke called attention to what was to become an important problem about representations of finite groups:

> Little is known, up to now, as to the nature of the irrationalities which—in our experience, so to say—occur as matrix coefficients of linear groups of finite order. In the following it will be proved that (with one restriction) each finite linear substitution group, in any number of variables, can always be transformed, so that the coefficients of the matrices [substitutions] are all *cyclotomic*, i.e. rational functions of roots of unity.

As a start, he proved, using the Jordan normal form, that the trace of each matrix belonging to a finite group of matrices over $\mathbb{C}$ is a sum of roots of unity, and hence is cyclotomic. The restriction was that, in a given finite group of $n \times n$ matrices, there exists some matrix having n distinct eigenvalues. Maschke's result led to further research by Burnside, Schur, and Brauer, to be discussed later. The main point for us, now, was that, in his proof, Maschke established a special case of the theorem on complete reducibility, and had the insight to hang on to the idea until he was able to prove a general result a year later.

Maschke began with the assumption that, in a representation $a \to A$ of a finite group G by $n \times n$ matrices, some fixed off-diagonal matrix coefficient vanishes, for all the matrices A. He was able to prove, from this not entirely natural hypothesis, that for some r, $1 < r < n$, the given representation is equivalent to a representation in which the matrices all have the form

$$a \to \begin{pmatrix} A_1' & B' \\ 0 & A_2' \end{pmatrix},$$

with A_1' an $r \times r$ submatrix, 0 an $(n-r) \times r$ submatrix, etc. He then stated:

> We agree to call a linear substitution group, that can be transformed in such a way that the variables of the transformed group can be divided into a number of systems, each with the property that the variables in the system are transformed only among themselves, *intransitive*, and will prove in all generality:
>
> *Each finite linear substitution group, in which one coefficient (not located on the principal diagonal) of all the substitutions is zero, is intransitive.*

Here is a modern version, stated for matrix representations, of what is known today as *Maschke's Theorem*.

THEOREM 4.1. *Let* $a \to A$ *be a matrix representation of a finite group* G, *of degree* n, *and assume that for some* $r < n$, *the matrices corresponding to the elements of the group all have the form*

$$\begin{pmatrix} A_1 & B \\ 0 & A_2 \end{pmatrix},$$

with A_1 *an* $r \times r$ *submatrix*, 0 *an* $(n-r) \times r$ *submatrix, etc. Then the representation is equivalent to one of the same form, with the submatrices* B *all equal to zero.*

A representation having the form given in the statement of the theorem is said to be *reducible*, with the *subrepresentations* $a \to A_1$ and $a \to A_2$ (it is easily shown that $a \to A_1$ and $a \to A_2$ are also representations of G). The conclusion of Maschke's Theorem is that the representation is equivalent to the *completely reduced* form

$$a \to \begin{pmatrix} A_1 & 0 \\ 0 & A_2 \end{pmatrix},$$

called today the *direct sum* of the subrepresentations $a \to A_1$ and $a \to A_2$. The full power of the theorem becomes apparent only when it is combined with other things, such as the concept of irreducible representation (see Burnside's Theorem 4.2 below).

Maschke's proof of his theorem was based on a result of E. H. Moore [**218**], which Moore stated as follows:

> *A finite group of n-ary linear homogeneous substitutions leaves absolutely invariant an n-ary positive Hermitian form. By proper linear transformations of the group, this invariant form is*
>
> $$x_1 \overline{x}_1 + \cdots + x_n \overline{x}_n.$$

This theorem, which Moore had presented to the Mathematics Club of the University of Chicago on July 10, 1896, was communicated to Klein, then one of the editors of the *Mathematische Annalen*, in September, 1896. Klein called Moore's attention to the fact that it had already been stated (without proof) in an article [**206**] by Alfred Loewy, published in 1896. Moore's paper appeared in the *Mathematische Annalen* two years later.

Moore's proof could not have been more direct. One starts with the positive definite Hermitian form[21] $H(x, \overline{x}) = \sum x_i \overline{x}_i$, and a finite group of matrices, viewed as linear substitutions of the variables $x_1, \ldots, x_n$. For each element $g = (g_{ij})$ belonging to the given finite group of matrices, let $g \cdot H$ denote the Hermitian form obtained from H by substituting $\sum_j g_{ij} x_j$ for each variable x_i. Let $\tilde{H} = \sum_g g \cdot H$; then it is easily verified that $\tilde{H}$ is also a positive definite Hermitian form, with the further property of being invariant under the linear substitutions (or matrices) g belonging to the given finite group. By a change of orthonormal basis, the form becomes $\sum x_i' \overline{x}_i'$, completing the proof.

Moore, and his colleague Maschke at the University of Chicago, shared an interest in polynomial invariants of finite groups,[22] but this was apparently not the motivation for Moore's theorem. Moore remarked, in the introduction to his paper:

> Klein first effected (*Mathematische Annalen* vol. 9, 1875) the complete determination of finite binary groups [groups of linear fractional transformations, discussed in §2] by the theorem (l.c. pp. 186-7) that the corresponding group of real quaternary collineations with positive

---

[21] For the definition of positive definite Hermitian form, see the discussion following Theorem 4.3.

[22] For example, Moore had edited a collection of papers read at the International Mathematical Congress held in conjunction with the World's Columbian Exposition, Chicago 1893, including one by Hilbert entitled *Über die Theorie der algebraischen Invarianten*. Courses on invariants were given at about this time at the University of Chicago. A set of lecture notes taken by a graduate student, E. E. DeCou, later to become department head at the University of Oregon, from a course on invariant theory given by Maschke in 1899, contained an account of Hilbert's Theorem (*Math. Annalen* vol. 36, 1890) on the finite generation of polynomial ideals, and applications of it to invariant theory.

determinant of the ellipsoid $x^2 + y^2 + z^2 - w^2 = 0$ into itself leaves invariant a point within the ellipsoid and so can be transformed into a finite group of real rotations around a fixed point. This result is easily derived from our theorem for $n = 2$. Indeed it was by the analytic phrasing in terms of binary groups of Klein's invariant point that I was led to the discovery of the universal invariant positive Hermitian form.

Maschke's proof of his theorem was based on the Loewy-Moore theorem; a modern version of it is given here.[23] The Loewy-Moore Theorem implies that, in any representation of a finite group G by linear substitutions, or, in modern terminology, by linear transformations $T(a), a \in G$, acting on a vector space V, there exists a positive definite Hermitian form $H(u, v)$ on V which is G-invariant, in the sense that

$$H(T(a)u, T(a)v) = H(u, v),$$

for all $u, v \in V$ and $a \in G$. The representation $a \to T(a)$ of G on the vector space V is reducible if there exists a proper subspace V_1 of V which is invariant under the linear transformations $T(a), a \in G$; this means that the matrices corresponding to the elements $a \in G$, taken with respect to a basis of V consisting of a basis of the subspace V_1, supplemented with other vectors, have the reduced form given in the statement of the theorem. The vector space V is the direct sum $V = V_1 \oplus V_1^\perp$, where $V_1^\perp$ is the subspace consisting of vectors v which are orthogonal to the vectors $u \in V_1$ with respect to the given G-invariant positive definite Hermitian form H, that is, $H(u, v) = 0$ for all vectors $u \in V_1$. The G-invariance of the form H implies that the orthogonal complement $V_1^\perp$ is also invariant under the linear transformations $T(a)$, as

$$H(T(a)v, u) = H(T(a^{-1})T(a)v, T(a^{-1})u) = H(v, T(a^{-1})u) = 0,$$

for all $a \in G, v \in V_1^\perp$, and $u \in V_1$. Here we have also used the facts that $T(a^{-1})T(a) = 1$, the identity transformation on V, and that V_1 is invariant under the linear transformations $T(a), a \in G$. The matrices corresponding to the elements of G, taken with respect to a basis of V consisting of a basis of V_1 supplemented by a basis of $V_1^\perp$, define a matrix representation equivalent to the given reducible representation, which has the required completely reduced form. This completes the proof of Maschke's Theorem.

The demonstration given above illustrates an important step in the evolution of representation theory, from the concept of a representation of a finite group by matrices, used by Maschke, Frobenius, and Schur in their proofs of Maschke's Theorem, to the concept of a representation of a finite group by linear transformations acting on a vector space. This step was taken during the 1920's by Hermann Weyl and Emmy Noether; for example, one finds the proof given above in Weyl's book ([**285**] (1928), Chapter III, §7) on the applications of group representations to quantum mechanics.

Burnside put his own stamp on the foundations of representation theory in a series of papers published between 1903 and 1905. These incorporated what he had learned from the papers of Moore and Maschke, and some new ideas of his own,

---

[23]Another proof, by Schur, is given later (see Chapter IV, Theorem 2.3); Schur's proof was suggested by a new proof of Maschke's Theorem by Frobenius, also sketched in Chapter IV, §2, which was close to Maschke's own proof.

leading to a clear analysis of the classification of representations of finite groups, in terms of the concepts of irreducible and completely reducible representations. Much of this material was included in the second edition of his book [**87**], published in 1911, which contained a comprehensive account of the theory of representations and characters of finite groups, based on a new set of principles, stated in the quotations below. In his approach, the factorization of the group determinant no longer occupied center stage.

As an illustration of Burnside's way of looking at things, and the influence on his work of the papers of Moore and Maschke discussed above, we shall follow the steps (in [**77**]) leading up to his new proof of the orthogonality relations for characters.[24]

His first proofs of the main properties of the irreducible characters, in [**72**], were based, as we have seen, on the Lie group and Lie algebra associated with the group algebra of G. In [**77**], he was ready to give a more direct approach, and remarked:

> It is hardly too much to say that the methods by which new ideas are introduced and new results obtained in any branch of mathematical science are in general indirect. In fact the historical development of a subject rarely follows the lines of the logical presentment that can be given when a sufficient number of results and their connexion with each other have become familiar.
>
> A case in point is the theory of group-characteristics and the allied subject of the representation of a group of finite order as a group of linear substitutions. In the remarkable series of memoirs, published in the *Berliner Sitzungsberichte*, in which Herr Frobenius developed this theory, the actual starting point is an algebraical theorem connected with permutable matrices; and the methods made use of in the subsequent discussion turn largely on the properties of matrices. In the paper on "Group-Characteristics" in which I have given an account of Herr Frobenius's theory the methods used, though quite distinct from his, are at least as indirect. They involve the consideration of a continuous group from whose properties those of the group-characteristics are deduced.

He continued:

> The present paper is an attempt to deal with the theory of the representation of a group of finite order as a group of linear substitutions and the properties of the group-characteristics from what may be described as a self-contained point of view. No considerations or ideas are introduced foreign to those involved in the conceptions of an abstract group of finite order and of a group of linear substitutions of finite order. Moreover, the theorems made use of are those which lie at the foundations of the subject. It will be seen that the order in which the various parts of the subject are considered differs materially from that in which they occur in Herr Frobenius's memoirs. First, the complete reducibility (defined below) of a group of linear substitutions of finite order is taken; secondly, the number of distinct irreducible representations; thirdly, the composition of the

---

[24]cf. Frobenius's version, discussed in Chapter II, §3.

irreducible representations; and, lastly, the group-characteristics and their properties.

The definition he referred to was the following one:

A group of linear substitutions in n variables is called *reducible* when it is possible to find n' ($< n$) linear functions of the variables which are transformed among themselves by every substitution of the group. I propose to call a group of linear substitutions *completely reducible* when it is possible to choose the variables in such a way that they fall into sets with the properties (i) that the variables in each set are transformed among themselves by every substitution of the group; and (ii) that the group of linear substitutions in each separate set is irreducible.

The first two steps of the program described in the paragraph quoted above, on complete reducibility and the number of irreducible representations, were published in a separate paper [**78**], submitted to *Acta Mathematica* in December, 1902. His result on complete reducibility carried Maschke's Theorem 4.1 one step farther.

THEOREM 4.2. *Every finite group of linear substitutions is either irreducible, or completely reducible.*

The theorem is equivalent to the statement that every representation of a finite group is either irreducible, or completely reducible. This meant that all that was required to describe an arbitrary representation of a finite group was a knowledge of the irreducible representations that occurred as *components*, in Burnside's terminology, when the representation was expressed as a direct sum, as we would say today, of irreducible representations.

Here is a shortened version of Burnside's proof. Let $a \to A$ be an arbitrary representation of a finite group G, of degree n. If the representation is irreducible, there is nothing to prove. Assume it is reducible, with a subrepresentation $a \to A_1$ of degree $n_1 < n$. If this representation is reducible, it has a subrepresentation of degree n_2 with $n_2 < n_1 < n$. Continuing this analysis, we arrive at some stage at an irreducible subrepresentation of the given representation. Thus we may assume that the given representation has an irreducible subrepresentation $a \to A_1$. By Maschke's Theorem, the representation $a \to A$ is equivalent to a representation

$$a \to \begin{pmatrix} A'_1 & 0 \\ 0 & A'_2 \end{pmatrix},$$

for some irreducible representation $a \to A'_1$ (equivalent to the given subrepresentation $a \to A_1$), and another representation $a \to A'_2$ of degree $n - n_1$. The argument is now repeated for the subrepresentation $a \to A'_2$. The result is that, after a finite number of steps, the given representation is equivalent to a representation in what Burnside called the completely reduced form

$$a \to \begin{pmatrix} A_1 & 0 & \cdots \\ 0 & A_2 & \cdots \\ \cdots & \cdots & \cdots \end{pmatrix},$$

with irreducible subrepresentations $A_1, A_2, \ldots$ in the blocks along the diagonal, and zeros elsewhere. This completes the proof.

The next steps in the program, on the number of distinct irreducible representations, and the analysis of the regular representation of a finite group, were carried out in several theorems, in [**78**], summarized below as parts of a single theorem. Its statement closely follows Burnside's. We first recall that the *regular representation* of a finite group G is the representation $a \to A$ which assigns to each element $a \in G$ the permutation matrix A describing the permutation of the group elements (taken in some order) by left multiplication by the element a.

THEOREM 4.3. *Let G be a finite group of order N.*

(*i*) *In the completely reduced form of the regular representation of G, the number of times that each irreducible component occurs is equal to its degree.*

(*ii*) *The number of distinct irreducible components of the regular representation of G is equal to the number of conjugacy classes of G.*

(*iii*) *Every irreducible representation of G is equivalent to a component of the regular representation.*

(*iv*) *Let the degrees of the distinct irreducible representations be $\nu_1, \nu_2, \ldots$. Then*

$$N = \sum \nu_i^2.$$

The statements (*i*)-(*iv*) are the main facts, as we know them today, about the complete reduction of the regular representation of a finite group. Part of the theorem, with primitive instead of irreducible representations, had been proved earlier, using characters, by Frobenius (Chapter II, Theorem 4.1). Burnside gave a direct, and quite remarkable, proof of it, without using characters, based instead on a classification of the different G-invariant Hermitian forms on the underlying vector space of a representation, using the Loewy-Moore Theorem as a start.

A modern version of his proof goes as follows. We first review the terminology. Let V be a vector space, and $a \to A$ a representation of G by linear transformations (or linear substitutions, as they were called at the time) acting on V. An *Hermitian form* on V is, by definition, a complex valued function $H(u,v)$ defined on pairs of vectors $u, v \in V$, which is additive in both variables, and satisfies the conditions: $H(u,v) = \overline{H(v,u)}$, where $\bar{z}$ denotes the complex conjugate of $z \in \mathbb{C}$; and $H(\alpha u, v) = \alpha H(u,v)$, for all vectors u and v and complex numbers α. The form is G-invariant if $H(Au, Av) = H(u,v)$ for $u, v \in V$, and all linear transformations A corresponding to elements of G. From the definition, $H(u,u)$ is a real number, for all vectors u. The form is *positive definite* if $H(u,u) \geq 0$ for all vectors u, and is zero only if $u = 0$.

Now let V be a vector space, and let $a \to A$ be a representation of G by linear transformations acting on V. Assume that, in the completely reduced form of the representation, there are distinct irreducible components $a \to A_1$, $a \to A_2, \ldots$, occurring n_1 times, n_2 times, $\ldots$ in the given representation, respectively. Burnside's main result stated that:

The number of linearly independent G-invariant Hermitian forms on V is $n_1^2 + n_2^2 + \ldots$.

His proof is long, and will not be given here, nor will we show exactly how it is used to prove the theorem above, as that can be done in other ways. We shall, however, give the underlying idea of the result, using some things that came later in the development of representation theory. Let $[u,v]$ be a fixed positive definite G-invariant Hermitian form on V, whose existence is guaranteed by the Loewy-Moore Theorem. Let $H(u,v)$ be an arbitrary G-invariant, not necessarily positive definite, Hermitian form on V. Using the theory of dual vector spaces, it is easily shown that there exists a linear transformation T from V to itself such that

$$H(u, v) = [Tu, v],$$

for all vectors $u, v \in V$. The fact that both forms are G-invariant implies that $AT = TA$ for all matrices A corresponding to elements of G. Thus the different invariant Hermitian forms correspond to elements of the *commuting algebra* or *centralizer* of the representation, and Burnside's result on the number of linearly independent ones follows from this fact, together with Schur's Lemma (see Chapter IV, §2, Theorem 2.5).

We now turn to Burnside's new proof of the orthogonality relations for characters, from [**77**]. It was based on two important constructions in representation theory. Let $a \rightarrow A = (a_{ij})$ be a representation of G. The first construction defines another representation $a \rightarrow \overline{A} = (\overline{a_{ij}})$, obtained by replacing each matrix entry of the first by its complex conjugate. Burnside called the second representation the *inverse* of the first; today it is called the *contragredient* representation. If χ is the character of the first representation, then the character of the inverse representation is given by

$$\overline{\chi}(a) = \chi(a^{-1}).$$

This follows because $\chi(a)$ is the sum of the eigenvalues of the matrix A, so $\overline{\chi(a)}$ is the sum of their complex conjugates, which are the eigenvalues of the matrix A^{-1} corresponding to a^{-1}. Here we have used the fact that the eigenvalues are roots of unity, and are inverted by taking the complex conjugate. The operation of taking the inverse (or contragredient) representation permutes the irreducible representations, up to equivalence. Burnside used the notation $g_1, g_2, \ldots$ for a set of distinct (inequivalent) irreducible representations, with g_1 the principal (or trivial) representation, and $g_{i'}$ denoting the inverse (up to equivalence) of the irreducible representation g_i, for each $i = 1, 2, \ldots$.

The second construction was apparently first defined by Frobenius [**135**], and was called by Frobenius the *composition* of representations. Consider two representations of G, of degrees p and q, which assign $a \in G$ to the linear substitutions $x_i' = \sum_j a_{ij}x_j$ and $y_k' = \sum_\ell a_{k\ell}'y_\ell$, respectively, for independent sets of variables x_i and y_k. It is not difficult to check that the correspondence that assigns $a \in G$ to the linear substitution on the products of the variables, $x_i'y_k' = \sum a_{ij}a_{k\ell}'x_jy_\ell$, defines a representation of G of degree pq. The $(ik, j\ell)$ entry of the matrix of the new representation, corresponding to $a \in G$, is $a_{ij}a_{k\ell}'$. Today we recognize it as the *Kronecker product* or *tensor product* of the matrices (a_{ij}) and $(a_{k\ell}')$, and the resulting representation of degree pq as the *tensor product* of the given representations. Let χ and χ' denote the characters of the representations $a \rightarrow (a_{ij})$ and $a \rightarrow (a_{k\ell}')$, respectively. Then the character of the tensor product representation is given by

$$a \rightarrow \sum_{i,k} a_{ii}a_{kk}' \rightarrow \chi(a)\chi'(a),$$

and answers the natural question as to whether the product of the characters of two representations of a finite group is itself the character of a representation.

Now let g_ig_j denote the tensor product of the irreducible representations g_i and g_j. While the representation g_ig_j is not necessarily irreducible, it will have a completely reduced form

$$g_ig_j = \sum_k d_{ijk}g_k,$$

with nonnegative integers d_{ijk}, which give the number of times g_k occurs as a component of the completely reducible representation $g_i g_j$.

THEOREM 4.4. *Let* $g_1, g_2, \ldots$ *be a set of distinct irreducible representations of* G, *with* g_1 *the principal representation, and* $g_{i'}$ *the contragredient representation of* g_i, *for each* i. *Let*

$$g_i g_j = \sum_k d_{ijk} g_k$$

describe the complete reduction of the tensor product representation $g_i g_j$. *Then* $d_{ii'1} = 1$, *and* $d_{ij1} = 0$ *if* $j \neq i'$.

Burnside's proof of the theorem was another application of Moore's Theorem on the existence of invariant Hermitian forms on the underlying space of a representation. If g_i and g_j are irreducible representations on the variables x_i and y_j, respectively, he observed that d_{ij1} is the number of G-invariant bilinear forms in the variables x_i and y_j. Moreover, for each i, he was able to show that $d_{ij1} = 1$ for one and only one value of j, and is zero for the other values of j. The proof was completed by pointing out that the Loewy-Moore Theorem implied the existence of a G-invariant bilinear form for the product representation $g_i g_{i'}$, so that $d_{ii'1} = 1$.

A modern version of this argument can be given as follows.[25] Let V_i and V_j be vector spaces affording irreducible representations of G. Let V_i^* denote the contragredient representation, this time defined as the dual space V_i^* of V_i, with the G-action given by $af(v) = f(a^{-1}v)$, for all $a \in G$, $f \in V_i^*$, and $v \in V_i$. It is not difficult to prove that this definition is equivalent to Burnside's definition of the inverse of a representation. It can further be proved that the vector space $V_i \otimes V_j$ is isomorphic to the vector space of linear maps from V_i^* to V_j in such a way that the space of G-invariants in $V_i \otimes V_j$ corresponds to the space of linear maps from V_i^* to V_j which intertwine the G-actions on V_i^* and V_j. It follows that d_{ij1} is the dimension of the space of G-homomorphisms from V_i^* to V_j, and is one or zero according as $i' = j$ or $i' \neq j$, by Schur's Lemma (see Chapter IV, §3). All this is implicit in Burnside's (much longer) argument.

Burnside's new proof of the orthogonality relations was a direct consequence of the preceding theorem. It is based on the observation that if χ is the character of a representation of G, then the number of times the principal representation occurs in the complete reduction of the representation is $\frac{1}{|G|} \sum_{a \in G} \chi(a)$. A modern proof of this fact is obtained as follows. Let $a \to A$ be the given representation, with character χ. Then the operator $\frac{1}{|G|} \sum_{a \in G} A$ is the idempotent projection from the underlying vector space of the representation to the subspace consisting of those elements fixed by all the linear transformations A corresponding to elements of the group. The dimension of this subspace is the number of times the principal representation occurs in a completely reduced form of the representation, and is given by the trace of the projection operator, which is $\frac{1}{|G|} \sum \chi(a)$. Letting χ^i and $\chi^{j'}$ denote the characters of the irreducible representations g_i and $g_{j'}$ respectively, the character of the product representation $g_i g_{j'}$ is $\chi^i \chi^{j'}$. Moreover, $\chi^{j'}(a) = \chi^j(a^{-1})$, for all $a \in G$, as noted above. Combining these observations with the preceding theorem, one has:

---

[25] For a more detailed account, see Curtis and Reiner ([**99**], §9, Exercise 11).

COROLLARY 4.1. *The integers d_{ij1} are given by:*

$$d_{ii'1} = \frac{1}{|G|} \sum_{a \in G} \chi^i(a)\chi^i(a^{-1}) = 1,$$

and

$$d_{ij'1} = \frac{1}{|G|} \sum_{a \in G} \chi^i(a)\chi^j(a^{-1}) = 0 \; if \, i \neq j.$$

These are the orthogonality relations for the irreducible characters. The Corollary is immediate from Theorem 4.4, and the preceding discussion.

Another contribution to the foundations of representation theory, and perhaps his most important one, was Burnside's characterization of irreducible representations of finite or infinite groups ([**80**], 1905). When a reference is made today to "Burnside's Theorem" without further qualification, it is usually to this result.

THEOREM 4.5. *Let $a \to A$ be a representation of a group G, of degree n. If the representation is irreducible, then there exists a set of n^2 linearly independent matrices $A_1, \ldots, A_{n^2}$ corresponding to elements of G. Conversely, if such a set of n^2 linearly independent matrices exists, then the representation is irreducible.*

The second statement of the theorem is easy to prove. The first statement, however, is more profound. It was applied, in the paper published next to it in the same journal [**81**], to an interesting finiteness problem in group theory (see Theorem 5.3 in the next section); this, in turn, was used by Schur as the point of departure for a study of groups of periodic linear transformations ([**254**], 1911).

Burnside's proof of his theorem is not given here; the result, at least for representations of finite groups, follows easily from Schur's theorem, published in the same year [**249**], on the orthogonality of coefficients of irreducible representations (see Chapter IV, §2). The deeper meaning, that it is really a general theorem about irreducible representations of algebras, was realized about a year later, by Frobenius and Schur, who extended it to show the linear independence of the coefficient functions associated with a finite set of inequivalent irreducible representations [**152**] (see Chapter IV, §3). The modern interpretation of Burnside's theorem, as an application of Wedderburn's theorems to irreducible representations of algebras, was explained by Emmy Noether ([**233**], 1929; see Chapter VI, §2).

5. The $p^a q^b$-Theorem and Other Applications of Character Theory

The theorem that all finite groups whose orders are of the form $p^a q^b$, for primes p, q, are solvable, had been proved for particular values of p, q, a, b by Jordan [**193**], Frobenius [**142**], and Burnside, in the first edition of his book [**69**], where the question of what happened in the general case was raised as an interesting problem for further research. Special cases of Frobenius's theorem on the structure of Frobenius groups and its permutation group version, obtained by Maillet and Burnside, also preceded the proof of the general theorem (see Chapter II, §5). The proofs of these earlier special cases were elementary in the sense that they required only basic and well understood facts from finite group theory and the theory of permutation groups. Frobenius's theorem ([**140**], 1901) and Burnside's general version of the $p^a q^b$-Theorem ([**79**], 1904), both proved using character theory, and apparently not accessible in any other way, were of a different nature altogether, and removed all doubts as to the significance of the theory of group characters as a powerful and, in some cases, unavoidable, method for obtaining new information about the structure of finite groups.

Burnside's paper [**79**] contained two proofs of:

THEOREM 5.1. *Every finite group whose order is of the form $p^a q^b$, with p, q primes, is solvable.*

The fact that finite groups of prime power order are solvable was proved by Sylow in 1872 [**270**]. On the other hand, the order of the nonabelian simple group of minimal order (A_5 of order 60), has 3 distinct prime factors. Burnside's Theorem showed that the orders of all nonabelian finite simple groups contain at least 3 distinct prime factors.

The starting point of Burnside's first proof was the following result about groups with an irreducible representation of prime power degree.

PROPOSITION 5.1. *Let G be a group of order $p^a s$, with p a prime, $a > 0$, and s relatively prime to p. Let $z \neq 1$ be an element belonging to the center of a Sylow p-subgroup, and let χ be an irreducible character of G of degree p^m. Then $\chi(z)$ is either zero or $p^m \alpha$, where α is a root of unity. In the latter case, the matrix corresponding to z in a representation whose character is χ has the form αI, where I is the identity matrix. Moreover, G has a proper normal subgroup containing z.*

Frobenius's proof of his theorem on the structure of Frobenius groups relied on the theory of induced characters. Burnside's proof of the proposition, and of the theorem stated above, are based on quite a different set of ideas, namely the application of the theory of algebraic numbers and Galois theory to cyclotomic fields containing the values of irreducible characters. The proof of the proposition is given here because it will turn out, as Burnside realized, that it contains the nucleus of a simple proof of the $p^a q^b$-Theorem.

The first step was a new proof of a result of Frobenius, originally proved in [**131**], §2, p. 9. For the statement, and throughout the rest of the discussion, the notation χ_i will be used for the value of an irreducible character at an element of the ith conjugacy class, h_i for the number of elements in the class, and χ_1 for the degree of the character, so the conjugacy class consisting of the identity element is taken as the first class.

LEMMA 5.1. *Let χ be an irreducible character of a finite group G. Then*

$$\frac{h_i \chi_i}{\chi_1}$$

is an algebraic integer, for each i.

Burnside's proof (which is the standard one used today) began with the formulas

$$h_j h_k \chi_j \chi_k = \chi_1 \sum_i a_{ijk} h_i \chi_i,$$

where a_{ijk} are the structure constants of the center of the group algebra. These are equivalent to Frobenius's definition of characters (see (7) in §3, Chapter II), and could also be viewed, as both Frobenius and Burnside understood, as the statement that

$$c_i \to \frac{h_i \chi_i}{\chi_1}$$

defines an irreducible representation of the center of the group algebra (see Chapter II, Theorem 4.2, and the proof of Theorem 3.5 in this chapter). In any case, the formulas imply that the elements $\frac{h_i \chi_i}{\chi_1}$, for $i = 1, 2, \ldots$, are a nontrivial solution of a system of homogeneous linear equations, so the determinant

$$\left| \delta_{ik} \frac{h_j \chi_j}{\chi_1} - a_{ijk} \right|$$

of the matrix of coefficients is zero. (The displayed entry of the determinant is the (i, k)-th entry of the coefficient matrix, and δ_{ik} is a Kronecker delta.) The structure constants a_{ijk} are rational integers, so the expansion of the determinant shows that, for each j, $\frac{h_j \chi_j}{\chi_1}$ satisfies a monic polynomial equation with integer coefficients, and is consequently an algebraic integer.

Now assume that χ is an irreducible character of degree p^m for some $m \geq 1$, as in the hypothesis of the proposition; then $|G|$ is divisible by $p^m = \chi_1$ by Theorem 3.5. Let z be a nontrivial element in the center of a Sylow p-subgroup of G. (Such an element exists because the center of a p-group is nontrivial, a result proved by Sylow in 1872 [**270**].) Let h_z be the number of elements in the conjugacy class of z. The centralizer of z contains a Sylow p-subgroup of G, so the index of the centralizer, which is h_z, is relatively prime to p. By the preceding lemma, $\frac{h_z \chi(z)}{\chi_1}$ is an algebraic integer. As h_z is relatively prime to p, and χ_1 is a power of p, it follows that

$$\frac{\chi(z)}{\chi_1}$$

is an algebraic integer. Burnside then reasoned as follows. Assume $\chi(z) \neq 0$. The character value $\chi(z)$ is a sum of p^m powers of ω, for a primitive p^mth root of unity ω, and the expression $\frac{\chi(z)}{\chi_1}$ belongs to the cyclotomic field $\mathbb{Q}(\omega)$. The elements of the Galois group of $\mathbb{Q}(\omega)$ over $\mathbb{Q}$ send ω onto the other primitive p^mth roots of unity, and $\frac{\chi(z)}{\chi_1}$ onto its conjugates. The product of these conjugates is fixed under the elements of the Galois group, and is therefore a rational number. It is, at the same time, an algebraic integer, so it is a rational integer different from zero. On the other hand, $\chi(z)$ is the sum of the p^m eigenvalues of the matrix Z corresponding to z in an irreducible representation of G whose character is χ, and these are roots of unity. It follows that the absolute value

$$\left| \frac{\chi(z)}{\chi_1} \right| = \left| \frac{\chi(z)}{p^m} \right|$$

is less than or equal to 1, with equality only if all the eigenvalues are equal, as it is easily shown from the graphical representation of the roots of unity. If not all the eigenvalues are equal, then

$$\left| \frac{\chi(z)}{\chi_1} \right| < 1,$$

and the same inequality holds for all the conjugates of $\frac{\chi(z)}{\chi_1}$, and for their product, contradicting the fact that their product is a nonzero rational integer.

At this point, it has been proved that, if $\chi(z) \neq 0$, one has $\chi(z) = p^m \omega^t$ for some integer t and that the matrix Z corresponding to z has only the one eigenvalue ω^t. Burnside already knew that in any representation $a \to A$ of a finite group by matrices with entries in $\mathbb{C}$, each matrix A corresponding to an element of the group is diagonalizable. These facts, taken together, imply that $Z = \omega^t I$, where I is the identity matrix. The set of all elements of G whose matrices in the given irreducible representation have the form αI, for some complex number α, is clearly a proper normal subgroup of G containing z, completing the proof of the proposition.

Using the proposition, Burnside completed his first proof of the $p^a q^b$-Theorem in several more steps, ending with an analysis of the possible degrees of the irreducible characters, using the fact that the sum of their squares is $|G| = p^a q^b$.

In a note dated February 9, 1904, Burnside remarked that he had arrived at a "materially simpler manner of establishing a rather more general result."

THEOREM 5.2. *If, in a group of finite order, the number of elements in any conjugacy class is a power of a prime, then the group cannot be simple.*

As he noted after the proof, the $p^a q^b$-Theorem follows immediately, by induction on the order of G. Indeed, let $|G| = p^a q^b$, with both a and b different from zero. Then the number of conjugates of a central element in a Sylow p-subgroup is a power of q, and G has a proper normal subgroup H, by Theorem 5.2. Both H and G/H are groups whose orders have the form $p^r q^s$, and are solvable by the induction hypothesis. It follows that G is solvable.

Burnside's proof of the preceding theorem, given below, is a wonderful example of how the ideas needed to establish an important general theorem are sometimes discovered by a careful reexamination of the proof of a rather special result, in this case Proposition 5.1.

Suppose, in a finite group G, the number of elements in a conjugacy class is a power of a prime, say $h_i = p^r$. Then $\frac{h_i \chi_i}{\chi_1}$ is an algebraic integer, for each irreducible character χ, by Lemma 5.1. By an argument similar to a step in the proof of the proposition, one may deduce that if χ_1 is not divisible by p, then χ_i/χ_1 is an algebraic integer. In that case, by the proof of the proposition, either $\chi_i = 0$, or $\chi_i = \chi_1 \omega$, for a root of unity ω.

Now consider the second orthogonality relation,

$$\sum_{\chi} \chi_1 \chi_i = 0,$$

where the sum is taken over the set of irreducible characters χ of G. If no χ_1 is one, except for the principal character, and if χ_i is zero whenever χ_1 is not divisible by p, the equation has the form

$$1 + p\xi = 0,$$

for some algebraic integer ξ, which is impossible. Hence, either some irreducible character degree other than the degree of the principal character is one, or some character value χ_i is equal to $\chi_1 \omega$, for a root of unity ω. In the first case, the commutator subgroup is a proper normal subgroup of G, while in the second case, G has a proper normal subgroup containing the ith conjugacy class, again by the reasoning used in the proof of the proposition. In either case, the group G is not simple, and the theorem is proved.

No proof of Frobenius's theorem (Chapter II, §6) has as yet been found which does not require the use of character theory. A proof of the $p^a q^b$-Theorem was obtained, however, three quarters of a century later, which is elementary, in the sense that character theory was not used (Bender [**12**], 1972). Like other so-called elementary proofs of deep theorems, such as Selberg's proof of the prime number theorem [**262**], the new proof of the $p^a q^b$-Theorem was neither shorter nor less demanding than Burnside's proof, given above. It was an important step in the search for new ways to analyze the structure of finite groups, in connection with the problem of classifying the finite simple groups (see Gorenstein [**163**], §4.8, for further discussion).

The period 1903-1905 was an extraordinarily productive one for Burnside. In those years, he published 12 papers on representations and characters of finite groups, and several others on finite groups and algebraic geometry. While the papers on groups of order $p^a q^b$, and the criterion for irreducibility (taken up in the preceding section), stand out as contributions of lasting importance from this period, the others were also significant forays into unexplored areas, and some of them are summarized here. These include an application of the irreducibility criterion to a question as to the finiteness of groups satisfying certain relations [**81**], and a continuation of Maschke's efforts, mentioned earlier, to prove that irreducible representations can be realized in cyclotomic fields [**82**]. Most of Burnside's work from this period, in updated form, and with notes and exercises showing his intent, was incorporated in the second edition of his book ([**87**], 1911).

The finiteness question was related to a general problem he had stated three years before [**76**]:

> A still undecided point in the theory of discontinuous groups is whether the order of a group may be not finite, while the order of every operation it contains is finite. A special form of this question may be stated as follows:
>
> Let $A_1, A_2, \ldots, A_m$ be a set of independent operations finite in number, and suppose they satisfy the system of relations given by
>
> $$S^n = 1,$$
>
> where n is a given finite integer, while S represents in turn any and every operation which can be generated from the m given operations A.
>
> Is the group thus defined one of finite order, and if so, what is its order?

He mentioned that the answer was known in the cases (i) $n = 2$ and m is any integer, and (ii) $n = 3$ and $m = 2$. In [**76**], he found the answer in two more cases: (iii) $n = 3$ and m is any integer, and (iv) $n = 4$ and $m = 2$, and added the intriguing comment,

> In the very interesting case where n is a prime p greater than 3 and m is 2, I have not been able to arrive at a definite result; but it is easy to show that, if the order of the group is finite, it is not less than p^{2p-3}.

Burnside returned to the finiteness problem, this time for groups of matrices (or linear substitutions), in 1905 [**81**]:

THEOREM 5.3. *Let G be a group of $n \times n$ matrices, with the property that for every matrix A belonging to G, the order of A is less than or equal to a finite number N. Then G is a group of finite order.*

His first observation was that the character, or trace function, $\chi(A)$, of each element $A \in G$, is the sum of the n eigenvalues of A, and that these are roots of unity of order not exceeding N, by the hypothesis of the theorem. It follows that the number of possible values of $\chi(A)$, for $A \in G$, is finite.

Now let $A = (a_{ij})$ and $B = (b_{ij})$ be elements of G. Then the formula

$$\sum a_{ij} b_{ji} = \chi(AB)$$

can be viewed as a system of equations for the entries of B, with coefficients taken from a given set of matrices $A \in G$, and with a finite number of possible choices of the expressions $\chi(AB)$, by the first step of the proof.

The main idea in the proof was to apply his irreducibility criterion, Theorem 4.5, which he had only just published. He first assumed that the given representation of G by the matrices A was irreducible. By the irreducibility criterion, there exist n^2 linearly independent matrices $A_1, \ldots, A_{n^2}$ in G. The coefficient matrix of the system of equations for $B \in G$ described above, with coefficients taken from the matrices $A_1, \ldots, A_{n^2}$ in turn, is invertible, so the solution B is uniquely determined. As the number of possibilities for $\chi(AB)$ is finite, it follows that the number of matrices belonging to G is finite, proving the theorem in this case.

To finish the proof, one may assume the given representation is reducible, so the matrices can be put in the form

$$A = \begin{pmatrix} A_1 & B \\ 0 & A_2 \end{pmatrix}.$$

The groups of matrices $\{A_1\}$ and $\{A_2\}$ satisfy the hypothesis of the theorem, and may be assumed finite, by induction on the degree of the representation. Then the kernels of the homomorphisms $A \to A_1$ and $A \to A_2$ both have finite index. The matrices belonging to their intersection have the form

$$\begin{pmatrix} I_1 & B \\ 0 & I_2 \end{pmatrix},$$

with identity matrices I_1 and I_2. Matrices of this form have infinite order unless $B = 0$. Therefore the kernels of the homomorphisms intersect trivially, and it follows that the group G is finite, completing the proof.

Burnside's finiteness problem discussed above, solved for groups of matrices in [81], (see also *Note J* in the second edition of his book [87]), attracted an increasing amount of attention in the latter part of the 20th century, perhaps because it was, like some other famous problems, easily stated and understood, but, as it turned out, very difficult to solve.[26]

In another paper published in 1905, entitled *On the complete reduction of any transitive permutation group; and on the arithmetical nature of the coefficients in its irreducible components*, Burnside carried Maschke's investigation of the subject mentioned in the second part of the title one step further (see the remarks at the beginning of §4). His new idea was to relate the problem to the field extension over $\mathbb{Q}$ containing the values of the characters:

> When the characteristics are not all rational, it is certain that the group cannot be exhibited in a form in which the coefficients are all rational; but it is possible that they may be rational functions of the characteristics. One of my objects here is to establish a condition

---

[26] Efim Zelmanov received a Fields Medal at the International Congress of Mathematicians in Zürich, 1994, for the solution of the *restricted Burnside problem*, a version of the general finiteness problem posed by Burnside (for a survey, and a report on the contributions of Zelmanov, see W. Feit, *On the work of Efim Zelmanov*, Proc. Int. Congress of Mathematicians, Zürich, Switzerland (1994)).

which shall be sufficient to ensure that an irreducible group of linear substitutions can be expressed so that the coefficients are rational functions of the characteristics, and to this I now proceed. It will be seen that the condition is satisfied for a very considerable class of cases. The quaternion group in two variables however shews that there are cases in which such a form of the group is not possible.

His main result was the following:

THEOREM 5.4. *Let $a \to A$ be an irreducible representation of a finite group G, and let χ be the character of the representation. If there exists a subgroup H of G such that the underlying vector space of the representation contains one and only one linear invariant for the subgroup H, then it is possible to find a representation equivalent to the given one in which the coefficients of all the transformations are rational functions of the character values $\chi(a), a \in G$.*

A crucial step in the proof was the observation that the given irreducible representation appears with multiplicity one in the complete reduction of the transitive permutation representation of G on the left cosets of H, by the hypothesis about the linear invariant of H. That hypothesis implies that the restriction of the character χ to the subgroup H contains the principal character of H with multiplicity one, so that, by Corollary 4.1 to the Frobenius Reciprocity Theorem (Chapter II, §4), χ occurs with multiplicity one in the character of G induced from the principal character of H. The latter is the character of the permutation representation of G on the cosets of H. Burnside gave his own proof of the result he needed, and repeated it in the second edition of his book (§207, Theorem II), without reference to the reciprocity theorem of Frobenius. A modern proof of a more general version of the theorem follows easily from a theorem of Janusz [**190**].

Within a year, Burnside and Schur obtained further results on the nature of the matrix coefficients of irreducible representations, which Burnside described in §233 of the second edition of his book [**87**] as follows:

> The general question as to the nature of the irrational quantities in terms of which the coefficients of any group of linear substitutions may be expressed has not yet received a complete answer. In every case that has been actually examined the coefficients may be expressed rationally in terms of the mth roots of unity, where m is the least common multiple of the orders of the operations of the group. Herr Schur has shown [see Chapter IV, §4], among other results, that this is certainly true for soluble groups; and the author has shown that, unless there is a number a, greater than unity, such that each multiplier of each operation of the group occurs a or a multiple of a times, it is the case.

The theory of polynomial invariants of finite groups had been one of the main topics in his correspondence with Baker (see the letters quoted in §1). He devoted a chapter in the second edition of his book to the subject and its connection with representation theory. His starting point was a powerful theorem of Molien [**216**]; we shall give Burnside's proof of it, from §227 of the second edition of his book, along with an example of how the theorem works in a special case.

Burnside's account of the matter started with a finite group G, and a representation of G by linear substitutions in the variables $x_1, \ldots, x_m$. The action of G on the variables $x_1, \ldots, x_m$ can be extended to an action of G as a finite group of automorphisms of the polynomial ring $\mathbb{C}[x_1, \ldots, x_m]$. The subspace of the polynomial ring consisting of homogeneous polynomials of degree n is carried to itself by the automorphisms corresponding to the elements of G, and affords a representation of G. It was a natural question to ask, for each n, how many times a given irreducible representation of G appears in a complete reduction of the representation of G on the polynomials of degree n. Molien's Theorem gave a generating function from which this information can be read off. The *polynomial invariants* of G are the polynomials in $\mathbb{C}[x_1, \ldots, x_m]$ which are fixed by all the automorphisms corresponding to elements of G. The space of homogeneous invariants of degree n affords the trivial representation of G, and Molien's Theorem gives a formula for the dimension of this space.

For the statement of Molien's Theorem, let $\omega_1(a), \ldots, \omega_m(a)$ be the eigenvalues of an element $a \in G$ in the given representation of G.

THEOREM 5.5. *Let χ be an irreducible character of G, and let N be the order of G. Then the coefficient of x^n in the expression*

$$\frac{1}{N} \sum_{a \in G} \frac{\chi(a^{-1})}{(1 - \omega_1(a)x) \cdots (1 - \omega_m(a)x)}$$

is the number of times the irreducible representation with character χ appears in the completely reduced form of the representation of G on the vector space V of homogeneous polynomials of degree n in the polynomial ring $\mathbb{C}[x_1, \ldots, x_m]$.

For each element $a \in G$, there exists a change of variables $x_i' = \sum_j p_{ij} x_j$ such that the linear substitution corresponding to a is the diagonal substitution $x_i' \to \omega_i(a) x_i'$. Now consider the term corresponding to $a \in G$ in the expression appearing in the statement of the theorem. Upon expanding the factors

$$\frac{1}{1 - \omega_i(a)x}$$

in geometric series, and multiplying the geometric series together, it is clear that the coefficient of x^n in the resulting product is

$$\sum_{i_1 + \cdots + i_m = n} \omega_1(a)^{i_1} \cdots \omega_m(a)^{i_m},$$

which is simply the trace of the a acting on the vector space V. Therefore the coefficient of x^n in the expression appearing in the statement of the theorem is

$$\frac{1}{N} \sum_{a \in G} \text{Trace } (a, V) \chi(a^{-1}).$$

By the orthogonality relations, this coefficient is the number of times the irreducible representation of G with character χ appears in the completely reduced form of the representation of G on V, completing the proof.

In particular, the coefficient of x^n in the expression

$$\frac{1}{N} \sum_{a \in G} \frac{1}{(1 - \omega_1(a)x) \cdots (1 - \omega_m(a)x)}$$

is the dimension of the space of homogeneous polynomial invariants of degree n in the vector space V.

As an illustration, Burnside considered, at the end of §227, the group G of order 10 generated by the matrices

$$\begin{pmatrix} \omega & 0 \\ 0 & \omega^{-1} \end{pmatrix}, \begin{pmatrix} 0 & 1 \\ 1 & 0 \end{pmatrix},$$

where $\omega^5 = 1$, viewed as a group of linear substitutions in the variables x_1, x_2. The generating function for the polynomial invariants of this group is given by the above formula, and is

$$\frac{1}{10}\left(\frac{1}{(1-x)^2} + \frac{2}{(1-\omega x)(1-\omega^{-1}x)} + \cdots\right),$$

which is easily shown to be

$$1 + x^2 + x^4 + x^5 + x^6 + x^7 + x^8 + x^9 + 2x^{10} + \cdots.$$

CHAPTER IV

Schur: A New Beginning

The combined efforts of Frobenius and Burnside built a solid foundation for the representation theory of finite groups; nevertheless, as gateways to the new subject, their accounts presented serious obstacles. To understand Frobenius's approach, one had to come to grips with the factorization of the group determinant; while to learn from Burnside, the classification of invariant Hermitian forms was required. An important step in the development of the subject was taken by Schur in 1905, when he gave a new introduction to the theory based on elementary facts from linear algebra, that opened the subject to a wide audience. Some of his ideas, such as those based on the result known as Schur's Lemma, are familiar to mathematicians everywhere.

Schur devoted the first part of his career to research in group theory and representations of finite and infinite groups. While he later branched out, with contributions to algebra, analysis, and number theory, he returned to groups and representations from time to time throughout his life. This chapter contains a section of biography, and an account of Schur's research on representation theory of finite groups. Section 3 is devoted to his joint work with Frobenius. The last two sections examine two more major undertakings begun during Schur's years as a Privatdozent in Berlin. The first, on projective representations of finite groups, was an interesting blend of representation theory and group theory, and contained ideas which were instrumental in starting a new branch of algebra, known today as the cohomology of groups. The second is an introduction to Schur's approach to permutation groups based on a study of the centralizer ring of a permutation representation, beginning with Schur's new proof, in 1908, of Burnside's theorem on permutation groups of prime degree. Schur's research on polynomial representations of the general linear group, and work of Jacques Deruyts that preceded it, are taken up in Chapter V.

1. Issai Schur in Berlin: 1894-1939

Issai Schur (1875-1941) was born in Mogilev, Russia, the son of a wholesale merchant, Moses Schur, and his wife Golde. When he was 13, he went to live with his sister and brother-in-law in Libau, Latvia, so that he could attend the Nicolai Gymnasium, where he became fluent in German. He achieved the highest mark on the final examination at his school, received a gold medal for his accomplishments there, and left in 1894 to enroll at the University of Berlin, as a student in mathematics and physics.

He was awarded the doctorate *summa cum laude* at the University of Berlin in 1901, with a dissertation entitled *Über eine Klasse von Matrizen, die sich einer gegebener Matrix zuordern lassen*. In the traditional brief autobiographical sketch at the end of his dissertation, he listed the professors of mathematics, physics,

Issai Schur (1875-1941)

and philosophy whose lectures he had attended, and the seminars in which he had participated: a mathematics colloquium organized by Schwarz, and the problem seminars of Planck in mathematical physics, and Blasius in experimental physics. He expressed special thanks to Professors Frobenius, Fuchs, Hensel, and Schwarz, and also to the University of Berlin Mathematics Club, known as the "Mapha" (Mathematisch-Physikalische Arbeitsgemeinschaft), whose meetings he had found stimulating and instructive.

With his dissertation [**247**], Schur turned a new leaf in representation theory. It contained a penetrating analysis of the representation theory of an important family of infinite groups, known today as the general linear groups $GL_n(\mathbb{C})$. Frobenius's work on the character theory of the symmetric group [**137**], which had only just been published (in 1900), and a few general results in representation theory due to Molien and Frobenius, were useful to him at several points in the dissertation, but the overall plan, and its implementation, required a new set of ideas (see Chapter V, where the dissertation, and its connection with the character theory of the symmetric group, are discussed, along with the work on algebraic forms of the Belgian mathematician Jacques Deruyts, who, it is now realized, had proved the main facts about polynomial representations of the general linear group almost a decade earlier).

Schur plunged into an active life of teaching and research, as Privatdozent at Berlin, from 1903-1911, and then with the title[1] of Professor from 1909-1913, lecturing in the standard cycles of introductory courses in algebra, analysis, and number theory. From time to time, he also gave advanced courses, out of the cycle, on the algebraic theory of quadratic forms, the theory of irrational numbers, and, in the academic year 1910-11, the new theory of integral equations and bounded bilinear forms in infinitely many variables, begun by David Hilbert and Erhard Schmidt in 1904-1905, ([**178**] and [**245**]), and to which he had contributed himself ([**253**], [**254**]), in 1909 and 1911. With the exception of the last two papers, his research during his first years on the Berlin faculty was mostly in the field of algebra and group theory, and included many of the major contributions to representation theory for which he is known today. During this period, he was in close touch with Frobenius, and their collaboration resulted in two important joint papers, [**151**] and [**152**], both published in 1906. The period was also marked by a happy marriage, in 1906, to a physician, Dr. Med. Regina Frumkin. They had two children: one, a son, Georg, named in honor of Frobenius, and a daughter, Hilde.

In 1913, Schur was appointed ausserordentlicher Professor at the University of Bonn, as the successor to Felix Hausdorff. Hausdorff was known for his research on the foundations of set theory and topology, so it is not surprising that Schur added set theory, for a short time, to his other interests. While he was at Bonn, Schur was named as a candidate for no fewer than 9 open positions on the Berlin faculty, mostly at the instigation of Frobenius.[2] Schur's name was also mentioned in a letter dated June 27, 1913, from Frobenius to the president of ETH Zürich, in response to an inquiry from Zürich.[3] In it, Frobenius discussed various candidates for the

---

[1]I am indebted to J. C. Jantzen for the following comments. " A book, *Mathematische Institute in Deutschland* 1800-1945, published by the Deutsche Mathematiker-Vereinigung, says that Schur got, in 1909, the title of a professor, but not a position as "ausserordentlicher Professor." Unfortunately the term "ausserordentlicher Professor" has been used to denote different things. It can mean a permanent position (with less pay and rights as an "ordentlicher Professor"), but it could also be an honorary title given to a Privatdozent. (At times one used the term "nicht-beamteter ausserordentlicher Professor" for the second type, sometimes "Titular-Professor"). The fact that Schur went in 1913 to Bonn as an ausserordentlicher Professor indicates that his position in Berlin in the preceding years had been only temporary."

[2]See Biermann [**14**], p. 139.

[3]The text of the letter appears in *Hermann Weyl und die Mathematik an der ETH Zürich* 1913-1930, by G. Frei and U. Stammbach, Birkhäuser Verlag, Basel (1992) pp. 14-15. I am indebted to Professor Stammbach for calling it to my attention, and for some quotations from the letter and other comments, in a letter dated October 26, 1992.

open position at Zürich, including Schur and Hermann Weyl. In connection with Schur, he stated: "Der ist viel zu gut für Zürich, und soll einmal mein Nachfolger in Berlin werden." [He is much too good for Zürich, and will at some time become my successor at Berlin]. He concluded, "Wenn Sie Weyl berufen, werden Sie eine vortreffliche und einwandfreie Wahl treffen." [If you appoint Weyl, you will make a splendid and unblemished choice.] The position was indeed offered to Weyl, who accepted it on July 5. In his letter (see the accompanying footnote), Stammbach commented, "People moved really fast in those days!".

Frobenius's efforts to have Schur return to Berlin were finally successful, and Schur was appointed ausserordentlicher Professor at Berlin, as the successor to Knoblauch, in 1916, and became ordentlicher Professor in 1919, following the death of Frobenius in 1917.

In 1918, soon after he rejoined the Berlin faculty, Schur, Konrad Knopp, Erhard Schmidt, and Leon Lichtenstein established a new journal, *Mathematische Zeitschrift*, published by Julius Springer in Berlin, with Lichtenstein the first editor. Its stated purpose was to serve pure mathematics, along with contributions of mathematical interest from theoretical physics and astronomy. Although the founding editors were all at Berlin, the journal soon became a leading international mathematics journal, and remains one today.

Schur was elected to membership in the Berlin Academy in 1922, to fill the vacancy caused by the death of Frobenius. By that time, he had branched out in his research, with contributions to integral equations, mentioned above, and to the theory of power series, and the roots of polynomial equations with integral coefficients. The formal statement proposing him for membership in the Academy, signed by 8 members including Erhard Schmidt, H. A. Schwarz, Schottky, and the Nobel Prize winners Planck, von Laue, and Nernst, gave greatest weight to his research on representation theory: the representations of the general linear group, the simplification of Frobenius's work on the factorization of the group determinant, the investigation of the algebraic number field in which the representations of a given finite group can be expressed, and the study of the representations of a finite group by linear fractional transformations, noting that Klein and his school had considered representations of finite groups by transformations in projective space.

Schur's inaugural address [**257**] (Antrittsrede) to the members of the Academy began with the acknowledgment of his debt to Frobenius, as Frobenius's pupil and collaborator. He continued with his own thoughts on representation theory, predicting its usefulness in other parts of mathematics:

> My first independent work deals with a problem from this discipline, namely the question as to the rational isomorphisms of the general linear group. I succeeded in obtaining the solution with the help of the theory of finite groups of linear substitutions recently established by Molien and Frobenius. Thereby I was led to penetrate more deeply into this theory and it was my good fortune to collaborate with Frobenius and other scholars in its further development, its simplification and its extension. What fascinated me so extraordinarily in these investigations was the fact that here, in the midst of the standstill that prevailed in other areas of the theory of forms, there arose a new and fertile chapter of Algebra which is also important for Geometry and Analysis and which was distinguished by great beauty and perfection.

The "rigid, artificial structure of Algebra," of which Kronecker once
spoke, gains here a particularly lively aspect and displays a multitude
of attractive and surprising details.[4]

University of Berlin, 1992

In his response, the Academy Secretary, Max Planck, recalled Frobenius's tes-
timony concerning Schur as his successor: "As only few mathematicians can, you
practise the great art of Abel, to formulate the problems correctly, recast them in a
suitable way, divide them skillfully, and then master them individually." After reviewing Schur's early accomplishments in representation theory, he then commented
on Schur's recent excursions into other areas of mathematics: "These achievements
led you subsequently to far-removed fields of application, and to the fundamental and extremely interesting utilization of algebraic methods for the solution of
analytic problems."

As Ordinarius [Professor] at Berlin, Schur drew large audiences to his lectures.
For example, in the Winter Semester 1930-1931, he was scheduled to lecture in the
second largest lecture hall at the university, with a capacity of about 500; but it was
still not large enough, and it was arranged for his assistant, Alfred Brauer, to give
parallel lectures in another room. Each student with any interest in mathematics
normally attended at least one of Schur's lecture courses, along with many students
whose main interest was in another field.[5] There was nothing relaxed or informal
in the presentation of the lectures. No notes were used, although it was known that
Schur kept a set of notes for the current lecture in the pocket of his jacket, and if

---

[4]Schur, I., *Antrittsrede*, Ges. Abh. II, Springer-Verlag, Berlin, 1973, pp. 413–415.
[5]See Biermann ([**15**], p. 219), and Ledermann ([**205**], §7) for recollections of Schur's teaching.

he needed something from them, such as the details of a calculation, he would turn aside, and surreptitiously find in his pocket the item he needed. In addition to the lectures, Schur held a seminar, with the details arranged by his assistant. Each student participating in the seminar gave a 50-minute lecture on a published paper selected by Schur, and was expected to speak without notes. Participants received a written report on the lecture from the assistant.

Representation theory of finite groups was first applied in physics by Eugene Wigner, at Göttingen, in two papers on the quantum mechanics of atomic spectra published in the *Zeitschrift für Physik* in 1926 and 1927, and by Hermann Weyl, in lectures given on related subjects in Zürich in 1927-28.[6] In another direction, Emil Artin, at Hamburg, took up the challenge of extending Dirichlet's theory of *L*-series associated with the characters of abelian groups to a theory of *L*-series associated with characters of general finite groups ([**7**], [**9**]; see Chapter VII, §4), in order to investigate the ideal theory in extensions of algebraic number fields with nonabelian Galois groups. In both of these important new applications of character theory, Schur's paper *Neue Begründung der Theorie der Gruppencharaktere* was a primary reference. Meanwhile, the representation theory of finite groups was absorbed in the problem of classification of modules over algebras, in the fundamental paper of Emmy Noether, *Hyperkomplexe Grössen und Darstellungstheorie* [**233**], published in 1929 (see Chapter VI, §2). While he knew of these developments, Schur did not participate in them directly. Nor was he caught up in the excitement generated by the publication of van der Waerden's book, *Moderne Algebra, unter Benutzung von Vorlesungen von E. Artin und E. Noether*, ([**276**], 1931), which offered what was at the time a new and sometimes revolutionary approach to algebra. Walter Ledermann recalled[7] that Heinz Hopf, who was a member of the Berlin faculty around 1930, told him of the "electrifying effect that van der Waerden's textbook on Modern Algebra produced in the mathematical world at that time." Hopf, Remak, and some of the other younger faculty members at Berlin, wanted to hold a seminar to study the new ideas it contained, including Noether's new approach to representation theory using modules. But in order to do this, it was necessary to obtain permission from Schur, the Professor in the field of Algebra. Hopf arranged an interview with Schur to take up the matter, with the result that permission was readily granted. But Schur himself did not attend the seminar.

The first lecture courses containing proofs of the theorems of Frobenius, Burnside, and Schur on representations of finite groups were given, as far as I know, by Emmy Noether, in Göttingen, beginning in 1924, and in a more complete form, in 1927-1928.[8] Her point of view was that semisimple hypercomplex systems, or algebras, as they are called today, are, by the Wedderburn structure theorems, direct sums of simple left ideals; their representations are completely reducible; and

---

[6]These were later expanded into books by Wigner ([**289**], 1931) and Weyl ([**285**], 1928). In his book, *The Computer*, Princeton 1972, Herman Goldstine wrote,

> Eugene Wigner was already at that time a central person among the young Göttingen pioneers. He was suspected to be familiar with some kind of black magic, called group theory. This was several years before the culmination of the so-called *"Gruppenpest"* when every paper on wave mechanics in order to be taken seriously had to start by stating the "group character" of the subject.

Goldstine, Herman, *The Computer*, Copyright©1972 by Goldstine, H. Reprinted by permission of Princeton University Press.

[7]See Ledermann, [**205**], §5.

[8]See T. Y. Lam [**203**].

from these facts, applied to the group algebra of a finite group over the field of complex numbers, the basic theorems on representations of finite groups, proved by Frobenius, Burnside, and Schur, follow (see Chapter VI, §2). In the Winter Semester, 1931-1932, Schur gave a course of lectures at Berlin,[9] out of the cycle, entitled *Gruppen linearer Substitutionen*, which contained all the main theorems of Frobenius, Burnside, and Schur on matrix representations of finite groups and their characters over the field of complex numbers, proved in a coherent way from the point of view of Schur's paper *Neue Begründung der Theorie der Gruppencharaktere*. This was followed by applications to finite groups that may well have appeared here for the first time in a course of lectures: Burnside's proof of his $p^a q^b$-Theorem, induced characters and the calculation of the irreducible characters of the alternating group A_5, Frobenius's calculation of the characters of the symmetric groups, with improvements and simplifications of his own, and a sketch of his work on the rational representations of the general linear group. The presentation is crystal clear, and, if given verbatim today, would make a fine graduate course.

Schur's research in the 1920's and the early 1930's covered a wide range of topics in complex analysis, number theory, and algebra. Some of the articles were the results of collaborations: with K. Knopp and G. Szegö in analysis, with L. Bieberbach on a continuation of Minkowski's work on the arithmetic theory of quadratic forms, and with A. Ostrowski and his former doctoral student, Richard Brauer, in algebra. He continued his work in representation theory, with a series of important papers simplifying and extending his work on the polynomial representations of the general linear group (see Chapter V, §2). His last published paper in representation theory ([**261**], 1933) introduced new methods in the theory of finite permutation groups based on a study of the ring of matrices commuting with the matrices in a given permutation representation. This idea was taken up in a powerful way by his former doctoral student, Helmut Wielandt (see §6; also Chapters IV and V of Wielandt's monograph on permutation groups [**288**], and its review in *Mathematical Reviews* by N. Ito (vol. 32, 1252)).

The dark cloud of Naziism descended over Europe when Hitler came to power in 1933. Jewish faculty members at German universities (including Schur) were dismissed from their positions. Some, including Emmy Noether, and Schur's doctoral student, Richard Brauer, were able to leave Germany for the United States. Schur himself decided to remain in Germany. In his article, *Issai Schur and his School in Berlin* [**205**], Walter Ledermann wrote:

> When the storm broke in 1933, Schur was 58 years of age and, like many German Jews of his generation, he did not grasp the brutal character of the Nazi leaders and their followers. It is an ironic twist of fate that, until it was too late, many middle-aged Jews clung to the belief that Germany was the land of Beethoven, Goethe, and Gauss rather than the country that was now being governed by Hitler, Himmler and Goebbels.[10]

Soon after his suspension from the university, Schur received an offer from the University of Wisconsin in Madison. He declined the offer, partly because he was

---

[9]I am indebted to Walter Ledermann for lending me his own set of lecture notes from this course to make a copy, and for reminiscences he shared with me of his student days in Berlin.

[10]Ledermann, W., "Issai Schur and his School in Berlin", *Bull. London Math. Soc.* **15** (1983), pp. 97–106. Used with permission.

DIE ALGEBRAISCHEN GRUNDLAGEN DER

DARSTELLUNGSTHEORIE DER GRUPPEN

Vorlesungen über Darstellungstheorie
von

DR. J. SCHUR

emeritierter Professor an
der Universität Berlin

gehalten auf Einladung des mathematischen Seminars
der Eidg. Techn. Hochschule Zürich

bearbeitet und herausgegeben
von

DR. E. STIEFEL

Assistent am mathematischen
Seminar der E.T.H.

1 9 3 6

Autographie: Graph.Anstalt Gebr. Frey & Kratz
Zürich.

Title page of the mimeographed notes from
Schur's lectures at the ETH Zurich, 1936

already fluent in two languages, and no longer felt energetic enough to give lectures in a third.[11]

During the period of his suspension, Schur was not allowed to come to the university, and he received students at his home.[12] Through the efforts of Erhard Schmidt, the suspension was cancelled in the Winter Semester, 1933–34, and he resumed his duties. After his reinstatement, members of the Nazi youth occasionally disrupted his lectures.[13]

Schur's period of reinstatement ended abruptly on August 31, 1935, when he was forced to resign. In 1936, as Professor Emeritus, he was invited to Zürich. He was a friend of Heinz Hopf, who probably arranged the visit.[14] While there, he gave a course of lectures at ETH Zürich. Notes on the lectures were prepared by E. Stiefel, then an assistant at the ETH, with the assistance of G. Polya and W. Frey, and published in Zürich with the title *Die Algebraischen Grundlagen der Darstellungstheorie der Gruppen*. The lectures were a revision of his lectures on groups of linear substitutions given at Berlin in 1931–32, whose contents were described above. The published version, from Zürich, was, to my knowledge, the first book entirely devoted to representations of finite groups.

Things began to unravel for Schur in the late 1930's. In 1938, he left the Berlin Academy, along with Goldschmidt and Norden. In ([17], pp. VII, VIII), Alfred Brauer gave a detailed account of the Schurs' plan to emigrate to Palestine (now Israel), the difficulties they encountered in carrying it out, and a moving and personal recollection of his last days with Schur in Berlin. In 1938, shortly before a scheduled appointment with the Gestapo, his wife arranged for Schur to be hospitalized—by then he had been diagnosed as suffering from arterial sclerosis[15]— and in January, 1939, to leave Germany, and join his daughter in Bern. A few weeks later, his wife was able to follow him, and soon afterward, they departed for Israel. There they were welcomed by Schiffer, whom Schur had known in Berlin, and Fekete, and Schur resumed his mathematical work. In 1940, he suffered a heart

---

[11]See A. Brauer [17], p. VI.

[12]W. Ledermann, personal communication, December, 1992.

[13]W. Ledermann, personal communication, December, 1992.

[14]His trip to Zürich did not pass unnoticed. His colleague, and former collaborator, Ludwig Bieberbach, reported the visit, in a letter dated 20 Feb. 1936, to the Reichsminister für Wissenschaft, Kunst, und Volksbildung, and stated that he would try to learn the contents of Schur's guest-lectures in Zürich. By then, Bieberbach's antisemitism had gained the upper hand, and he had become associated with the Nazis (see Biermann, [15], §8.2.3). In 1934 Bieberbach had published an article entitled *Persönlichkeitstruktur und mathematisches Schaffen* in the journal *Forschungen und Fortschritte*, on the influence of nationality, blood, and race on the creative style. It was reviewed in *Nature* [169] by G. H. Hardy. After examining Bieberbach's distinction between the "J-type" and the "S-type" among mathematicians, Hardy concluded with the comment: "It is not reasonable to criticise too closely the utterances, even of men of science, in times of intense political or national excitement. There are many of us, many Englishmen and many Germans, who said things during the War which we scarcely meant and are sorry to remember now. Anxiety for one's own position, dread of falling behind the rising torrent of folly, determination at all costs not to be outdone, may be natural if not particularly heroic excuses. Prof. Bieberbach's reputation excludes such explanations of his utterances; and I find myself driven to the more uncharitable conclusion that he really believes them true." Hardy, G. H., "The *J*-type and the *S*-type among mathematicians," *Collected Papers of G. H. Hardy*, London Mathematical Society, Vol. 7, pp. 610–611. Used with permission.

[15]This diagnosis had been reported to the German authorities in 1936, according to a note in Schur's *Personal-Akten* [Personal Papers] in the *Archiv der Berlin Universität*.

attack while giving a lecture, and died a few weeks later on his 66th birthday, January 10, 1941.

In 1956, on the occasion of Schur's 80th birthday, the editors of *Mathematische Zeitschrift* dedicated Volume 63 of the journal to him. It contained a brief statement prepared by the editors entitled *Issai Schur zum Gedächtnis*, and articles dedicated to him by several of his friends and former students and colleagues. The collection includes a landmark paper on representation theory by his former student, Richard Brauer, entitled *Zur Darstellungstheorie der Gruppen endlicher Ordnung*, to which we shall return in Chapter VII, §3.

A list of the titles of 22 dissertations supervised by Schur can be found in his *Gesammelte Abhandlungen*, Bd. III, Springer-Verlag, Berlin, 1973, on pp. 479-480. Almost all were in algebra or number theory, and are listed below, with the date of the doctoral degree, name of the student, and title of the dissertation.

13.10.1921, Heinz Prüfer, *Unendliche Abelsche Gruppen von Elementen endlicher Ordnung.*

14.8.1922, Dora Prölsz, *Über Zahlkörper, die aus dem Körper der rationalen Zahlen durch Adjunktion von Wurzelausdrücken hervorgehen.*

15.12.1922, Felix Pollaczek, *Über die Kreiskörper der ℓ-ten und ℓ²-ten Einheitswurzeln.*

7.3.1923, Maximilian Herzberger, *Über Systeme hyperkomplexer Grössen.*

12.3.1924, Hildegard Ille, *Zur Irreduzibilität der Kugelfunktionen.*

9.5.1925, Karl Dörge, *Über die ganzen rationalen Lösungspaare von algebraischen Gleichungen in zwei Variablen.*

16.3.1926, Richard Brauer, *Über die Darstellung der Drehungsgruppe durch Gruppen linearer Substitutionen.*

12.10.1928, Udo Wegner, *Über die ganzzahligen Polynome, die für unendlich viele Primzahlmoduln Permutationen liefern.*

19.12.1928, Alfred Brauer, *Über diophantische Gleichungen mit endlich vielen Lösungen.*

19.12.1928, Arnold Scholz, *Über die Bildung algebraischer Zahlkörper mit auflösbarer Galoisscher Gruppe.*

7.2.1931, Robert Frucht, *Über die Darstellung endlicher Abelscher Gruppen von Kollineationen.*

9.5.1932, Wilhelm Specht, *Eine Verallgemeinerung der symmetrischen Gruppe.*

25.7.1932, Bernhard Neumann, *Die Automorphismgruppe der freien Gruppen.*

25.7.1932, Hans Rohrbach, *Die Charaktere der binären Kongruenzgruppen mod* p^2.

8.2.1935, Helmut Wielandt, *Abschätzungen für den Grad einer Permutationsgruppe von vorgeschriebenem Transitivitätsgrad.*[16]

2. Schur's New Foundations of Character Theory

Burnside, and shortly afterwards, Schur, both gave new introductions to Frobenius's theory of characters and representations of finite groups, which they believed were more direct or simpler (see Chapter III, §4, for Burnside's contributions). In his paper *Neue Begründung der Theorie der Gruppencharaktere* ([**249**], 1905), Schur put it this way:

---

[16]Schur. I., Gesammelte Abhandlungen, Bd. III, Springer-Verlag, Berlin, 1973. Used with permission.

The following work contains a thoroughly elementary introduction to the theory of group characters, established by Frobenius. It can also be characterized as the study of representations of finite groups by linear homogeneous substitutions.

> An elementary presentation of this theory has indeed been given recently by Burnside [citing [**77**] and [**78**]; see Chapter III §4]. Burnside, however, makes use of material far removed from the subject, namely the concept of Hermitian forms. I consider it not superfluous, therefore, to give a new account of Frobenius's theory, that operates with still simpler methods.

Schur's introduction to the subject contained short and incisive new proofs of the main results, organized in a beautiful way to take advantage, as much as possible, of the result known today as Schur's Lemma, which, as Schur mentioned, "had also played an important role in Burnside's presentation"([**249**], p. 409; Ges. Abh. I, p. 146). Schur and Frobenius were together in Berlin at the time the paper was written, and remarks and footnotes show that Frobenius took a keen interest in it, and contributed some of his own ideas. Its thrust was continued in a joint paper with Frobenius [**151**], published the following year, that contained another basic new result for the foundations of representation theory, extending Burnside's theorem (Chapter III, Theorem 4.5), and also based on Schur's Lemma.

This section contains a fairly complete account of the results in Schur's paper [**249**], following, to a great extent, Schur's own proofs and explanations; its crisp approach stands up well in comparison with contemporary expositions of the same material. It begins with the concept of a *representation* $r \to A(r)$ of a finite group G by $f \times f$ matrices $A(r), r \in G$, with entries in $\mathbb{C}$, and such that $A(rs) = A(r)A(s)$, for all $r, s \in G$, as in Frobenius's paper [**133**] (see Chapter II, §4). The number f is called the *degree* of the representation. Following Frobenius, he introduced independent variables x_r, indexed by the elements $r \in G$, and the matrix, which he called a *group matrix*

$$X = \sum_{r \in G} A(r)x_r,$$

associated with the given representation.

Let 1 denote the identity element of G; then one clearly has $A(1)X = X$, for a group matrix X as above. Therefore the determinant $|X|$ is identically zero, that is, $|X| = 0$ for all values of the x_r, if and only if the determinant of $A(1)$ is zero. It follows that, if $|X| \neq 0$, then $A(1)$ is the identity matrix E, and all the matrices $A(r)$ are invertible, for $r \in G$, as each element $r \in G$ satisfies the equation $r^g = 1$, where $g = |G|$, so $A(r)^g = A(1) = E$. After this interlude, it is natural to confine our attention, as we shall in what follows, to representations $r \to A(r)$ satisfying the conditions that $A(1)$ is the identity matrix, and that all the matrices $A(r), r \in G$, are invertible.

A group matrix X associated with a representation $r \to A(r)$ is called *equivalent* to the group matrix X' whenever $X' = S^{-1}XS$ for an invertible constant matrix S; in that case, X' is associated with the representation $r \to A'(r) = S^{-1}A(r)S$, and the representations $r \to A(r)$ and $r \to A'(r)$ are said to be equivalent. A group matrix X is defined to be *reducible* if it is equivalent to a group matrix of the form

$$\begin{pmatrix} X_1 & 0 \\ U & X_2 \end{pmatrix},$$

with square submatrices X_1 and X_2. When this occurs, the submatrices X_1 and X_2 are group matrices associated with representations. An *irreducible* group matrix is one that is not reducible; reducible and irreducible representations are defined similarly. A group matrix X is also reducible if it is equivalent to a matrix of the form

$$\begin{pmatrix} X_1 & V \\ 0 & X_2 \end{pmatrix},$$

as is shown by applying the similarity transformation defined by the matrix

$$\begin{pmatrix} 0 & E \\ E' & 0 \end{pmatrix},$$

where E and E' are identity matrices.

Representations of groups were, for Schur, always *matrix* representations; the idea that matrices defined linear transformations acting on vectors in a vector space was never adopted by him, although this point of view was embraced, somewhat later, by his contemporaries Hermann Weyl and Emmy Noether (see Chapter VI, §2).

Schur stated the result known today as Schur's Lemma as follows.

THEOREM 2.1 (Schur's Lemma). *(i) Let X and X' be irreducible group matrices, of degrees f and f', respectively. Let P be a constant matrix with f rows and f' columns, such that*

$$XP = PX'.$$

Then either $P = 0$, or X and X' are equivalent and P is an invertible square matrix.

(ii) Let X be an irreducible group matrix of degree f. Then each constant matrix P commuting with X has the form aE_f, with $a \in \mathbb{C}$ and E_f the $f \times f$ identity matrix.

In modern terminology, the first part of the theorem states that if $r \to A(r)$ and $r \to B(r)$ are two irreducible representations of G, and P is a matrix such that $A(r)P = PB(r)$ for all $r \in G$, then either $P = 0$ or the representations $r \to A(r)$ and $r \to B(r)$ are equivalent, and P is a (necessarily square) invertible matrix. The second part states simply that the only matrices which commute with all the matrices $A(r)$, in an irreducible representation $r \to A(r)$, are multiples of the identity matrix by elements of $\mathbb{C}$.

Schur's proof began with the elementary fact that there exist invertible $f \times f$ and $f' \times f'$ matrices A and B such that, if $P \neq 0$, then

$$APB = Q,$$

and Q has the form

$$\begin{pmatrix} E_m & 0 \\ 0 & 0 \end{pmatrix}$$

with E_m the $m \times m$ identity matrix, and the rest of the entries equal to zero. (This follows, for example, by reducing P to Q by performing a sequence of elementary row and column operations, and interpreting each step as multiplication on the left or right by an elementary matrix.) The next step was to set $X_1 = AXA^{-1}, X_1' = B^{-1}X'B$, and observe that

$$X_1 Q = QX_1'.$$

Upon putting X_1 and X_1' in the block forms

$$X_1 = \begin{pmatrix} X_{mm} & Y \\ Z & W \end{pmatrix}, X_1' = \begin{pmatrix} X'_{mm} & Y' \\ Z' & W' \end{pmatrix}$$

with X_{mm} and X'_{mm} both $m \times m$ submatrices, it follows from the preceding equation that $Z = 0$ and $Y' = 0$. Therefore, if either $m < f$ or $m < f'$, then X or X' is reducible. As $P \neq 0$, the conclusion is that $m = f = f'$, and P is an invertible matrix such that

$$P^{-1}XP = X'.$$

For the proof of (ii), let a be an eigenvalue of P, so that

$$|P - aE_f| = 0.$$

Then

$$X(P - aE_f) = (P - aE_f)X,$$

and by part (i), $P - aE_f = 0$, completing the proof.

The principal new result in the paper was the following set of orthogonality relations for the coefficients of an irreducible representation.

THEOREM 2.2. *Let $r \to A(r) = (a_{ij}^r)$ be an irreducible representation, of degree f, of a finite group G. Then, for all $i, j, k, \ell = 1, \ldots, f$,*

$$\sum_{r \in G} a_{ij}^{r^{-1}} a_{k\ell}^r = \frac{|G|}{f} e_{i\ell} e_{jk},$$

where $e_{ab} = 1$ or 0 according as $a = b$ or $a \neq b$. If $r \to B(r) = (b_{ij}^r)$ is a second irreducible representation, not equivalent to the first, then

$$\sum_{r \in G} a_{ij}^{r^{-1}} b_{k\ell}^r = 0,$$

for all i, j, k, ℓ.

In a footnote, Schur remarked that, although the relations had not appeared previously in the literature, they were implicit in the known orthogonality relations for characters. There was a fundamentally new idea in his proof, however, which gives an efficient proof of the theorem, and incidentally supersedes all previous proofs of the orthogonality relations for characters. It has been in constant use ever since, and consists of an ingenious construction (of a matrix V defined below) which brings Schur's Lemma into play in a decisive way.

Here is the main idea. Let $U = (u_{ij})$ be an $f \times f$ matrix, and let

$$V = \sum_{r \in G} A(r^{-1}) U A(r).$$

Using the fact that $r \to A(r)$ is a representation, so that $A(rs) = A(r)A(s)$ for all elements $r, s \in G$, one has

$$A(s^{-1})VA(s) = \sum_{r \in G} A(s^{-1}r^{-1})UA(rs),$$

for each element $s \in G$. The elements rs, for fixed s, run through the elements of G because G is a group. Therefore the right side of the equation above is equal to V. Upon multiplying on the left by $A(s)$, and noting that $A(s)A(s^{-1}) = A(1) = E_f$ (the identity matrix), one obtains

$$VA(s) = A(s)V,$$

for all elements $s \in G$. Part (ii) of Schur's Lemma can now be applied, with the result that $V = vE_f$ for some complex number v. The entries of V are linear combinations of the entries of U, so

$$v = \sum_{a,b} c_{ab} u_{ab},$$

for some elements $c_{ab} \in \mathbb{C}$. From the definition of V, and the fact that $V = vE_f$, one has

$$\sum_{r \in G} \sum_{j,k} a_{ij}^{r^{-1}} u_{jk} a_{k\ell}^r = e_{i\ell} \sum_{ab} c_{ab} u_{ab},$$

for all choices of i and ℓ. By taking $u_{jk} = 1$ and all other entries of U equal to 0, it follows that

$$\sum_{r \in G} a_{ij}^{r^{-1}} a_{k\ell}^r = e_{i\ell} c_{jk},$$

and it remains to evaluate c_{jk}. This is done by taking $i = \ell$, and summing the preceding formula on i. This yields the formula

$$\sum_{r \in G} \sum_i a_{ki}^r a_{ij}^{r^{-1}} = f c_{jk}.$$

Finally, use the relation $A(r)A(r^{-1}) = E_f$ to obtain

$$|G| e_{kj} = f c_{jk}.$$

This completes the proof of the first set of relations.

The proof of the second set of relations is similar to the proof of the first. Starting from an $f \times f'$ matrix U (where f' is the degree of the representation $r \to B(r)$), let

$$V = \sum_{r \in G} A(r^{-1}) U B(r),$$

and proceed as in the first part of the proof to obtain $A(r)V = VB(r)$ for all elements $r \in G$. As the representations $r \to A(r)$ and $R \to B(r)$ are assumed not to be equivalent, part (i) of Schur's Lemma implies that $V = 0$. Then, as above, one has

$$\sum_{r \in G} \sum_{j,k} a_{ij}^{r^{-1}} u_{jk} b_{k\ell}^r = 0,$$

for all i, ℓ, and all choices of elements u_{jk}. Upon specializing the elements u_{jk} as above, taking all but one of them equal to zero, the second set of orthogonality relations follows. This completes the proof of the theorem.

COROLLARY 2.1. *Let*

$$r \to A(r) = (a_{ij}^r), r \to B(r) = (b_{k\ell}^r), \ldots$$

be a finite set of irreducible representations of G of degrees $f, f', \ldots$, no two of which are equivalent. Then the coefficients $a_{ij}^r, b_{k\ell}^r, \ldots$ form a set of $f^2 + (f')^2 + \ldots$ linearly independent functions on G, in the sense that a relation

$$\sum_{i,j} u_{ij} a_{ij}^r + \sum_{k,\ell} v_{k\ell} b_{k\ell}^r + \cdots = 0,$$

with constant coefficients $u_{ij}, v_{k\ell}, \ldots$, can only hold for all $r \in G$, if all the coefficients are zero.

For the proof, one multiplies the relation given in the statement of the corollary by $a_{pq}^{r^{-1}}$, for fixed p, q, and sums over $r \in G$. One has, by the theorem, the result that

$$\frac{|G|}{f} u_{qp} = 0,$$

and hence $u_{qp} = 0$ for all p, q. Similarly, all other coefficients are equal to zero.

The preceding corollary implies Burnside's Theorem (Chapter III, Theorem 4.5), for the case of representations of finite groups. Burnside's Theorem also applies to representations of infinite groups, and this situation is not covered by the preceding argument. Frobenius and Schur considered this anomaly, and shortly afterwards, found a way to prove a more general version of the corollary which, in particular, accounted for representations of infinite groups [**151**]. Their idea for extending the proof of the corollary to a more general situation was an important milestone in the passage from representations of finite groups to representations of algebras, culminating in Emmy Noether's paper [**233**], (1929).

No reworking of the basic results about group representations could be considered satisfactory if it failed to give a simple proof of Frobenius's theorems on the factorization of the group determinant (see Chapter II, §3). Schur's first step in this direction was also based on the preceding theorem and corollary.

COROLLARY 2.2. *The determinant of an irreducible group matrix $X = (x_{ij})$ is an irreducible polynomial in the variables $x_r, r \in G$. Two irreducible group matrices are equivalent if and only if their determinants are equal.*

By the previous corollary, the entries x_{ij} of an irreducible group matrix X of degree f are a set of f^2 linearly independent functions of the variables $x_r, r \in G$. Let u_{ij} be a set of f^2 independent variables. One can replace the variables x_r by linear combinations of the u_{ij} in such a way that the entries in the group matrix X are given by $x_{ij} = u_{ij}$ for all i, j. Therefore a nontrivial factorization of the determinant $|x_{ij}|$, viewed as a polynomial in the variables x_r, implies a nontrivial factorization of the polynomial $|u_{ij}|$ in the independent variables u_{ij}. The latter was known to be impossible, as Schur noted, proving the irreducibility of the determinant $|X|$. Now let $X = (x_{ij})$ and $X' = (x'_{ij})$ be two inequivalent irreducible group matrices. By the corollary above, one can specialize the variables x_r so that the matrix entries x_{ij} and $x'_{k\ell}$ can be chosen arbitrarily. It follows that $|X| \neq |X'|$. On the other hand, if X is equivalent to X', then clearly $|X| = |X'|$, completing the proof of the corollary.

Schur next turned his attention to reducible group matrices (or reducible representations). Maschke's theorem (Chapter III, Theorem 4.1), as Burnside had already shown (see Chapter III, §4), was crucial for understanding how reducible representations can be analyzed in terms of irreducible ones. Schur gave a new self-contained proof, which demonstrated that the existence of invariant Hermitian forms, proved by Loewy and Moore, and applied by Maschke in the proof of his theorem, was not required for the proof. In a footnote (Ges. Abh. Bd. I, p. 152) he remarked that he was led to the elementary proof given below by a simplified version of Maschke's proof involving invariant Hermitian forms, communicated to him by Frobenius.

THEOREM 2.3. *Let X be a group matrix for a finite group G, which is equivalent to a group matrix of the form*

$$Y = \begin{pmatrix} X_1 & 0 \\ U & X_2 \end{pmatrix}.$$

Then it is also equivalent to a group matrix of the form

$$Z = \begin{pmatrix} X_1 & 0 \\ 0 & X_2 \end{pmatrix}.$$

It had to be proved that there exists an invertible constant matrix P such that

$$Z = P^{-1}YP.$$

Schur translated the problem into the language of representations as follows. Let $Y = \sum_{r \in G} A(r)x_r$, and let the representation $r \rightarrow A(r)$ associated with the group matrix Y have the form

$$A(r) = \begin{pmatrix} B(r) & 0 \\ D(r) & C(r) \end{pmatrix},$$

where $r \rightarrow B(r)$ and $r \rightarrow C(r)$ are the representations associated with the group matrices X_1 and X_2, respectively. His next step was the observation that it was sufficient to find a matrix P of the form

$$P = \begin{pmatrix} E_1 & 0 \\ F & E_2 \end{pmatrix},$$

with E_1 and E_2 identity matrices, such that

$$P^{-1}A(r)P = \begin{pmatrix} B(r) & 0 \\ 0 & C(r) \end{pmatrix},$$

for all $r \in G$. The necessary and sufficient condition for this to occur is that

$$D(r) + C(r)F = FB(r),$$

for all $r \in G$. He then proceeded to solve these equations for a matrix F directly, using only the facts that $r \rightarrow A(r), B(r), C(r)$ are representations of G, so that $A(rs) = A(r)A(s)$ for all $r, s \in G$ etc..

From the equations $A(rs) = A(r)A(s)$ etc., one has

$$D(rs) = D(r)B(s) + C(r)D(s),$$

for all $r, s \in G$. Multiplying this equation on the right by $B(s^{-1})$, and summing over $s \in G$, he obtained

$$\sum_{s \in G} D(rs)B(s^{-1}) = |G|D(r) + \sum_{s \in G} C(r)D(s)B(s^{-1}),$$

for all $r \in G$, using the fact that $B(s)B(s^{-1}) = E_1$. The key step in the proof was to show that the matrix

$$F = \frac{1}{|G|} \sum_{s \in G} D(s)B(s^{-1})$$

does indeed satisfy the equations stated in the paragraph above. To prove this, he rewrote the preceding formula, with s replaced by $r^{-1}s$. The result is that

$$\sum_{s \in G} D(s)B(s^{-1}r) = \sum_{s \in G} D(s)B(s^{-1})B(r) = |G|FB(r) = |G|D(r) + C(r)|G|F,$$

for all $r \in G$. By cancelling $|G|$, one has

$$FB(r) = D(r) + C(r)F$$

for all $r \in G$, completing the proof.[17]

Schur remarked that he was led to the preceding elementary proof from the following communication he had received from Frobenius, containing a "simplification and sharpening" of Maschke's proof. By the theorem of Moore and Loewy (Chapter III, §4), there exists a positive definite Hermitian form $f = \sum h_{ij} x_i \bar{x}_j$ left invariant by the finite group of matrices

$$A(r) = \begin{pmatrix} B(r) & 0 \\ D(r) & C(r) \end{pmatrix},$$

corresponding to the elements $r \in G$ in the given reducible representation. This means that, for all $r \in G$, one has

$$A(r)'HA(r) = H,$$

where $H = (h_{ij})$ is the matrix of the Hermitian form f, and $A(r)'$ denotes the conjugate transpose of the matrix $A(r)$. (The change of variables required to obtain an invariant form whose matrix is the identity, as at the conclusion of the statement of Moore's Theorem, would alter the form of the matrices $A(r)$, so H is not assumed to be the identity in what follows.) Putting H in the form

$$\begin{pmatrix} J & K \\ L & M \end{pmatrix}$$

corresponding to the form of the matrices $A(r)$, one has the equations

$$C(r)'LB(r) + C(r)'MD(r) = L, \text{ and } C(r)'MC(r) = M,$$

for all $r \in G$. Frobenius then used the fact that in the matrix of a positive definite Hermitian form, the determinant of each principal submatrix is different from zero, so in particular, $|M| \neq 0$, and M is an invertible matrix. Upon eliminating $C(r)'$ from the preceding equations, it follows that the matrix $F = M^{-1}L$ satisfies the equation

$$FB(r) + D(r) = C(r)F,$$

for all $r \in G$, and the proof is completed as in the discussion given above.

---

[17]In modern terminology, the set of rectangular matrices U with f rows and e columns, where e and f are the degrees of the subrepresentations $r \to B(r)$ and $r \to C(r)$, respectively, form a two-sided G-module $\mathcal{U}$, with left and right actions by elements of G defined by $rU = C(r)U$ and $Us = UB(s)$, satisfying the associativity condition $(rU)s = r(Us)$, for $r, s \in G$ and $U \in \mathcal{U}$. The matrices $D(r) \in \mathcal{U}$, in Schur's proof above, define a map $r \to D(r)$ from G to $\mathcal{U}$, and hence a linear map $a \to D(a) \in \mathcal{U}$ from the group algebra $\mathbb{C}G$ to $\mathcal{U}$. The formulas

$$D(rs) = D(r)B(s) + C(r)D(s),$$

for $r, s \in G$ imply that $D(ab) = aD(b) + D(a)b$, for all elements $a, b \in \mathbb{C}G$. Linear maps from an algebra over $\mathbb{C}$ (in this case the group algebra $\mathbb{C}G$) to a two-sided module for the algebra satisfying the preceding condition are called *crossed homomorphisms* or *derivations* (see Cartan and Eilenberg, *Homological Algebra* [91] (1956), p. 168). Examples of crossed homomorphisms are the maps $a \to Fa - aF$, for a fixed matrix $F \in \mathcal{U}$; these are called *principal crossed homomorphisms* or *inner derivations*. Schur's proof of Maschke's theorem amounts to a proof of the fact that every crossed homomorphism of the group algebra $\mathbb{C}G$ of a finite group G is principal, or that the first cohomology group $H^1(\mathbb{C}G, \mathcal{U})$ is zero (see Cartan-Eilenberg, [91], Prop. 4.1, p. 170).

Maschke's theorem implies that every representation of a finite group is either irreducible, or completely reducible, in the sense that it is equivalent to a representation of the form

$$r \to \begin{pmatrix} A_1(r) & 0 & \dots \\ 0 & A_2(r) & \dots \\ \dots & \dots & \dots \end{pmatrix},$$

for $r \in G$, with irreducible subrepresentations $r \to A_i(r)$ along the diagonal, and zeros elsewhere (see the proof of Theorem 4.2 in Chapter III). Further investigation of the decomposition of reducible representations into irreducible ones was based on the following uniqueness theorem, stated, as Schur did, in terms of group matrices.

THEOREM 2.4. *Let X be a group matrix of a finite group which is equivalent to the group matrices*

$$\begin{pmatrix} X_1 & \dots \\ \dots & \dots \\ \dots & X_m \end{pmatrix} \quad and \quad \begin{pmatrix} X_1' & \dots \\ \dots & \dots \\ \dots & X_{m'}' \end{pmatrix},$$

with X_i and X_j' irreducible, for all i and j, and all entries equal to zero except in the diagonal blocks. Then $m = m'$ and the group matrices $X_1, X_2, \dots$ are equivalent, apart from order, to the group matrices $X_1', X_2', \dots$.

The determinants of the two group matrices in question are equal, and are products of the determinants $|X_1|, |X_2|, \dots$, and $|X_1'|, |X_2'|, \dots$, respectively. As these are irreducible monic polynomials in the variables $x_r, r \in G$ by Corollary 2.2, the theorem follows at once from the uniqueness of factorization of polynomials in several variables over $\mathbb{C}$.

The preceding theorem implies that the number of times a given irreducible group matrix X_i occurs in the completely reduced form of a group matrix X is uniquely determined; Schur called it the *index* of X_i in X. Today we would call it the *multiplicity* of a given irreducible representation in the completely reduced form of a representation; it counts the number of components of the representation which are equivalent to the given irreducible representation.

The following two theorems gave Schur's version of Frobenius's main results on the group determinant of a finite group (Chapter II, Theorems 3.4 and 4.1) and Burnside's analysis of the completely reduced form of the regular representation of a finite group G, which Burnside had derived using his classification of G-invariant Hermitian forms (Chapter III, Theorem 4.3).

THEOREM 2.5. *Let $X = (x_{ij})$ be a group matrix of a finite group G, let $e_1, \dots, e_\ell$ be the indices of the inequivalent irreducible group matrices $X_i, i = 1, \dots, \ell$, occurring in a completely reduced form of X, and let $f_1, \dots, f_\ell$ be their degrees. Then:*

(i) $f_1^2 + \cdots + f_\ell^2$ is the number of linearly independent functions x_{ij} of the variables $x_r, r \in G$; and

(ii) $e_1^2 + \cdots + e_\ell^2$ is the number of linearly independent constant matrices P commuting with X.

(iii) The number of linearly independent matrices Q commuting with X and with all matrices P as in part (ii) is equal to ℓ.

A knowledgeable reader today will immediately understand the theorem in terms of representations of semisimple algebras, and their commuting algebras, over an algebraically closed field. But this somewhat high-powered approach is unnecessary—and was not available to Schur! He simply noted that the first part of the theorem follows at once from Corollary 2.1, while the second part is a fairly easy consequence of Schur's Lemma. Indeed, for the proof of part (ii), use the fact that X is equivalent to a group matrix, in completely reduced form, with the first irreducible group matrix X_1 repeated e_1 times along the diagonal, the second repeated e_2 times, etc.. Put a matrix commuting with this matrix in corresponding block form, write out the commutativity conditions, and apply Schur's Lemma. It is a nice exercise to carry this out. Part (iii) follows from the proof of part (ii).

Before stating the next theorem, Schur recalled that the matrix $X = (x_{rs^{-1}})$, with $|G|$ rows and columns indexed by the elements of G, was the group matrix associated with the regular representation of G. He let $X_1, \ldots, X_k$ denote the inequivalent irreducible group matrices occurring in a completely reduced form of X, and denoted their indices and degrees by $e_1, \ldots, e_k$ and $f_1, \ldots, f_k$, respectively. Letting Θ denote the determinant of X, and Φ_i the determinant of X_i, for $i = 1, \ldots, k$, it follows from the preceding discussion that

$$\Theta = \Phi_1^{e_1} \cdots \Phi_k^{e_k},$$

and that this equation gives the factorization of Θ as a product of irreducible polynomials.

THEOREM 2.6. *The numbers e_i and f_i are equal, for $i = 1, \ldots, k$. Moreover, the number k is equal to the number of conjugacy classes in G.*

From the factorization given above, one has

$$e_1 f_1 + \cdots + e_k f_k = |G|.$$

By part (i) of the preceding theorem, it follows that

$$f_1^2 + \cdots + f_k^2 = |G|.$$

Part (ii) of the preceding theorem yields a further relation as soon as the matrices $Y = (y_{r,s})$ commuting with X are known. A calculation shows that Y commutes with X if and only if $y_{p,q} = y_{rp,rq}$ for all $r, p, q \in G$. Put $y_r = y_{r,1}$ for each $r \in G$ (where 1 is the identity element of G, as usual). It follows that the commuting matrices Y are all matrices of the form $Y = (y_{q^{-1}p})$, and hence

$$e_1^2 + \cdots + e_k^2 = |G|.$$

Combining the formulas, one has

$$(e_1 - f_1)^2 + (e_2 - f_2)^2 + \cdots = 0.$$

This proves that $e_i = f_i$ for each i, and establishes the result that Frobenius had called the "Fundamental Theorem in the Theory of the Group Determinant" (Chapter II, Theorem 3.4).

In order to prove the second statement, that k is the number of conjugacy classes in G, Schur reasoned as follows. A matrix $Z = (z_{p,q})$ commutes with X and all matrices Y commuting with X if and only if

$$z_{p,q} = z_{rp,rq} = z_{pr,qr},$$

for all $p, q, r \in G$, by the calculations used in the first part of the proof. Setting $z_{r,1} = z_r$, the conditions become

$$z_{p,q} = z_{q^{-1}p} = z_{pq^{-1}}.$$

Replacing p by pq in these equations, it follows that a matrix Z as above commutes with X and all matrices Y commuting with X if and only if $z_p = z_{q^{-1}pq}$ for all $p, q \in G$, so that the number of linearly independent matrices Z is equal to the number of conjugacy classes in G. But it is also equal to k, by part (iii) of the preceding theorem, completing the proof.

Schur defined the *character* of an arbitrary representation $r \to A(r) = (a_{ij}^r)$ as the complex valued function χ on G

$$\chi(r) = \text{ Trace } (A(r)) = \sum_i a_{ii}^r.$$

The value of a character at the identity element $1 \in G$ was called the *degree* of the character; it coincides with the degree of the corresponding representation. He noted that characters are class functions, so that $\chi(s^{-1}rs) = \chi(r)$, for all $r, s \in G$, and that the characters of equivalent representations coincide, because of the property of the trace function that Trace $(AB) =$ Trace (BA) for all square matrices A and B. The characters of the irreducible representations were called by Schur the *simple* [einfach] characters; we shall continue to refer to them as the irreducible characters. The orthogonality relations for the irreducible characters, which were not easy for Frobenius and Burnside to prove (see Chapter II, Theorem 3.2, and Chapter II, Theorem 4.1), followed directly, in Schur's approach, from the orthogonality relation for coefficients of irreducible representations, Theorem 2.2.

THEOREM 2.7. *Let ψ and χ be irreducible characters of a finite group G. Then*

$$\sum_{r \in G} \chi(r^{-1})\chi(r) = |G|,$$

and

$$\sum_{r \in G} \psi(r^{-1})\chi(r) = 0 \text{ if } \psi \neq \chi.$$

For example, in order to prove the first relation, let $r \to A(r) = (a_{ij}^r)$ be the irreducible representation whose character is χ, of degree f. Then

$$\sum_{r \in G} \chi(r^{-1})\chi(r) = \sum_{r \in G} \sum_i \sum_\ell a_{ii}^{r^{-1}} a_{\ell\ell}^r = \frac{|G|}{f} \cdot f = |G|,$$

by the first formula in Theorem 2.2. This proves the first statement of the theorem. The second follows from the last part of Theorem 2.2, along with the observation that the irreducible representations with characters ψ and χ are inequivalent, if $\psi \neq \chi$.

From the orthogonality relations, it follows that two irreducible representations of a finite group are equivalent if and only if they have equal characters. Schur proved that the same criterion applied to reducible representations. It is a fundamental result.

THEOREM 2.8. *Let $r \to A(r)$ and $r \to B(r)$ be two representations of a finite group G. The representations are equivalent if and only if they have the same character.*

It has already been noted that the characters of equivalent representations coincide. For the proof of the converse statement, Schur used the concept of index (or multiplicity) with which a given irreducible representation occurs in a reducible representation. He pointed out that, by the remarks following Maschke's Theorem, it follows that two representations are equivalent if and only if they have the same indices, for each irreducible representation. The proof of the theorem is completed by showing that the indices of a given representation $r \to A(r)$ are uniquely determined by the character χ of the representation. Let $r \to A_1(r), r \to A_2(r), \ldots$ be a full set of inequivalent irreducible representations, with characters $\chi^{(1)}, \chi^{(2)}, \ldots$, respectively, and let $e_1, e_2, \ldots$ denote the indices of the irreducible representations in the given representation $r \to A(r)$. This means that the representation $r \to A(r)$ is equivalent to a completely reduced representation containing the irreducible representation $r \to A_1(r)$ exactly e_1 times, the representation $r \to A_2(r)$ exactly e_2 times, etc.. Because the trace function of a matrix is equal to the sum of the diagonal elements, it follows that the character χ of the representation $r \to A(r)$ is given by the expression

$$\chi(r) = e_1 \chi^{(1)}(r) + e_2 \chi^{(2)}(r) + \cdots ,$$

for $r \in G$. From the orthogonality relations, one has

$$\sum_{r \in G} \chi(r^{-1}) \chi^{(j)}(r) = |G| e_j,$$

for $j = 1, 2, \ldots$. This proves that the indices are uniquely determined by the character, and completes the proof of the theorem.

In his second paper on representations of finite groups [136], published in 1899, Frobenius had also obtained a criterion for two not necessarily irreducible representations $r \to A(r)$ and $r \to B(r)$ of a finite group G to be equivalent. It is not surprising that the result involved the factorization of the group determinant. His result was that the representations are equivalent if the group matrices $X = \sum A(r) x_r$ and $Y = \sum B(r) x_r$ have equal determinants. His proof was rather sketchy, and apparently was based on the assumption that the representations were completely reducible, as the difficulties connected with Maschke's Theorem (also published in 1899) were not addressed. In his first paper on representations [133], he had proved that the determinant of a group matrix $|X|$ is a product of the irreducible factors Φ of the group determinant $|x_{pq^{-1}}|$ of G, so that $|X| = \Phi_1^{e_1} \Phi_2^{e_2} \cdots$. Assuming that the representation associated with X is completely reducible, it follows that the exponents $e_1, e_2, \ldots$ are the indices of the irreducible representations occurring in it, and the result follows from the uniqueness of factorization of polynomials.

3. Joint Work of Frobenius and Schur: 1905-1906

Frobenius was drawn into Schur's circle of interests during the preparation of Schur's paper *Neue Begründung der Theorie der Gruppencharaktere*, to which he had contributed some suggestions. The outcome of their collaboration was the publication in the following year of two joint papers, which are important not only for the results they contain, but for their extension of the scope of representation theory in several new directions. One, entitled *Über die reellen Darstellungen der endlichen Gruppen* [151], contained a many-sided exploration of the question as to whether a given irreducible representation of a finite group (over $\mathbb{C}$) is or is not equivalent to a representation by matrices with entries in the real field $\mathbb{R}$. The second, entitled *Über die Äquivalenz der Gruppen linearer Substitutionen* [152], was devoted to an extension of Burnside's Theorem (Chapter III, Theorem 4.5).

Although he continued to publish papers every year until his death in 1917, these two joint papers were Frobenius's last contributions to the general theory of

representations.[18] The next year, as he had done several times before, he turned
to another subject. He had become intrigued by Oskar Perron's discovery that, in
Frobenius's words from the introduction to [147], the "characteristic determinant
and its subdeterminants" of matrices with positive real entries "have some remarkable properties". He published three papers on the subject himself, from 1908-1912
([147], [148], [150]), containing simplified proofs, and significant extensions, of
Perron's results, known today as the Perron-Frobenius Theorems on the properties
of eigenvalues of matrices with nonnegative real entries.

The first joint paper [151], along with a paper by Schur published in the same
year [250], examined representations of finite groups by matrices with entries in
fields other than $\mathbb{C}$, namely $\mathbb{R}$ in the first case, and an algebraic number field, in
the second. This problem had already been considered by Maschke and Burnside,
from a different point of view, namely to determine minimal subfields K of $\mathbb{C}$ with
the property that a given irreducible representation of a finite group, over $\mathbb{C}$, is
equivalent to a representation with entries in the field K (see Chapter III, §4). It
is but a short step to consider representations of a finite group over arbitrary fields
(such as finite fields, for example) but this step was not taken until later, by Emmy
Noether and Richard Brauer (see Chapter VI).

The joint paper on real representations of finite groups began by recalling
some results obtained by Alfred Loewy, in Freiburg, published as the second of two
papers by him in the *Transactions of the American Mathematical Society* ([207],
[208], 1903).[19] Loewy proved that a finite or infinite irreducible group of real
matrices,[20] viewed as a matrix group over $\mathbb{C}$, satisfies one of two conditions: (i) it
remains irreducible as a matrix group over $\mathbb{C}$; or (ii) it is completely reducible as a
representation over $\mathbb{C}$, and is the direct sum of an irreducible matrix group (a_{ij}),
and a second irreducible matrix group consisting of the complex conjugates of the
first, $(\bar{a}_{ij})$. In case (ii) holds, the complex matrix groups (a_{ij}) and $(\bar{a}_{ij})$ may or
may not be equivalent, but, in any case, neither is equivalent to a group of real
matrices.

Frobenius and Schur began by turning Loewy's analysis around. They assigned
an irreducible matrix group over $\mathbb{C}$ (or an irreducible representation of a finite or
infinite group), to one of three *types*, according as it is: (I) equivalent to a group of
real matrices; (II) equivalent to its complex conjugate, but not equivalent to a real
representation; or (III) not equivalent to its complex conjugate matrix group. They
showed that a completely reducible group is equivalent to a real matrix group if and

---

[18]There were two more papers on applications of character theory. One, on a theme he had
first developed in his 1896 series of papers, gave another connection between character theory and
the number of solutions of certain equations in a finite group ([146], 1907). The other gave a new
derivation, using character theory, of the classification of crystal classes ([149], 1911). The latter
seems to have been as close as either Frobenius or Schur came to acknowledging the potential
usefulness of character theory of finite groups for applications to chemistry and physics.

[19]Both papers were presented to the American Mathematical Society, the first in September,
1902, by E. H. Moore, the second in February, 1903, by E. W. Brown, and read, in the absence of
the author, by H. Maschke.

[20]A group of matrices G with entries in $\mathbb{R}$ is irreducible if there does not exist an invertible
matrix P, with real entries, for which the matrices in $P^{-1}GP$ all have the form

$$\begin{pmatrix} A & 0 \\ B & C \end{pmatrix}.$$

only if the number of irreducible summands of the second type is even, and each irreducible summand of the third type occurs as often as its complex conjugate.

They next turned their attention to representations of finite groups, noting that a representation (over $\mathbb{C}$) of a finite group G is uniquely determined, up to equivalence, by its character χ. Their main result was that it is possible to determine the type of an irreducible representation by its character.

THEOREM 3.1. *Let χ be the character of a complex irreducible representation of a finite group G. Then*

$$\sum_{r \in G} \chi(r^2) = c|G|,$$

with $c = 1, -1,$ or 0 according as the representation is of type I, II, or III.

The expression $|G|^{-1} \sum \chi(r^2)$ is called today the *Frobenius-Schur indicator* of the character χ. The preceding theorem about it contains a surprising amount of information, in a form that lends itself to applications; for example, Frobenius used the theorem several times in his elegant derivation, based on character theory, of the 32 crystal classes [149]. A column giving the Frobenius-Schur indicator has been included in the character tables of all the simple groups considered in the *Atlas of finite groups* [96], whose stated purpose was "to convey every interesting fact about every interesting finite group."

An outline of their proof of the preceding theorem follows. It provides a good illustration of Frobenius's and Schur's virtuosity with matrix theory, and brings out interesting conceptual features of the proof, which still figure in modern proofs.

They began by interpreting concepts related to bilinear forms and Hermitian forms in terms of matrices. A matrix H of an Hermitian form is defined by the condition $H' = \overline{H}$, where H' denotes the transpose of H, and $\overline{H}$ the matrix whose entries are the complex conjugates of the entries of H. The *Hermitian form* associated with H is the complex-valued function on pairs of complex column vectors X, Y given by $X'H\overline{Y}$. It has the property that $X'H\overline{X}$ is real for all X, and is called *positive definite* provided that $X'H\overline{X} \geq 0$, and is zero only if $X = 0$. An arbitrary matrix U defines a *bilinear form* $X'UY$ on pairs of complex vectors X and Y. The bilinear form defined by U is *symmetric* if $U' = U$, and *alternating* if $U' = -U$. In case U has real entries, the bilinear form on pairs of real vectors X and Y is *positive definite* provided that $X'UX \geq 0$, and is zero only if $X = 0$. The action of an invertible matrix R on a Hermitian form H produces another Hermitian form whose matrix is $R'H\overline{R}$. The form defined by H is *G-invariant* for a finite group G of invertible matrices R provided that $R'H\overline{R} = H$, for all elements $R \in G$. Similar remarks apply to bilinear forms. The action of an invertible matrix R on a bilinear form with matrix U produces another bilinear form with matrix $R'UR$. The form is *G-invariant* provided that $R'UR = U$ for all elements $R \in G$. Finally, a real or complex matrix S is defined to be *orthogonal* whenever $S'S = SS' = E$, where E denotes the identity matrix. Frobenius and Schur proved their theorem by finding connections between these concepts and ideas from representation theory, such as Schur's Lemma.

LEMMA 3.1. *A real representation $r \to R$ of a finite group G is equivalent to a representation by real orthogonal matrices. Conversely, an irreducible representation of G by orthogonal matrices is equivalent to a real representation.*

Both statements of the lemma follow from the fact that a representation $r \to R$ of G leaves invariant a positive definite Hermitian form, by the theorem of Loewy and Moore (Chapter III, §4). If H is the matrix of the form, then $R'H\overline{R} = H$, for all matrices R corresponding to elements of G. In case the matrices are all real, the preceding formula, and the fact that $H' = \overline{H}$, show that $K = H + H'$ is a positive definite real symmetric matrix such that $R'KR = K$ for all R. By an orthogonal change of coordinates the matrix K can be transformed to the identity matrix. The same change of coordinates transforms the matrices R to real orthogonal matrices, proving the first statement of the lemma.

The proof of the second statement was more subtle. The assumption that the matrices R are orthogonal, combined with the G-invariance of the Hermitian form with the matrix H, yield:

$$H = R'H\overline{R},\, RH = H\overline{R} \text{ and } H\overline{R}' = R'H,$$

for all the matrices R corresponding to elements of G. For the matrix $H' = \overline{H}$, one has

$$H' = \overline{R}'H'R \text{ and } \overline{R}H' = H'R;$$

consequently,

$$R(HH') = (RH)H' = H(\overline{R}H') = (HH')R,$$

for each R. As the representation $r \to R$ is irreducible, Schur's Lemma implies that $HH' = cE$, for some complex number c. Moreover, $H' = \overline{H}$ and hence $H\overline{H} = cE$. Then $H'H\overline{H} = cH'$, and multiplication by c carries one positive definite Hermitian form to another. It follows that c is a positive real number. Replacing H by $(c)^{-\frac{1}{2}}H$ produces another positive definite Hermitian form, whose matrix (also denoted by H) satisfies

$$HH' = H'H = E \text{ and } H\overline{H} = \overline{H}H = E.$$

As E and H are matrices of positive definite Hermitian forms, so is their sum, and hence $E + H$ is an invertible matrix. Then

$$r \to (E + H)^{-1}R(E + H) = S$$

is a representation equivalent to the given one, and it will be shown that the matrices S have real entries. As $H\overline{H} = \overline{H}H = E$, one has $H = (E + H)(E + \overline{H})^{-1}$. Combining this formula with $H^{-1}RH = \overline{R}$ from the first part of the proof, it follows that

$$(E + H)^{-1}R(E + H) = (E + \overline{H})^{-1}\overline{R}(E + \overline{H}),$$

for all the matrices R corresponding to elements of G. Therefore the matrices S in the equivalent representation satisfy $S = \overline{S}$ and are real, completing the proof.

The proof of their main result (Theorem 3.1) began with the observation that an irreducible representation $r \to R$ of a finite group G is of type I, II, or III according as its character χ satisfies one of the conditions: χ is real valued, and the representation is equivalent to a real representation; χ is real valued, but the representation is not equivalent to a real representation; or χ is not a real valued character.

In each of the first two cases, the representations $r \to R$, and $r \to (R^{-1})'$, called today the *contragredient* of the first representation, are equivalent, because they have equal characters. This followed, as they explained, from the fact that the eigenvalues of $(R^{-1})'$ are the inverses of the eigenvalues of R, and hence their complex conjugates, because the eigenvalues of R are roots of unity, for a representation $r \to R$ of a finite group. Thus the trace of the matrix $(R^{-1})'$ is $\chi(r^{-1}) = \overline{\chi(r)} = \chi(r)$, for $r \in G$, in case the character χ has real values, so the characters of the representation $r \to R$ and its contragredient are equal. The equivalence of the representation $r \to R$ and the contragredient representation $r \to (R^{-1})'$ means that there exists an invertible matrix F such that

$$FRF^{-1} = (R^{-1})',$$

or

$$R'FR = F$$

for all matrices R corresponding to elements of G. This means that there is a nonzero G-invariant bilinear form.

Conversely, if there exists a nonzero G-invariant bilinear form for an irreducible representation $r \to R$ of a finite group G, then for all $r \in G$,

$$FR = (R^{-1})'F$$

where F is the matrix of the bilinear form. By Schur's Lemma (Theorem 2.1) and the assumption that $F \neq 0$, it follows that the representation $r \to R$ is equivalent to its contragredient. Then its character χ satisfies $\chi(r) = \chi(r^{-1})$ for all $r \in G$, and is real valued.

Frobenius and Schur pushed this promising line of investigation one step farther by observing that any two nonzero G-invariant bilinear forms for an irreducible representation $r \to R$ are multiples of each other by a constant, by another application of Schur's Lemma (left here as an exercise). They applied this remark to the following situation. Let F be the matrix of a nonzero G-invariant bilinear form for an irreducible representation $r \to R$, so that $F = R'FR$ for all $r \in G$. Then $F' = R'F'R$ for all R, and consequently, $F' = cF$ for some element $c \in \mathbb{C}$. From this, one has also $F = cF'$, so that $F = c^2F, c^2 = 1$, and $c = \pm 1$. This means that the bilinear form is either symmetric or alternating, and has an invertible matrix, by the preceding remarks. If the form is symmetric, then it is carried to the identity matrix by a change of coordinates, and the representation $r \to R$ is equivalent to a representation of G by orthogonal matrices. By the preceding lemma, this implies that the representation is equivalent to a real representation. On the other hand, if the given representation is equivalent to a real representation, it follows easily from Lemma 3.1 that there exists a G-invariant symmetric bilinear form, and hence the given G-invariant bilinear form is also symmetric, as the two are constant multiples of one another. This completes the proof of the following interesting result, which was an important step towards the proof of their theorem.

PROPOSITION 3.1. *Let $r \to R$ be an irreducible representation of a finite group G. Then the representation is of type I, II, or III according as the corresponding one of the following statements holds. There exists a nonzero G-invariant symmetric bilinear form for the representation; there exists a nonzero G-invariant alternating bilinear form for the representation; or there does not exist a nonzero G-invariant bilinear form for the representation.*

Using the preceding result, they completed the proof of their theorem as follows. Let $r \to R = (r_{ij})$ be an irreducible representation of the finite group G of degree

d, let $U = (u_{ij})$ be a $d \times d$ matrix whose entries are independent variables, and let

$$F = \sum_R R'UR.$$

If the given representation $r \to R$ is of type III, then F vanishes for all choices of the variables $u_{ij} \in \mathbb{C}$, otherwise the matrix F defines a nonzero G-invariant bilinear form, contrary to Proposition 3.1. It follows that

$$\sum_{r \in G} \sum_{i,j} r_{ih} u_{ij} r_{jk} = 0,$$

for all h, k and all choices of $u_{ij} \in \mathbb{C}$. Then

$$\sum_{r \in G} r_{ih} r_{jk} = 0,$$

for all h, i, j, k. As $\sum_j r_{ij} r_{jk}$ is the (i, k) entry in the product matrix R^2, the preceding formulas imply that

$$\sum_{r \in G} R^2 = 0,$$

and hence

$$\sum_{r \in G} \chi(r^2) = 0,$$

whenever χ is the character of an irreducible representation of type III.

Now let $r \to R = (r_{ij})$ be an irreducible representation of type I or II; then $F = cF'$, with $c = \pm 1$ according as the representation is of type I or II, by the discussion preceding the statement of Proposition 3.1. Then

$$\sum_{r \in G} \sum_{i,j} r_{ih} u_{ij} r_{jk} = c \sum_{r \in G} \sum_{ij} r_{ik} u_{ij} r_{jh},$$

for all h, k and all choices of $u_{ij} \in \mathbb{C}$. Upon specializing the variables u_{ij}, one obtains

$$\sum_{r \in G} r_{ih} r_{jk} = c \sum_{r \in G} r_{ik} r_{jh},$$

for all h, i, j, k. Specialization of the indices yields the formulas

$$\sum_{r \in G} r_{k\ell} r_{\ell k} = c \sum_{r \in G} r_{kk} r_{\ell\ell},$$

for all k, ℓ. It follows that

$$\sum_{r \in G} \chi(r^2) = c \sum_{r \in G} \chi(r)\chi(r),$$

where χ is the character of the given irreducible representation. As the representation is of type I or II, one has $\chi(r) = \chi(r^{-1})$ by the previous discussion. Combining this remark with the orthogonality relation $\sum_r \chi(r^{-1})\chi(r) = |G|$, the formula stated above becomes

$$\sum_{r \in G} \chi(r^2) = c|G|,$$

for the character of a representation of type I or II. This, together with the remarks about representations of type III given earlier, completes the proof of the theorem.

The second joint paper of Frobenius and Schur [151] extended the scope of representation theory in another direction, through the consideration of infinite groups of matrices not assumed to be completely reducible. They began by generalizing the notion of a *group* of linear substitutions to include any finite or infinite set $\mathfrak{G}$ of linear substitutions (or matrices) with the property that whenever $A \in \mathfrak{G}$ and $B \in \mathfrak{G}$, then their product $AB \in \mathfrak{G}$. It was not assumed that the matrices belonging to $\mathfrak{G}$ were invertible, or that $\mathfrak{G}$ contained the identity matrix. Two groups[21] $\mathfrak{G}$ and $\mathfrak{G}'$ were defined to be equivalent if $\mathfrak{G}' = P^{-1}\mathfrak{G}P$, for some invertible matrix P. A mapping from one group $\mathfrak{G}$ to another group $\mathfrak{H}$, given by $R \to H_R$, with $R \in \mathfrak{G}$ and $H_R \in \mathfrak{H}$, was called a *homomorphism* in case $H_R H_S = H_{RS}$ for all $R, S \in \mathfrak{G}$, and an *isomorphism* if the mapping was one-to-one. They defined a group $\mathfrak{G}$ to be *reducible* if it is equivalent to a group $\mathfrak{G}'$ all of whose elements have the form

$$\begin{pmatrix} R & 0 \\ S & T \end{pmatrix},$$

and *irreducible* if $\mathfrak{G} \neq 0$, and is not reducible. Their generalization of a matrix group was important for two reasons; it included algebras of matrices (sets of matrices closed under the operations of taking sums and products, and scalar multiples, of elements), later to become a focus of representation theory; and, as simple examples show that Maschke's Theorem no longer applies to reducible groups in the extended sense, their work applied to representations which fail to be completely reducible (see Theorem 3.3 below).

Their main theorem, for groups of matrices in the extended sense, was as follows.

THEOREM 3.2. *Let $\mathfrak{A}, \mathfrak{B}, \ldots$ be irreducible groups of matrices, which are all homomorphic images of a given group $\mathfrak{G}$, and let an element $R \in \mathfrak{G}$ correspond to the matrices*

$$A(R) = (a_{ij}(R)) \in \mathfrak{A}, B(R) = (b_{k\ell}(R)) \in \mathfrak{B}, \ldots.$$

If no two of the groups $\mathfrak{A}, \mathfrak{B}, \ldots$ are equivalent, then there exists no nontrivial linear relation

$$\sum_{i,j} p_{ij} a_{ij}(R) + \sum_{k,\ell} q_{k\ell} b_{k\ell}(R) + \cdots = 0,$$

with constant coefficients $p_{ij}, q_{k\ell}, \ldots$ which holds among the matrix coefficients $a_{ij}(R), b_{k\ell}(R), \ldots$ for all $R \in \mathfrak{G}$.

The preceding theorem, for the case of one irreducible group, was Burnside's Theorem (Chapter III, Theorem 4.5), as they noted, along with the remark that Burnside's Theorem was "of fundamental significance for the theory of groups of linear substitutions." Their theorem, for the case of irreducible representations of finite groups, follows from Schur's orthogonality theorem, Corollary 2.1. But their proof for irreducible representations of groups in the extended sense required a different argument, based on a subtle application of Schur's Lemma, which they used first to give a new proof of Burnside's Theorem, before tackling the general case. We shall not include the details here, as the theorem follows easily from Emmy Noether's approach to the representation theory of semisimple algebras based on the Wedderburn theorems (Chapter VI, §2). We shall however, discuss their application of Theorem 3.2 to representations that are not necessarily completely reducible.

A. Loewy was apparently the first to consider representations of groups in the extended sense that might fail to be completely reducible. In the first of two papers

---

[21]In what follows, we shall use Frobenius and Schur's concept of group in the extended sense.

published in the *Transactions of the American Mathematical Society* in 1903, he
obtained a forerunner of what is known today as the *Jordan-Hölder theorem for
groups with operators* (see Chapter VI, §2). A group in the extended sense $\mathfrak{G}$ is
equivalent to a group of matrices of the form

$$\mathfrak{G}' = \begin{pmatrix} G_{11} & 0 & 0 & \ldots & 0 \\ G_{21} & G_{22} & 0 & \ldots & 0 \\ \cdot & \cdot & \cdot & \ldots & \cdot \\ G_{p1} & G_{p2} & \cdot & \ldots & G_{pp} \end{pmatrix},$$

with each set of matrices G_{ii}, for fixed i, either irreducible, or all equal to zero. In
[**208**], Loewy proved that the set of inequivalent irreducible groups $\mathfrak{G}_1, \mathfrak{G}_2, \ldots$ of
the form G_{ii}, for some i, in a matrix group $\mathfrak{G}'$ as above, and the number of times
each is equivalent to a diagonal component G_{ii} in $\mathfrak{G}'$, are uniquely determined by
the given matrix group $\mathfrak{G}$. That means, in particular, that the data is independent
of the choice of the matrix group of the form $\mathfrak{G}'$ equivalent to the given matrix
group $\mathfrak{G}$.

After recalling the statement of Loewy's theorem, and adopting the terminology
irreducible components [irreduziblen Bestandteile] (also called *composition factors*
in current literature) for the irreducible representations of the form G_{ii} occurring
in a matrix group $\mathfrak{G}'$, as above, Frobenius and Schur proved the following result,
which extends, and gives another proof of, Loewy's theorem.

THEOREM 3.3. (*i*) *Two isomorphic irreducible groups $\mathfrak{A}$ and $\mathfrak{B}$ are equivalent
if and only if all pairs of corresponding matrices $A \in \mathfrak{A}$ and $B \in \mathfrak{B}$ have equal
traces.*

(*ii*) *Two isomorphic matrix groups $\mathfrak{G}$ and $\mathfrak{H}$, in m and n variables, contain
the same set of irreducible components, with each one occurring with the same
multiplicity, if and only if all pairs of corresponding matrices have equal traces.*

For representations of finite groups, the first part of the theorem follows directly
from the orthogonality relation for the characters of irreducible representations
(Theorem 2.8). For irreducible representations of groups in the extended sense,
the orthogonality relations are not available, but Theorem 3.2 can be used instead.
Denoting the trace of a matrix A by $\chi(A)$, one has to prove that if $\chi(A) = \chi(B)$
for all pairs of corresponding matrices A and B, then the groups $\mathfrak{A}$ and $\mathfrak{B}$ are
equivalent. If they are not equivalent, then the formula $\chi(A) - \chi(B) = 0$ is a
nontrivial linear relation among the coefficients of corresponding matrices in $\mathfrak{A}$ and
$\mathfrak{B}$, and this contradicts Theorem 3.2, completing the proof of part (*i*).

For the proof of part (*ii*), one may assume that the groups $\mathfrak{G}$ and $\mathfrak{H}$ are equivalent to groups $\mathfrak{G}'$ and $\mathfrak{H}'$, respectively, which have the form described in connection
with Loewy's uniqueness result. In particular, it is assumed that the blocks of
matrices along the diagonal G_{ii} and H_{jj}, respectively, are either all equal to zero,
or generate irreducible matrix groups. Let $\mathfrak{R}_1, \mathfrak{R}_2, \ldots, \mathfrak{R}_\ell$ be a set of inequivalent
irreducible matrix groups obtained in this way. An element $G \in \mathfrak{G}$ corresponds to
elements $R_1 \in \mathfrak{R}_1, R_2 \in \mathfrak{R}_2, \ldots$ so that the trace is given by

$$\chi(G) = \sum_{i=1}^{\ell} p_i \chi(R_i),$$

where p_i is the number of diagonal matrix groups $\{G_{ii}\}$ equivalent to $\mathfrak{R}_i$, for $i =
1, \ldots, \ell$. If G corresponds to the element $H \in \mathfrak{H}$, then the trace of H is given by a

similar formula,

$$\chi(H) = \sum_{i=1}^{\ell} q_i \chi(R_i).$$

Now assume that $\chi(G) = \chi(H)$, as in the hypothesis of part (ii) of the theorem. Then one has the formula

$$\sum_i (p_i - q_i)\chi(R_i) = 0.$$

If any of the differences $p_i - q_i$ are different from zero, then the preceding formula is a nontrivial linear relation among the coefficients of corresponding elements of the matrix groups $\Re_1, \Re_2, \ldots$. The existence of such a relation is ruled out by Theorem 3.2. It follows that $p_i = q_i$ for $i = 1, \ldots, \ell$. This completes the proof of the theorem.

4. The Schur Index

While Frobenius and Schur were working on their joint paper on real representations of finite groups, Schur began an investigation of his own of representations of finite groups by matrices with coefficients in algebraic number fields ([250], 1906). The theory of algebraic number fields was one of the main subjects of research at the time in Germany, building on the achievements of Eisenstein, Kronecker, and Kummer in Berlin, and Gauss, Dirichlet, Dedekind, and Hilbert in Göttingen.[22] Algebraic number theory had already been applied to representations of finite groups in a few situations by Frobenius and Burnside, for example in the proof that the degrees of the irreducible characters divide the order of the group, and in the proof of Burnside's $p^a q^b$-Theorem. Maschke and Burnside had raised the question as to whether every irreducible representation of a finite group G was equivalent to a representation by matrices with entries in the cyclotomic field of $|G|$th roots of unity, and answered it affirmatively in some special cases ([212], [82]; see also Chapter III, §5). But Schur was the first to undertake a systematic study of representation theory in the context of algebraic number fields. This section contains an account of the new methods he developed for this purpose, and in particular the numerical invariant attached to each irreducible character, called today the *Schur index* of the character with respect to a given subfield F of $\mathbb{C}$.

Finding the appropriate numerical invariant to organize a complicated situation is always a significant step in the evolution of a theory. Schur obtained several definitive results about the index in his article [250] which demonstrated its importance, and settled the question of Maschke and Burnside about rationality of representations in cyclotomic fields, for solvable groups. The problem of Maschke and Burnside was eventually settled by Brauer for all finite groups ([33]; see Chapter VII, §4, Theorem 4.4).

As a start, Schur introduced the following notation and terminology. Let $\chi^{(1)}, \ldots, \chi^{(k)}$ denote the distinct irreducible characters of a finite group G, where k is the number of conjugacy classes in G, and let $r \to A^{(i)}(r)$ be an irreducible representation of G whose character is $\chi^{(i)}$, for $1 \leq i \leq k$. He called a representation $r \to R$ (for example one of the irreducible representations $r \to A^{(i)}(r)$) *rational* in a subfield F of $\mathbb{C}$, or, to use a modern abbreviation, an *F-rational representation*, provided that it is equivalent to a representation by matrices with coefficients in F.

---

[22]For a survey, see Hilbert's monumental report *Die Theorie der algebraischen Zahlkörper* ([177], 1897).

As he remarked, Frobenius had proved that each irreducible representation of G is rational in an algebraic number field $F = \mathbb{Q}(\mu)$.[23] In view of this, Schur proposed to find the properties characterizing algebraic number fields F in which a given irreducible representation is rational, and in particular, to determine the least degree (as an extension field of $\mathbb{Q}$) of such a field F. He continued:

> This problem can be generalized somewhat. Let E be a given domain of rationality [i.e. a subfield of $\mathbb{C}$]. It can then be asked, in which algebraic extension fields $E(\mu)$ can a given irreducible representation $r \to R^{(i)}$ be rational?
>
> Such a field $E(\mu)$ must first of all contain all the numbers $\chi^{(i)}(r)$. If the k numbers $\chi^{(i)}(r)$ generate a number field $E(\chi)$, and χ satisfies an irreducible [polynomial] equation of degree ℓ over E, then the degree n of the field $E(\mu)$ (in relation to E) is divisible by ℓ. If the least degree n of such a field is $m\ell$, then I will call the number m the *index* of the irreducible representation $r \to R^{(i)}$, or of the character $\chi^{(i)}$, with respect to the field E.
>
> The number m is evidently completely determined, if besides the field E and the multiplication table for the elements of the group G, the k numbers $\chi^{(i)}(r)$ are known. The exact determination of m with the use of this data appears however to present considerable difficulties.
>
> The following work contains some contributions towards the solution of this problem.[24]

The number m defined above is now called the *Schur index* of the irreducible character $\chi^{(i)}$ with respect to the field E. Schur's remark about the difficulty of computing it has turned out, over the course of a century, to be an understatement. In recent work on the problem, computer calculations of character tables have played an important role (see Feit [**119**], 1983).

Let $r \to A(r)$ be an F-rational representation of a finite group G. The representation is called *irreducible in F* (or simply irreducible, in the context of F-rational representations) if it is not equivalent to an F-rational representation of the form

$$r \to \begin{pmatrix} B(r) & 0 \\ 0 & C(r) \end{pmatrix},$$

for F-rational subrepresentations $r \to B(r)$ and $r \to C(r)$. A group matrix $X = \sum A(r)x_r$ is called rational in F, or rational and irreducible in F, if the representation $r \to A(r)$ has these properties. In this setting, Schur's Lemma becomes:

LEMMA 4.1. *Let X and X' be two group matrices, rational and irreducible in a field F, of degrees f and f' respectively. Let P be a constant matrix with coefficients in F, with f rows and f' columns, such that $XP = PX'$; then either $P = 0$, or X and X' are equivalent, $f = f'$, and P is an invertible $f \times f$ matrix.*

---

[23]An algebraic number field is a finite extension of the rational field (see Chapter I, §1). Here, and in what follows, Schur assumed familiarity with the *primitive element theorem* of Galois theory, namely that each finite extension of a subfield of $\mathbb{C}$ is generated by a single element.

[24]Schur, I., *Arithmetische Untersuchungen über endliche Gruppen linearer Substitutionen*, S'ber Akad. Wiss. Berlin (1906); Ges. Abh. I, Springer-Verlag, Berlin, 1973.

The proof is exactly the same as the proof of Theorem 2.1. One has only to check that the auxiliary matrices used in the proof can be taken with coefficients in F.

Schur's two main results analyzed the character (or trace function) of a representation, rational and irreducible in a field F, in terms of the irreducible characters of the given finite group, and their Schur indices. Nothing like this had been obtained before. For the proof of the first result, he made an ingenious use of matrices commuting with an irreducible F-rational group matrix and the F-rational version of Schur's Lemma, combined with several new applications of Galois theory to representation theory.

LEMMA 4.2. *Let X be an irreducible, F-rational group matrix, for an arbitrary subfield F of $\mathbb{C}$, and let P be a constant matrix, with coefficients in F, such that $XP = PX$. Then the characteristic polynomial $|xE - P|$ of P is a power of an irreducible polynomial $\varphi(x)$ with coefficients in F, and $\varphi(P) = 0$.*

Let E denote the identity matrix, and let

$$|xE - P| = \varphi(x)\varphi_1(x)\cdots$$

be a factorization of the characteristic polynomial of P as a product of irreducible monic polynomials with coefficients in F. Then $\varphi(P), \varphi_1(P), \ldots$ are matrices with entries in F commuting with X. By the Cayley-Hamilton Theorem, P satisfies its characteristic equation, so $\varphi(P)\varphi_1(P)\cdots = 0$, and it follows that each of the matrices $\varphi(P), \varphi_1(P), \ldots$ is not invertible (this is a nice exercise in linear algebra). The preceding version of Schur's Lemma implies that $\varphi(P) = \varphi_1(P) = \cdots = 0$. Therefore, if ω is an eigenvalue of P, one has $\varphi(\omega) = \varphi_1(\omega) = \cdots = 0$, and it follows that no two of the irreducible polynomials $\varphi(x), \varphi_1(x), \ldots$ are relatively prime. Therefore they are all equal, completing the proof.

Now let $r \to A(r)$ be an arbitrary F-rational, irreducible representation of G, and let $\xi(r) = \text{Trace}(A(r)), r \in G$ be its character, or trace function. As F is a subfield of $\mathbb{C}$, the representation $r \to A(r)$ can be viewed as a $\mathbb{C}$-representation, which of course is not necessarily irreducible. Its character can be expressed in terms of the irreducible characters $\chi^{(i)}$:

$$\xi(r) = \sum_{i=1}^{k} r_i \chi^{(i)},$$

with nonnegative integer coefficients r_i. In this setting, Schur proved that the characters $\chi^{(i)}$ appearing with nonzero coefficients in the above formula for ξ are conjugate under the action of a Galois group of an extension field of F, and appear in ξ with equal multiplicities r_i.

His proof began with an application of the F-rational version of Schur's Lemma given above. He observed that the matrices

$$Y = \sum_{r \in G} A(r) y_r,$$

commute with all the matrices $A(s), s \in G$, provided that the elements $y_r \in F$ satisfy the condition $y_r = y_{s^{-1}rs}$ for all $r, s \in G$. This follows by checking that, because of the condition on the coefficients y_r, one has $A(s)^{-1}YA(s) = Y$, for all $s \in G$. By the preceding lemma, the characteristic polynomial $|xE - Y|$ of such a matrix Y is a power of an irreducible polynomial with coefficients in F. On the other hand, as he noted, the eigenvalues of Y can be calculated by an argument similar to one used by Frobenius in the proof of Theorem 3.3.

LEMMA 4.3. *Let* $Y = \sum A(r)y_r$, *as above. Then its characteristic polynomial factors as follows:*

$$|xE - Y| = \prod_{i=1}^{k} (x - \frac{\eta_i}{f_i})^{r_i f_i},$$

where $\eta_i = \sum_r \chi^{(i)}(r)y_r$, f_i *is the degree of the character* $\chi^{(i)}$, *and* r_i *is the multiplicity of* $\chi^{(i)}$ *in the expression for* ξ *as a linear combination of the characters.*

For the proof of the lemma, Schur referred to Frobenius [**132**], p. 1361. We give here a modern proof. The condition on the elements y_r implies that $c = \sum y_r r$ belongs to the center of the group algebra of G. Therefore, $A^{(i)}(c) = \sum A^{(i)}(r)y_r$ commutes with all the matrices $A^{(i)}(s)$ in any one of the irreducible $\mathbb{C}$-representations $r \to A^{(i)}(r)$. By Schur's Lemma 2.1,

$$A^{(i)}(c) = \alpha^{(i)}(c)E_{f_i}$$

for some complex number $\alpha^{(i)}(c)$, and the $f_i \times f_i$ identity matrix E_{f_i}. Computing the trace of both sides yields the following formula[25] for the eigenvalue $\alpha^{(i)}(c)$:

$$\alpha^{(i)}(c) = \frac{1}{f_i} \sum_{r \in G} \chi^{(i)}(r)y_r = \frac{\eta_i}{f_i}.$$

As the irreducible representation $r \to A^{(i)}(r)$ appears in a completely reduced form of the representation $r \to A(r)$ with multiplicity r_i, and the eigenvalue $\alpha^{(i)}(c)$ with multiplicity f_i in $A^{(i)}(c)$, it follows that the characteristic polynomial of Y has the factorization given in the statement of the lemma.

The two lemmas stated above imply that the characteristic polynomial $|xE - Y|$ is a power of an irreducible polynomial $\varphi(x)$ with coefficients in F, and that the expressions $\eta_i = \sum \chi^{(i)}(r)y_r$ for which $r_i \neq 0$ are roots of $\varphi(x)$. Schur realized that this meant that the elements η_i are conjugates under the action of the Galois group of a Galois extension field of F (see Chapter I, §1, and the references given there), and hence was led to the important result that the characters $\chi^{(i)}$ themselves, for which $r_i \neq 0$, are conjugates under the Galois group. More precisely, let L be a Galois extension field of F containing the values of the irreducible characters $\chi^{(i)}, 1 \leq i \leq k$, and let $\mathcal{H}$ be the Galois group. His new concept was to call two characters $\chi^{(i)}$ and $\chi^{(j)}$ *conjugate with respect to the field* F if there exists an element $\sigma \in \mathcal{H}$ such that $\sigma(\chi^{(i)}(r)) = \chi^{(j)}(r)$ for all elements $r \in G$.[26] Notice that conjugate characters have the same degree.

Schur's first theorem can be stated as follows.

THEOREM 4.1. *(i) The character* ξ *of an F-rational, irreducible representation* $r \to A(r)$ *of* G *has the form*

$$\xi(r) = m(\chi(r) + \chi_1(r) + \cdots + \chi_{t-1}(r)),$$

---

[25]This result was first proved by Frobenius, in another way (see Theorem 4.2 in Chapter II).

[26]At this point, it is necessary to check that conjugates, with respect to a field F, of a character χ are indeed characters. This follows, for example, from the fact that the values of χ satisfy the equations $h_j h_k \chi_j \chi_k = f \sum h_{i'jk}\chi_i$. If χ' is a conjugate of χ with respect to F, then these equations are also satisfied by the values of χ', proving that χ' is also a character (see Chapter II, §3).

for all elements $r \in G$, *where* $\chi, \chi_1, \ldots, \chi_{t-1}$ *are the distinct irreducible characters which are conjugates, with respect to* F, *of an irreducible character* $\chi = \chi^{(i)}$, *appearing with positive multiplicity* m *in* ξ.

(*ii*) *The number of equivalence classes of irreducible,* F-*rational representations of* G *is equal to the number of equivalence classes of irreducible characters* $\chi^{(i)}, 1 \leq i \leq k$, *under conjugation with respect to the field* F.

The first statement of the theorem follows directly from the preceding two lemmas, the form of the elements η_i, and the fact that conjugate characters have the same degree.

For the proof of the second statement, let $r \to B(r)$ be another F-rational irreducible representation, not equivalent to the representation $r \to A(r)$ from part (*i*), and let f, f' be the degrees of the representations $r \to A(r)$ and $r \to B(r)$ respectively. By part (*i*), the character ξ' of the second representation is given by a formula,

$$\xi(r) = m'(\chi'(r) + \chi_1'(r) + \cdots),$$

for $r \in G$, where $\chi', \chi_1', \ldots$ are the conjugates, with respect to F, of another irreducible character $\chi' = \chi^{(j)}$. Although the orthogonality relations do not hold in general for the characters of irreducible F-rational representations, Schur brought them into the picture by another clever application of the F-rational form of Schur's Lemma. He let

$$P = \sum_{r \in G} A(r^{-1}) U B(r),$$

for an $f \times f'$ matrix U whose coefficients are independent variables. Then

$$A(s)P = PB(s)$$

for all $s \in G$, and hence $P = 0$ by Lemma 4.1, because the representations $r \to A(r)$ and $r \to B(r)$ are not equivalent. Upon substituting the coefficients $a_{ij}^{r^{-1}}$ of $A(r^{-1})$ and $b_{h\ell}^r$ of $B(r)$, and suitably chosen entries of U, in the formula for P, one obtains

$$\sum_{r \in G} a_{ij}^{r^{-1}} b_{h\ell}^r = 0,$$

for all choices of i, j, h, ℓ. As $\xi(r^{-1}) = \sum a_{ii}^{r^{-1}}$ and $\xi'(r) = \sum b_{hh}^r$, it follows that

$$\sum_{r \in G} \xi(r^{-1}) \xi'(r) = 0.$$

Combining this formula with the orthogonality relations (Theorem 2.7) for the irreducible characters $\chi^{(i)}$, one sees that the characters $\chi, \chi_1, \ldots$ and $\chi', \chi_1', \ldots$ appearing in the expressions for ξ and ξ' are all distinct.

In order to complete the proof of part (*ii*), it remains to show that each irreducible character $\chi^{(\ell)}$ occurs in a formula for the character of some irreducible F-rational representation, as in part (*i*). For this, Schur considered the regular group matrix $X = (x_{pq^{-1}})$, whose character ρ was known to be given by the formula

$$\rho = \sum_{i=1}^{k} f_i \chi^{(i)}.$$

In other words, each irreducible character occurs in the regular character as many times as its degree. But the group matrix X is clearly F-rational, and is consequently, by the F-rational version of Maschke's Theorem, the direct sum of irreducible, F-rational group matrices. Evidently, the character of one of them must contain the given irreducible character $\chi^{(\ell)}$ as a summand. This completes the proof of part (ii) of the theorem.

COROLLARY 4.1. *Let m be the multiplicity, assumed positive, of an irreducible character χ in the trace function of an irreducible F-rational representation, as in the preceding theorem. Let $\tilde{\xi}$ be the character, or trace function, of an arbitrary F-rational representation, and assume that χ appears with positive multiplicity r as a summand of $\tilde{\xi}$. Then r is a multiple of m.*

Let $\tilde{X}$ be an F-rational group matrix whose character is $\tilde{\xi}$. From the proof of the preceding theorem, Schur deduced that χ appears with positive multiplicity in just one irreducible F-rational group matrix X occurring as a summand of $\tilde{X}$. Then χ appears with multiplicity m in the character of X, by the preceding theorem. If X occurs q times as a summand of $\tilde{X}$, then χ occurs qm times as a summand of $\tilde{\xi}$, so that $r = qm$ as required.

COROLLARY 4.2. *The multiplicity m, assumed positive, of a character χ in the trace function of an irreducible F-representation divides the degree $\chi(1)$ of the character χ.*

One simply applies the preceding corollary to the regular representation of G, which is clearly an F-rational representation whose character contains χ with multiplicity equal to its degree $\chi(1)$.

Schur's second main result was the assertion that if χ occurs in the character ξ of an irreducible F-rational representation with positive multiplicity m, then m is, in fact, the Schur index, as defined earlier, of the irreducible character χ, with respect to the field F.

THEOREM 4.2. *Let χ be an irreducible character of a finite group G, and let χ appear with positive multiplicity m in the character of an irreducible F-rational representation, as in Theorem 4.1. Let $F(\chi)$ be the algebraic extension of F generated by the character values $\chi(r), r \in G$, and let $t = (F(\chi) : F)$; then t is the number of distinct conjugates of the character χ with respect to F. Let Z be an irreducible group matrix whose character is χ. Then the minimal degree of an extension field of F in which the irreducible group matrix Z is rational, is mt. Moreover, if Z is rational in an algebraic extension field $F(\mu)$ of degree q over F, then q is divisible by mt.*

Schur's proof used all the tools from linear algebra and field theory that were available to him at the time, and is outlined below. In particular, the theory of elementary divisors and the rational canonical form of a matrix, which were by then well understood, played a prominent role. As in the case of other results about group representations, further research led to proofs of the two preceding theorems based on the theory of algebras ([100], §74A).

Schur began with an irreducible, F-rational group matrix $X = \sum A(r)x_r$, whose character is given by the formula

$$\xi = m(\chi + \chi_1 + \cdots + \chi_{t-1}),$$

as in Theorem 4.1. The degree of the group matrix X is $\xi(1) = mtf$, where f is the degree of the character χ, and the degree of the field extension $F(\chi)$ over F is the number of conjugate characters, which is t. The most general constant matrix commuting with X is given by

$$V = \sum_{r \in G} A(r^{-1}) U A(r),$$

where $U = (u_{ij})$ is a matrix of degree mtf with variable coefficients. By the formula for its character ξ, the representation $r \to A(r)$, viewed as a representation over $\mathbb{C}$, is a direct sum of mt irreducible representations of degree f. Referring to his proof of Theorem 2.5, Schur reasoned that there exist matrices V commuting with X with mt distinct eigenvalues, each occurring with multiplicity f.[27] The characteristic polynomial of such a matrix is the fth power of a polynomial with mt distinct roots, whose coefficients are polynomials in the variables u_{ij}. The condition that this polynomial have distinct roots is given by the nonvanishing of a polynomial in the variables u_{ij}. This condition holds for suitably chosen rational values of the variables u_{ij}. It follows that there exists a matrix V commuting with X with coefficients in F, whose characteristic polynomial $|xE - V|$ is the fth power of a polynomial with coefficients in F, having mt distinct roots. Upon applying Lemma 4.2 to this situation, Schur concluded that

$$|xE - V| = (\varphi(x))^f,$$

for an irreducible polynomial $\varphi(x)$ of degree mt, with coefficients in F, and that $\varphi(V) = 0$, so that the elementary divisors of V are linear, and V is diagonalizable.

Let

$$\varphi(x) = x^{mt} - a_1 x^{mt-1} - \cdots - a_{mt} = (x - \mu_1) \cdots (x - \mu_{mt}),$$

and put

$$M = \begin{pmatrix} a_1 & \cdots & a_{mt} \\ 1 & \cdots & \cdots \\ \cdots & \cdots & \cdots \\ \cdots & 1 & 0 \end{pmatrix}, \overline{M} = \begin{pmatrix} \mu_1 & \cdots & \cdots \\ \cdots & \cdots & \cdots \\ \cdots & \cdots & \mu_{mt} \end{pmatrix}.$$

Then the matrices M and $\overline{M}$ have the same minimal polynomial and are similar. By the theory of the rational canonical form (see [97], Chapter 7, §25), the matrix

$$T = \begin{pmatrix} M & \cdots & \cdots \\ \cdots & \cdots & \cdots \\ \cdots & \cdots & M \end{pmatrix},$$

with f copies of M along the diagonal and zeros elsewhere, is similar to V over the field F, and hence there exists an invertible matrix Q with entries in F such that

$$Q^{-1} V Q = T.$$

Set

$$Q^{-1} X Q = X' = \sum_{r \in G} B(r) x_r;$$

---

[27]To put it in modern terminology, the subspace of the representation space consisting of eigenvectors for a fixed eigenvalue of a commuting transformation V are invariant subspaces for the given representation. For a commuting transformation V with the maximum number of distinct eigenvalues, the subspaces defined above are irreducible, and give a direct decomposition of the representation as a sum of mt irreducible representations. The dimensions of these spaces are the multiplicities of the eigenvalues.

then the group matrix X' commutes with T. Put $B(r) = (B_{ij}(r)), r \in G$, with square submatrices $B_{ij}(r), 1 \leq i, j \leq f$. As the matrices $B(r)$ commute with T, it follows that the submatrices $B_{ij}(r)$ commute with M, for all $r \in G$. From this fact, Schur argued, it follows that the commuting matrices $B_{ij}(r)$ can be expressed as polynomials $g_{ij}^{(r)}(M)$ in M, for a set of polynomials $g_{ij}^{(r)}(x)$ with coefficients in F. A modern proof of this fact is to observe that the matrix M acts on its underlying vector space over F as a cyclic linear transformation, because it has only one elementary divisor, namely $\varphi(x)$ (see [**97**], loc. cit.). Let v be a generator of the cyclic vector space, and let B denote an arbitrary linear transformation commuting with M. Then $Bv = f(M)v$, for some polynomial $f(x) \in F[x]$, and one has $BM^j v = M^j Bv = M^j f(M)v = f(M)M^j v$ for $j = 1, 2, \ldots$. As the vector space is cyclic relative to M, it is generated by the vectors $M^j v, j = 0, 1, 2, \ldots$, and hence $B = f(M)$, as required.

As noted above, the matrices M and $\overline{M}$ are similar, so there exists an invertible matrix L such that

$$L^{-1}B(r)L = (\overline{B}_{ij}(r)), \ r \in G,$$

with $\overline{B}_{ij}(r) = g_{ij}^{(r)}(\overline{M})$ for all i, j. For each root μ_ℓ of the polynomial $\varphi(x)$, Schur defined $C^{(\ell)}(r) = (g_{ij}^{(r)}(\mu_\ell))$, for $r \in G$, and reasoned that $r \to C^{(\ell)}(r)$ is a representation of G of degree f, which is rational in a field $F(\mu_\ell)$ of degree mt over F. This is not difficult to prove, using the facts that $r \to (g_{ij}^{(r)}(M))$, and hence $r \to (g_{ij}^{(r)}(\overline{M}))$, are representations. Letting $Z^{(\ell)} = \sum C^{(\ell)}(r)x_r$ denote the corresponding group matrices, for $\ell = 1, \ldots, mt$, one has

$$L^{-1}X'L = Z^{(1)} \oplus \cdots \oplus Z^{(mt)},$$

a direct sum of group matrices all of degree f. As the group matrix X has mt irreducible components, the uniqueness theorem (Theorem 2.4) implies that the group matrices $Z^{(\ell)}$ are irreducible, and the sum of their characters is $m(\chi + \chi_1 + \cdots + \chi_t)$. It follows that an arbitrary irreducible group matrix Z with character χ must be equivalent to one of the group matrices $Z^{(\ell)}$, and is consequently rational in a field $F(\mu_\ell)$ of degree mt. The fact that mt is the minimal degree of a field extension in which Z is rational follows from the second statement of the theorem.

For the proof of the second assertion of the theorem, let $r \to D(r)$ be an irreducible representation over $\mathbb{C}$, of degree f, whose character is χ, and assume that the coefficients of the matrices $D(r)$ belong to a field $F(\nu)$ of degree q over F. Then, for each $r \in G$, one has

$$D(r) = (g_{ij}^{(r)}(\nu)),$$

for some polynomials $g_{ij}^{(r)}(x) \in F[x]$. Let

$$\psi(x) = x^q - b_1 x^{q-1} - \cdots - b_q \in F[x]$$

be the minimal polynomial of ν over the field F, and let

$$N = \begin{pmatrix} b_1 & \cdots & b_q \\ 1 & \cdots & \cdots \\ \cdots & \cdots & \cdots \\ \cdots & 1 & 0 \end{pmatrix}$$

be the companion matrix of the irreducible polynomial $\psi(x)$, so that $\psi(N) = 0$. As ν and N have the same minimal polynomial, there is an isomorphism of rings (actually fields in this case) from $F[\nu]$ to $F[N]$, given by $h(\nu) \to h(N), h(x) \in F[x]$. From this, it follows easily that $r \to E(r) = (g_{ij}^{(r)}(N))$ is a representation of G, of degree qf, which is rational in the field F.

Schur's idea was to prove that the F-representation $r \to E(r)$ is equivalent to a direct sum of copies of the F-rational irreducible representations $r \to A(r)$ of degree mtf considered in the proof of Theorem 4.1. From this it will follow that qf is divisible by mtf, and hence that q is divisible by mt.

His proof of the final step followed easily from the classification of characters of irreducible F-representations, obtained in Theorem 4.1. Let $\nu_0 = \nu, \nu_1, \ldots, \nu_{q-1}$ be the roots of $\psi(x)$, and set

$$\chi_j(r) = \sum_{i=1}^{f} g_{ii}^{(r)}(\nu_j),$$

for $r \in G$, and $j = 0, \ldots, q-1$. Then $\chi_0, \chi_1, \ldots$ are all characters of G, because they are, by construction, conjugates with respect to the field F of the character $\chi_0 = \chi$. From this information, it follows that the irreducible F-representations occurring as components of the representation $r \to E(r)$ are all equivalent to the irreducible F-representation $r \to A(r)$, and this completes the proof of the Theorem.

Before tackling the problem of the rationality of representations of a solvable group G in the cyclotomic field of $|G|$th roots of unity, Schur illustrated his new theory with several interesting and subtle special cases. The first was an example to show that the field $F(\mu)$ of minimal degree mt in which a given irreducible group matrix Z is rational, according to the preceding theorem, is not necessarily unique. He considered the irreducible matrix group G of order 8 generated by the matrices

$$A = \begin{pmatrix} i & 0 \\ 0 & -i \end{pmatrix}, B = \begin{pmatrix} 0 & 1 \\ -1 & 0 \end{pmatrix}.$$

He asserted, without giving the details of his proof, that the index of the character χ of this representation with respect to the field of rational numbers $\mathbb{Q}$ is 2, and that the representation is rational not only in the field $\mathbb{Q}(i)$, but also in each field $\mathbb{Q}(\sqrt{-n})$ whenever n is a positive integer which can be expressed as a sum of three squares. The values of the character χ lie in the rational field $\mathbb{Q}$, so the number of conjugates t of the character, with respect to the field $\mathbb{Q}$, is equal to 1, and the example defines fields $\mathbb{Q}(\mu)$ and $\mathbb{Q}(\mu')$ in which the given representation is rational, both of minimal degree (in this case 2) over $\mathbb{Q}$, which intersect in the field $\mathbb{Q}(\chi) = \mathbb{Q}$. For a proof that the representation is rational in the field $\mathbb{Q}(\sqrt{-3})$ the reader is referred to [116], §11.

Another special case he considered was the set of possibilities for the index in case the field F is the real field $\mathbb{R}$. As all the representations of a given finite group are rational in the field $\mathbb{C} = \mathbb{R}(i)$ of degree 2 over $\mathbb{R}$, it follows that mt divides 2, for a character χ of index m and number of conjugates t, with respect to $\mathbb{R}$. The possibilities are:

$$t = 1, m = 1; \ t = 1, m = 2; \ t = 2, m = 1.$$

As Schur pointed out, these correspond exactly to the three possibilities for an irreducible real matrix group identified by Loewy [208], and used as the jumping-off point for Frobenius and Schur's paper [151] on real representations of finite groups (see §3). The possibilities for a given real irreducible matrix group were: (i) it is irreducible as a matrix group over $\mathbb{C}$; (ii) it splits as a direct sum of two equivalent irreducible matrix groups over $\mathbb{C}$, and (iii) it splits as a direct sum of two inequivalent irreducible matrix groups over $\mathbb{C}$. It is a nice exercise to check that these possibilities do indeed match the choices for m and t given earlier.

Burnside and Schur both considered the problem of relating the Schur index of an irreducible character χ of G to properties of the subgroups of G. In the smallest interesting case, of the two nonisomorphic nonabelian groups of order 8, the character tables are the same, but the irreducible characters of degree two (each group has just one) have different Schur indices, with respect to the rational field. The fact that one Schur index is one (for the dihedral group) and the other is two (for the quaternion group) can be explained by finding minimal extensions of the rational field in which the irreducible representations of degree two are rational. This, in turn, can be done using the generators and relations of the two groups, which follow from a knowledge of their subgroups. In §3 of his paper [250], Schur raised the question of which finite groups G have a character χ whose Schur index m, with respect to the rational field $\mathbb{Q}$, is equal to $\chi(1)$, the maximal permissible value for m, in view of Corollary 4.2, and related the problem to the subgroup structure, using some work of Burnside [83] published in 1905. An example of this phenomenon was given by Burnside in the second edition of his book ([87], 1911). G. J. Janusz commented on it as follows:

> There is an exercise that has been one of my favorites. In §238, Ex. 8, he asks the reader to prove that a certain group of order 63 has a representation of degree three with Schur index three. His language is, of course, that the reader should show "it is not possible to represent the group in a form in which the coefficients are rational functions of the characteristics.". This gives a noncyclic group of odd order that is a subgroup of the multiplicative group of a division ring. In the 1950s, Herstein conjectured (in print) that odd order subgroups of a division ring are cyclic and, with that as motivation, Amitsur ([6], 1955) went on to classify all finite subgroups of division rings. Burnside's group of order 63 is the smallest odd order (nonabelian) subgroup of a division ring. Good thing Herstein didn't know about Burnside's example.[28]

5. Projective Representations of Finite Groups

Soon after completing his dissertation on polynomial representations of $GL_n(\mathbb{C})$, Schur turned to the problem of classifying representations of a finite group G by collineations of projective space $\mathbb{P}^{n-1}$. The points in projective space $\mathbb{P}^{n-1}$ are the lines through the origin in complex n-space $\mathbb{C}^n$, and collineations in $\mathbb{P}^{n-1}$ are the functions from points to points in $\mathbb{P}^{n-1}$ defined by linear substitutions associated with $n \times n$-matrices with entries in $\mathbb{C}$. The collineation of $\mathbb{P}^{n-1}$ defined by a matrix A coincides with the collineation defined by a scalar multiple αA of A, for $\alpha \neq 0$ in $\mathbb{C}$. A *representation of G by collineations* of $\mathbb{P}^{n-1}$ is defined as a function $r \to A(r)$ from G to invertible $n \times n$-matrices, such that, for all $r, s \in G$,

$$A(r)A(s) = a_{r,s}A(rs),$$

with $a_{r,s}, r, s \in G$ a set of nonzero complex numbers, called a *factor set* associated with the given representation. Such a representation is called today a *projective*

---

[28]Letter to the author dated February 3, 1996. Used with permission.

representation of G,[29] and does indeed assign to each element $r \in G$ a collineation $\tilde{A}(r)$ of $\mathbb{P}^{n-1}$, so that $\tilde{A}(rs) = \tilde{A}(r)\tilde{A}(s)$ for all $r, s \in G$. Representations for which $a_{r,s} = 1$, for all $r, s \in G$, will be called *ordinary representations*, for the sake of clarity.

Schur recalled that Molien and Frobenius had shown that the problem of determining all ordinary representations of a finite group G was equivalent to the problem of expressing the group matrix of G in terms of irreducible submatrices, and that the first and essential step towards the solution of this problem was the factorization of the group determinant, and hence the calculation of the characters of G (see §2). His aim in the new investigation was to use the group matrix and group characters to classify the representations of G by fractional linear substitutions, or as we have seen, by projective representations of G. His work on this subject was contained in three papers ([**248**], 1904), ([**251**], 1907), and ([**255**], 1911). They are a pleasure to read, and with some changes and additions, have become a standard item in the literature on representation theory. This section contains a brief introduction to them, closely following Schur's approach.

He first defined the concepts of degree, equivalence, and irreducibility (or primitivity) of projective representations exactly as for ordinary representations. Thus a projective representation is irreducible if it is not equivalent to a representation of the form

$$r \to \begin{pmatrix} A(r) & 0 \\ 0 & B(r) \end{pmatrix},$$

with $r \to A(r)$ and $r \to B(r)$ projective representations of lower degree.[30] He noticed that two projective representations of the same degree, $r \to A(r)$ and $r \to B(r)$, define the same representation of G by collineations in projective space whenever there exists a set of nonzero constants $c_r, r \in G$, such that $A(r) = c(r)B(r)$ for all $r \in G$, and that in this case, it follows from the definitions that the factor sets $a_{r,s}$ and $b_{r,s}$ of the two projective representations satisfy the equations

$$a_{r,s} = \frac{c_r c_s}{c_{rs}} b_{r,s},$$

for all $r, s \in G$. When this occurs, he called the two projective representations *associates* of one another.

Schur realized that the study of ordinary representations of finite groups, and their subgroups and quotient groups, inevitably leads to projective representations. This comes about in the following way. Let $r \to A(r)$ be an irreducible ordinary representation of a finite group G, let Z be a subgroup of the center of G, and let H denote the quotient group G/Z. Then, in modern language, there is a short exact

---

[29]In his first paper on the subject [**248**], Schur began with representations of G by *fractional linear substitutions* [gebrochene lineare Substitutionen]

$$x_j = \frac{a_{j1}^r y_1 + \cdots + a_{j,n-1}^r y_{n-1} + a_{jn}^r}{a_{n1}^r y_1 + \cdots + a_{n,n-1}^r y_{n-1} + a_{nn}^r}, \; j = 1, \ldots, n-1,$$

for invertible $n \times n$ matrices $A(r) = (a_{ij}^r)$ corresponding to elements $r \in G$, and composed according to the multiplication of pairs of elements in G. As Schur observed, this definition is easily shown to be equivalent to the definition of projective representations $r \to A(r) = (a_{ij}^r)$ given above.

[30]Frobenius first defined primitive representations of a finite group in terms of the factorization of the group determinant (see Chapter II, §4), and proved that the concept was equivalent to the notion of irreducibility given above in his second paper on representations [**136**].

sequence

$$1 \to Z \to G \to H \to 1,$$

with $Z \to G$ the inclusion map, and $G \to H$ the natural homomorphism. In this situation, Schur called the group G the group H extended by the group Z [durch die Gruppe Z ergänzte Gruppe von H]; we shall use the modern terminology, and call G a *central extension of H by Z*. Let $u_h, h \in H$, be a set of coset representatives of Z in G, with the elements of the coset $u_h Z$ all mapped to $h \in H$ by the homomorphism $G \to H$. Then one has

$$u_h u_k = z_{h,k} u_{hk}$$

for all $h, k \in H$, for some set of elements $z_{h,k} \in Z$. For each pair of elements h, k in H, $A(z_{h,k})A(r) = A(r)A(z_{h,k})$ for all $r \in G$, because the elements $z_{h,k}$ belong to the center of G. As the representation $r \to A(r)$ is irreducible, the preceding equation implies that $A(z_{h,k}) = a_{h,k}I$ for some complex number $a_{h,k}$ and the identity matrix I, by Schur's Lemma 2.1. Letting $A'(h) = A(u_h)$ for each $h \in H$, it follows that

$$A'(h)A'(k) = a_{h,k}A'(hk),$$

for all $h, k \in H$, and that $h \to A'(h)$ is an irreducible projective representation of H. In this situation, the projective representation $h \to A'(h)$ is said to be *lifted* to the ordinary representation $r \to A(r)$ of G.

At this point, he raised the interesting question whether, for a given finite group H, there exists a central extension G of H with the property that all irreducible projective representations of H can be lifted to irreducible ordinary representations of G. When this occurs, we shall say that the extension G of H has the *projective lifting property*; Schur called a group G with this property *sufficiently extended* [hinreichend ergänzt]; and called a sufficiently extended group G of minimal order a *representation group* [Darstellungsgruppe] of H. In the first two papers [248], [251], Schur proved that representation groups exist for all finite groups, constructed and characterized them in several ways, and proved that the kernel Z of the natural homomorphism $G \to H$ from a representation group G is a uniquely determined abelian group, for which he gave a set of generators and relations. He called the kernel of the homomorphism $G \to H$ the *Multiplicator* of H. It is called today the *Schur multiplier* of H, and is, in modern terminology, the second cohomology group $H^2(H, \mathbb{C})$ (see the remarks and the footnote at the beginning of the proof of Theorem 5.1).

In the second and third papers [251], [255], he found representation groups and gave the structure of the Schur multipliers for several infinite families of finite groups. These included the groups $PSL_2(q)$, $SL_2(q)$, and $PGL_2(q)$ defined by 2×2 matrices with entries in the finite field of q elements, for a prime power q, and the symmetric and alternating groups. Among other things, he proved that the group $SL_2(q)$, which is by definition a central extension of the group $PSL_2(q)$, is in fact a representation group of $PSL_2(q)$, except when $q = 4$ or 9, and that, with these exceptions, $SL_2(q)$ is its own representation group. It follows that the irreducible projective representations of $PSL_2(q)$ are obtained from ordinary irreducible representations of $SL_2(q)$ in the nonexceptional cases, and to complete the picture, he derived the character tables of the groups $SL_2(q)$ for all values of q. All this concrete information has been a valued resource up to the present time,[31] and demonstrated that Schur, like Frobenius and Burnside before him, not only built general theories, but took an interest in special cases, along with the sometimes difficult calculations necessary to understand them.

The route from the idea of projective representations to the Schur multiplier and representation groups is not straightforward, and we explore, with Schur, some of the steps along the way.

Let $r \to A(r)$ be a projective representation of a finite group G, with factor set $a_{r,s}$. Then

$$A(r)A(s) = a_{r,s}A(rs),$$

for all $r, s \in G$. It follows from these equations and the associative law $(rs)t = r(st)$ that one has the *factor set identity*:

$$a_{r,s}a_{rs,t} = a_{r,st}a_{s,t},$$

for all $r, s, t \in G$.

Schur's first observation was that, conversely, an arbitrary set of nonzero complex numbers $a_{r,s}$, indexed by pairs of elements $r, s \in G$, and satisfying the factor set identity, is indeed a factor set of some projective representation. He proved this by defining a generalized group matrix

$$X = (a_{rs^{-1},s}x_{rs^{-1}}),$$

with indeterminates $x_r, r \in G$, writing it in the form

$$X = \sum A(r)x_r,$$

and checking that $A(r)A(s) = a_{r,s}A(rs)$, for $r, s \in G$. As an exercise, the reader can check that there exists an irreducible projective representation with the given factor set.

He then had the idea of considering factor sets by themselves, that is, sets of nonzero complex numbers $a_{r,s}$, for $r, s \in G$, satisfying the factor set identity, entirely apart from any connection with projective representations. He divided the set of factor sets into equivalence classes, denoted here by $\{a_{r,s}\}$, with two factor sets $a_{r,s}$ and $b_{r,s}$ belonging to the same equivalence class provided that

$$a_{r,s} = \frac{c_r c_s}{c_{rs}} b_{r,s},$$

for some set of nonzero complex numbers $c_r, r \in G$, as in the definition of associated projective representations. It is easily verified that the product of two factor sets $a_{r,s}b_{r,s}$ satisfies the factor set identity, and is a factor set. In terms of this operation, one obtains a well-defined operation of multiplication of classes of factor sets, defined by

$$\{a_{r,s}\}\{b_{r,s}\} = \{a_{r,s}b_{r,s}\}.$$

THEOREM 5.1. *For a given finite group G, the set M of classes of factor sets, with the operation of multiplication defined as above, is a finite abelian group.*

The proof that M is an abelian group is left as an exercise for the reader. The steps leading up to the definition of the group M were precursors of ideas in the

---

[31]The Schur multipliers are given for all the simple groups taken up in the *Atlas of finite groups* [**96**], where their importance for contemporary research in representation theory is explained.

development of the cohomology of groups.[32] We shall give a simplified version of Schur's proof that M is finite, based on an idea stated in a footnote to Schur's own proof of the theorem. Let $n = |G|$. Let $a_{r,s}$ be a factor set, and for each $r \in G$, put

$$d_r = \prod_{s \in G} a_{r,s}.$$

Then one checks, using the factor set identity, that one has

$$a_{r,s}^n = \frac{d_r d_s}{d_{rs}},$$

for $r, s \in G$. From this it follows that the class $\{a_{r,s}\}$ has finite order dividing n, in the group M.

For a factor set $a_{r,s}$ and elements d_r as above, let d_r' be a fixed set of solutions of the equations $(d_r')^n = d_r$, for $r \in G$. (Note that in this construction, we have used the fact that $\mathbb{C}$ is algebraically closed.) Now, put

$$a_{r,s}' = \frac{(d_r')^{-1}(d_s')^{-1}}{(d_{rs}')^{-1}} a_{r,s},$$

for each $r, s \in G$. Then $a_{r,s}'$ is a factor set in the class $\{a_{r,s}\}$, and it follows from the definition of the elements d_r' and the calculation of $a_{r,s}^n$ given above, that $(a_{r,s}')^n = 1$ for all $r, s \in G$. This proves that each equivalence class of factor sets contains a representative whose values are nth roots of unity. It follows that there are only finitely many equivalence classes, and hence M is a finite group. This completes the proof.

The abelian group M, today called the *Schur multiplier* of H, is not related, in any obvious way, to central extensions

$$1 \to Z \to G \to H \to 1.$$

Schur did find such a relationship, however, and it provided the connection between the Schur multiplier and representation groups described earlier (see Theorem 5.3). For a central extension as above, he began with a set of coset representatives u_h, $h \in H$, of Z in G, satisfying the equations

$$u_h u_k = z_{h,k} u_{hk},$$

with $z_{h,k} \in Z$, for all $h, k \in H$, as before, and this time noticed that the elements $z_{h,k} \in Z$ satisfy the identity:

$$z_{h,k} z_{hk,\ell} = z_{h,k\ell} z_{k,\ell},$$

for all $h, k, \ell \in H$. This means that, for each character ψ of the abelian group Z (as defined in Chapter I, §1), the complex numbers $r_{h,k} = \psi(z_{h,k})$ satisfy the factor set identities, and hence define an element $\{r_{h,k}\}$ of the Schur multiplier M.

THEOREM 5.2. *Let G be an arbitrary central extension of H, with kernel Z, and let $[G, G]$ denote the commutator group of G. Then $[G, G] \cap Z$ is isomorphic to the subgroup of the Schur multiplier of M represented by the factor sets of the form $r_{h,k} = \psi(z_{h,k})$, for all characters ψ of Z. This subgroup coincides with M if and only if the group G has the projective lifting property.*

---

[32]For some historical remarks about the Schur multiplier in the context of the cohomology of groups, see [**211**], Chapter IV, Notes.

For the proof of the theorem, Schur needed to take a closer look at the characters of Z and their relation to linear characters (that is characters of degree one) of the group G containing Z. He used the fact that the finite abelian group Z and its character group are isomorphic.[33] This means that the characters of Z can be indexed by the elements of Z, in such a way that if ψ_z and ψ_w are characters indexed by the elements z and w in Z, then $\psi_z \psi_w = \psi_{zw}$. With reference to the discussion preceding the statement of the theorem, two characters ψ_z and ψ_w define the same element of the Schur multiplier if and only if the factor sets $\psi_z(z_{h,k})$ and $\psi_w(z_{h,k})$ are equivalent, that is:

$$\psi_z(z_{h,k}) = \frac{\varphi(h)\varphi(k)}{\varphi(hk)} \psi_w(z_{h,k})$$

for some set of nonzero complex numbers $\varphi(h), h \in H$. This is equivalent to the statement that

$$\varphi(h)\varphi(k) = \psi_{zw^{-1}}(z_{h,k})\varphi(hk),$$

for all $h, k \in H$. From this, it follows easily that if one sets

$$\chi(r) = \chi(au_h) = \psi_{zw^{-1}}(a)\varphi(h),$$

for each element $r \in G$ expressed in the form au_h, with $h \in H$ and $a \in Z$, then

$$\chi(rs) = \chi(r)\chi(s)$$

for all elements $r, s \in G$, and the character $\psi_{zw^{-1}}$ extends to a linear character of G. It follows that the image M' of the homomorphism from characters of Z to elements of the Schur multiplier M is isomorphic to the factor group Z/W, where W is the subgroup of Z consisting of elements w whose corresponding characters ψ_w can be extended to linear characters of G.

The crucial point, as Schur recognized, was that the subgroup $M' \cong Z/W$ can be obtained in another way. This comes about as follows. From elementary character theory, it is evident that $[G, G] \cap Z$ consists precisely of those elements $z \in Z$ with the property that $\chi(z) = 1$ for all linear characters χ of G. As the restriction of a linear character χ of G to the subgroup Z has the form $\chi|_Z = \psi_w$, for some element $w \in W$, by the definition of the subgroup W of Z, it follows that $[G, G] \cap Z$ can be identified with the subgroup of Z consisting of elements z such that $\psi_w(z) = 1$ for all elements $w \in W$. It is not difficult to prove that this subgroup is isomorphic to Z/W, completing the proof of the first part of the theorem.

The proof of the second part of the theorem follows easily from the remarks about the projective lifting property, given earlier, and is left as an exercise.

The main result of Schur's first paper [**248**] was the following theorem.

THEOREM 5.3. *Let H be a finite group, and M the Schur multiplier of H. Then there exists a central extension G of H, with kernel M, having the projective lifting property. The group G has minimal order among all central extensions of H having the projective lifting property, and is a representation group.*

It follows from Theorem 5.2 that any central extension G of H with the projective lifting property has order at least $|H||M|$, so the second statement of the theorem follows from the first. Schur's proof of the first statement is interesting,

---

[33]The isomorphism is not unique, as it depends on expressing Z as a direct product of cyclic groups, and this can be done in more than one way.

and is sketched below. It was an application of the relatively new idea of defining groups by generators and relations, which Schur attributed to Dyck [**115**].

Schur began, as in the proof of Theorem 5.2, by recalling that the structure of a central extension

$$1 \to Z \to G \to H \to 1$$

is determined by a set of coset representatives $\{u_h\}_{h \in H}$ of Z in G satisfying the relations

$$u_h u_k u_{hk}^{-1} = z_{h,k} \in Z,$$

and

$$z_{h,k} z_{hk,\ell} = z_{h,k\ell} z_{k,\ell}$$

for all $h, k, \ell \in H$. He had the idea of using these relations to define a group by generators and relations, namely the group $\mathcal{E}$ with generators $\{U_h\}_{h \in H}$ and $\{Z_{h,k}\}_{h,k \in H}$, satisfying the defining relations

$$U_h U_k U_{hk}^{-1} = Z_{h,k}, U_h Z_{k,\ell} = Z_{k,\ell} U_h, \text{ and } Z_{h,k} Z_{hk,\ell} = Z_{h,k\ell} Z_{k,\ell},$$

for all $h, k, \ell \in H$.

Then $\mathcal{E}$ is an infinite group, and is a central extension

$$1 \to \mathcal{Z} \to \mathcal{E} \to H \to 1,$$

with $\mathcal{Z}$ the finitely generated abelian group contained in the center of $\mathcal{E}$ generated by the elements $\{Z_{h,k}\}_{h,k \in H}$. The main structure theorem for finitely generated infinite abelian groups had conveniently been published a few years earlier by Frobenius and Stickelberger [**153**]; it stated that such a group is the direct sum of a free group and its torsion group, and that the latter is a finite group. Schur applied the theorem to the kernel $\mathcal{Z}$ of the central extension he had constructed, and deduced that $\mathcal{Z}$ is the direct sum of its torsion subgroup $\mathcal{T}$ and a free subgroup $\mathcal{F}$. It follows that $\mathcal{E}/\mathcal{F}$ is a finite group, and that

$$1 \to \mathcal{T} = \mathcal{Z}/\mathcal{F} \to \mathcal{E}/\mathcal{F} \to H \to 1$$

is a central extension of H. It remains to prove that the kernel $\mathcal{T}$ is isomorphic to the Schur multiplier M of H, and that the central extension $\mathcal{E}/\mathcal{F}$ has the projective lifting property. Proofs of these steps are not included here; for them, the reader is referred to Schur's paper [**248**], or to a modern proof of the theorem (Huppert [**183**], Kap. V, §23, Hauptsatz 23.5).

From the two preceding theorems, the following result, mentioned earlier, follows directly.

COROLLARY 5.1. *Let G' be an arbitrary representation group of the finite group H. Then the kernel of the homomorphism from G' to H is isomorphic to the Schur multiplier of H.*

6. Permutation Groups and Centralizer Rings

From around 1900, many of the papers of Frobenius and Schur on finite groups and representations contain references to, and in some cases immediately follow up, related work of Burnside. An investigation, by Schur, in the theory of permutation groups, began in this way, with a paper by Schur ([**252**], 1908) containing a new proof of Burnside's theorem on permutation groups of prime degree ([**73**], 1900; Chapter III, Theorem 3.8):

A transitive permutation group G on p symbols, for a prime p, is either solvable or doubly transitive.

Schur stated that this theorem was "one of the most important new results in finite group theory." Burnside himself gave several proofs of the theorem: the first, cited above, using character theory, and the second in 1906, without characters; he also gave a proof of the theorem, using characters, in the second edition of his book ([**87**], §250). Schur's proof was a further simplification of Burnside's second proof; it contained the essential ideas of a new method in the theory of permutation representations, involving the centralizer ring of a permutation representation, to which he returned, in connection with some other problems raised by Burnside, in his last major work on representation theory ([**261**]; 1933).

An outline of the part of Schur's first paper containing the ideas underlying his new method, and some discussion of its continuation, are given below. Schur began with some general remarks about a group G of permutations of the set $\{1, 2, \ldots, n\}$, represented as a group of $n \times n$ permutation matrices. He considered bilinear forms

$$F = \sum_{i,j=1}^{n} a_{ij} x_i x_j$$

which are invariant under the action of the permutation matrices R corresponding to the elements of G, and noted that a bilinear form F is G-invariant if and only if the matrix $A = (a_{ij})$ of the form satisfies the condition

$$AR = RA$$

for all permutation matrices R corresponding to elements of G.[34] The set of all matrices A with this property is closed under addition and multiplication, and forms a ring, today called the *centralizer ring* of the permutation representation. The centralizer ring always contains the identity matrix E, the matrix J all of whose entries are equal to 1, and all their linear combinations $aE + bJ$, for $a, b \in \mathbb{C}$. The first step in Schur's new proof of Burnside's theorem was:

LEMMA 6.1. *A transitive permutation group G is doubly transitive if and only if the centralizer ring of the corresponding representation of G by permutation matrices consists of the matrices $aE + bJ$, for $a, b \in \mathbb{C}$, and only these.*

The proof of the lemma is not difficult, and was not included in [**252**]. Concerning the lemma, however, he stated, in his second paper on permutation groups: "The criterion is identical with one often used by Frobenius and Burnside, that a transitive permutation group G is doubly transitive if and only if the permutation representation $r \to R$ has exactly two irreducible components." For Frobenius's proof of this fact, see Chapter II, §4, Proposition 4.1.

After these preliminary remarks, Schur turned to the proof of Burnside's theorem. He put $n = p$, a prime number, and noted that, by the theorem of Cauchy-Sylow, G contains a p-cycle, which may be assumed to have the form

$$\ell = (0\,1\,\ldots\,(p-1)).$$

---

[34] In [**151**] Frobenius and Schur proved (see the proof of Proposition 3.1) that a bilinear form is G-invariant, for an arbitrary representation $r \to R$ of a finite group G, if and only if the matrix A of the form satisfies the condition $R'AR = A$, for all matrices R. (Here R' denotes the transpose of the matrix R.) If $r \to R$ is a representation of G by permutation matrices, then $R' = R^{-1}$ for each matrix R, and the condition becomes $AR = RA$.

Then the matrices A in the centralizer ring commute, in particular, with the matrix corresponding to ℓ:

$$L = \begin{pmatrix} 0 & 0 & \cdots & 1 \\ 1 & 0 & \cdots & 0 \\ & \cdots & & \\ 0 & \cdots & 1 & 0 \end{pmatrix}.$$

Therefore, from linear algebra, all the matrices in the centralizer ring are polynomials in L, and have the form

$$A = \varphi(L) = a_0 E + a_1 L + \cdots + a_{p-1} L^{p-1},$$

for polynomials

$$\varphi(x) = a_0 + a_1 x + \cdots + a_{p-1} x^{p-1} \in \mathbb{C}[x].$$

By Lemma 6.1, if the permutation group G is not doubly transitive, there exist matrices in the centralizer ring which are not of the form $aE + bJ$. As the centralizer ring contains all matrices $A = \varphi(L)$ as above, with $\varphi(x) \in \mathbb{Q}[x]$, and $J = E + L + \cdots + L^{p-1}$, it follows that there exists a matrix $A = \varphi(L)$ belonging to the centralizer ring such that $\varphi(x) \in \mathbb{Z}[x]$, and not all the coefficients $a_1, \ldots, a_{p-1}$ are equal to each other.

Schur then applied Gauss's theory of the cyclotomic field generated by a primitive pth root of unity ρ (see Chapter I, §3) to the polynomial $\varphi(\rho)$, where $\varphi(x) \in \mathbb{Z}[x]$ is the polynomial obtained in the preceding paragraph. As not all the coefficients $a_1, \ldots, a_{p-1}$ are equal, $\varphi(\rho)$ is not a rational number, and satisfies an irreducible equation of degree $e = \frac{p-1}{f}$. As the periods form a basis of the field generated by $\varphi(\rho)$ (see Chapter I, Theorem 2.1), it follows that there exists a polynomial $\psi(x) \in \mathbb{Z}[x]$ such that $\psi(\varphi(\rho))$ is the Gaussian period

$$\psi(\varphi(\rho)) = \rho + \rho^c + \cdots + \rho^{c^{f-1}},$$

where the integer c has exponent $f \pmod{p}$. It follows that, in $\mathbb{Z}[x]$, there is an identity of the form

$$x + x^c + \cdots + x^{c^{f-1}} = \psi(\varphi(x)) + \lambda(x)(x + \cdots + x^{p-1}),$$

for some polynomial $\lambda(x) \in \mathbb{Z}[x]$. As $LJ = J$, one has $\lambda(L)J = tJ$, where t is the sum of the coefficients of $\lambda(x)$. Upon substituting L in the identity, Schur deduced that

$$H = L + L^c + \cdots + L^{c^{f-1}} = \psi(A) + tJ = \psi(\varphi(L)) + tJ,$$

and belongs to the centralizer ring.

He finished the proof of Burnside's theorem with a close study (not given here) of the matrices commuting with H, and was able to prove that the permutations in G all have the form $x \to ax + b$, for x, a, b in the finite field $\mathbb{Z}_p$, and $a \neq 0$. The group of all such transformations is easily shown to be solvable, and as G is isomorphic to a subgroup, it follows that G is solvable, completing the proof.

In his second paper on permutation groups [261], Schur mentioned another theorem of Burnside:

THEOREM 6.1. *Let $p^a, a > 1$, be a composite power of a prime p. If a permutation group G on a set of p^a elements contains a cycle of order p^a, then G is either doubly transitive or imprimitive.*[35]

Burnside's proof of the theorem, using character theory, was given in §252 of the second edition of his book [**87**]. He conjectured, at the end of §252, that the same result held for an arbitrary permutation group G of degree n, containing a regular abelian subgroup of order n.[36] In this generality, the conjecture is false. Schur gave a counterexample in §6 of his paper. Other counterexamples were obtained by Dorothy Manning (see H. Wielandt ([**288**], §25) and P. Neumann [**227**] for further discussion and references to the literature). The main object of his paper was to establish a general theory of centralizer rings of permutation representations, and apply it to prove the deep result that the conjecture is true if the abelian group is cyclic:

THEOREM 6.2. *Let G be a permutation group on a set of n elements, for a positive integer n which is not a prime. If G contains a cycle of order n, then G is either doubly transitive or imprimitive.*

A modern account of Schur's theory of centralizer rings (or S-rings, as they are often called in the literature), and proofs of Theorem 6.2 and other related theorems obtained subsequently, are contained in the monograph on permutation groups by Helmut Wielandt [**288**].

Burnside's proof of Theorem 6.1 involved some interesting, but intricate, assertions about sums of roots of unity, and attracted further attention. In a paper *On Burnside's Method* [**197**], W. Knapp summarized the situation as follows:

> However it is a strange fact that Burnside's proof of [Theorem 6.1] published in the second edition of his book ([**87**], §252) has been generally accepted but is not at all convincing since the crucial argument in the last step is false, as noticed (apparently first) by Peter M. Neumann (see [**227**] where explicit counterexamples to Burnside's relevant assertion about sums of p^m-th roots of unity are given). ... So it happens that the first correct proof of [Theorem 6.1] has been given by Issai Schur [**261**].[37]

The point of Knapp's article was to show that Burnside's ideas in [**87**], §251, are sufficient to give a proof of Theorem 6.1 "along the same line of arguments as in his book."

---

[35] A permutation group G on a set X is *imprimitive* if X is the disjoint union of subsets $B_i, i = 1, \ldots, s, s > 1$, called *blocks*, with the property that for each block B_i, the elements $g \in G$ either fix the block, or map it onto another block.

[36] A transitive permutation group G acting on a set X is called *regular* if the stabilizer in G of each element of X is trivial. For example, the regular representation of a finite group, by left multiplications, has this property.

[37] Knapp, W., "On Burnside's method," *J. Algebra* **175** (1995), pp. 644–660. Used with permission of Academic Press, Inc.

CHAPTER V

Polynomial Representations of $GL_n(\mathbb{C})$

At the conclusion of his article *Hypercomplex Numbers, Lie Groups, and the Creation of Group Representation Theory* [173], T. Hawkins wrote:

> It is clear that no less than four lines of mathematical investigation were leading to group representation theory: (1) the structure theory of hypercomplex numbers (Molien, Cartan); (2) the theory of Lie groups and algebras (Cartan, Burnside); (3) Klein's form-problem (Maschke); (4) number theory and group determinants (Dedekind, Frobenius).[1]

In this chapter, we follow a fifth line of investigation leading to group representation theory: invariants and polynomial representations of the general linear group $GL_n(\mathbb{C})$. This direction was first taken by the Belgian mathematician Jacques Deruyts (1862-1945), in his memoir *Essai d'une théorie générale des formes algébriques* [105], published in Liège in 1892,[2] and a few years later by Schur, in his doctoral dissertation ([247], 1901). In these articles, Deruyts and Schur, independently, and by different methods, solved the problem of constructing and classifying the polynomial representations of $GL_n(\mathbb{C})$, starting a new branch of group representation theory, known today as the representation theory of linear groups and algebraic groups. While this subject is not, strictly speaking, part of the representation theory of finite groups, modern finite group representation theory is inextricably entangled with it. An example of this occurs in Schur's dissertation, where the polynomial representations of $GL_n(\mathbb{C})$ are investigated by relating them to representations of the symmetric group.

1. Invariant Theory and Representations of GL_n: Jacques Deruyts

Deruyts's work was presented in terms of the classical theory of algebraic forms, which was in vogue during the latter part of the 19th century, owing to the efforts of Aronhold, Cayley, Clebsch, Gordan, Hilbert, and Sylvester,[3] and it was by no means a straightforward task to understand its connection to representation theory. This connection was fully explained only recently, by J. A. Green [166].[4]

---

[1]Hawkins, T., "Hypercomplex numbers, Lie groups, and the creation of group representation theory," Archive for History of Exact Sciences **8** (1972), pp. 243–287. Used with permission of Springer-Verlag.

[2]I am indebted to J. A. Green for calling my attention to the work of Deruyts, and for sending me reprints of his articles ([166], [167]) along with other information about Deruyts.

[3]See Parshall and Rowe [236], Chapter 3, for an historical account of the work of Sylvester, with references to some of the literature on the subject.

[4]The fact that Deruyts's memoir contained proofs of the main results about the polynomial representations of the general linear group was suggested to J. A. Green by J. Towber, around 1979.

Jacques Deruyts was born in Liège, and spent most of his life there. He received his doctorate from the University of Liège in 1883, and began publishing articles on the theory of forms and other subjects soon afterwards. His academic career began at the University of Liège with his appointment as assistant to Professor Pérard, in experimental physics. After occupying some intermediate-level positions in mathematics, he was appointed professor of geometry in 1893, and remained on the faculty at Liège until his retirement.[5] The paper of interest to us here ([105], 1892), published in the *Mémoires de la Société des Sciences de Liège*, was, to some extent, a reorganization of earlier publications, with new proofs; it also contained new material. Referring to Deruyts's memoir, J. A. Green wrote:

> His language is the language of invariant theory, and he makes little use of matrices. But we can look back on Deruyts's work and find a wealth of methods which, to our eyes, are pure representation theory; some of these methods are still unfamiliar today. One can speculate that the representation theory of linear groups could have taken a different course, or at least have matured more rapidly, if Schur and Weyl had taken up the ideas which are lying just below the surface of Deruyts's memoir![6]

Our sketch of its contents, in modern terminology, from the point of view of group representations, is based on the article of J. A. Green cited above. A fuller account of the theory of polynomial representations of GL_n, based on connections with the representations of the symmetric group, and with proofs of most of the main results, is given in the next section.

We begin with some notation. Let n be a fixed positive integer and let G be the general linear group $GL_n(\mathbb{C})$.[7] An *algebraic form* of degree r is a polynomial

$$f(a, x) = \sum \frac{r!}{r_1! \cdots r_n!} a_{r_1, \ldots, r_n} x_1^{r_1} \cdots x_n^{r_n},$$

homogeneous of degree r, in n independent variables $x_1, \ldots, x_n$ over $\mathbb{C}$. The coefficients $a_{r_1, \ldots, r_n}$ are also viewed as independent indeterminates over $\mathbb{C}$, and will sometimes be denoted by $a_{(r)}$, for an n-tuple $(r) = (r_1, \ldots, r_n)$ with $r_1 + \cdots + r_n = r$. A *linear substitution* (see Chapter II, §4) corresponding to an invertible matrix $s = (s_{ij})$ in $G = GL_n(\mathbb{C})$, is the operation of replacing $x_1, \ldots, x_n$ by new variables $X_1, \ldots, X_n$ given by

$$x_i = \sum_{j=1}^n s_{ij} X_j.$$

When this operation is applied to the variables in a form $f(a, x)$, as above, one obtains a new form $f(A, X)$, of degree r, in the new variables $X_1, \ldots, X_n$, whose coefficients $A_{(p)}$, for $(p) = (p_1, \ldots, p_n), \sum p_j = r$, are determined by the equation

$$f(a, x) = f(A, X),$$

---

[5] The obituary article *Notice sur Jacques Deruyts* [161], by L. Godeaux, contains a biography and a report on Deruyts's mathematical work.

[6] Green, J. A., "Classical invariants and the general linear group," *Progress in Mathematics*, **95** (1991), pp. 247–272. Used with permission of Birkhauser, Boston.

[7] In Green's article [166], it is shown that all the main results are valid with $\mathbb{C}$ replaced by an arbitrary field of characteristic zero.

and are given in terms of the original coefficients $a_{(r)}$ by formulas

$$A_{(p)} = \sum_{(r)} T_{(r),(p)}(s) a_{(r)},$$

where the expressions $T_{(r),(p)}(s)$ are homogeneous polynomials of degree r in the entries s_{ij} of the matrix s. If one views the polynomials $T_{(r),(p)}(s)$ as entries of a matrix S whose rows and columns are indexed by the n-tuples (p) with $\sum p_i = r$, then it is easily checked that the correspondence $s \to S$ is a representation of G, as defined in Chapter II, §4, with the property that the entries of the matrix S are polynomials in the entries of the matrix $s \in G$. Such a representation is called a *polynomial representation* of G. The main results of Deruyts's memoir, on the application of linear substitutions to algebraic forms, can be interpreted as results about polynomial representations of G.

Let $|s|$ denote the determinant of an element $s \in G$. A polynomial $J(a)$ in the coefficients $a_{(r)}$ is called an *invariant* of *weight* w (for the given form $f(a,x)$), if the relation

$$J(A) = |s|^w J(a)$$

holds for all $s \in G$. A polynomial $J(a,x)$, which involves the coefficients $a_{(r)}$ and the variables x_i is called a *covariant* of weight w if

$$J(A,X) = |s|^w J(a,x),$$

for all $s \in G$.

In his memoir, Deruyts made a thorough study of the covariants associated with a system of linear forms, obtained as follows. Consider a linear form

$$a \cdot x = a_1 x_1 + \cdots + a_n x_n$$

in the variables $x_1, x_2, \ldots$. A linear substitution corresponding to $s \in G$ transforms the form $a \cdot x$ into a new linear form $A \cdot X$ such that

$$a \cdot x = A \cdot X.$$

The coefficients $A_1, \ldots, A_n$ of the new form are given in terms of $a_1, \ldots, a_n$ by the relations

$$A_k = \sum_\ell s_{\ell k} a_\ell,$$

for $1 \le k \le n$. Deruyts considered a family of linear forms $a_{(j)} \cdot x^{(j)} = \sum_i a_{ji} x_{ij}$, for $1 \le j \le p$, with variable sets $x^{(j)} = (x_{1j}, \ldots, x_{nj})$ and coefficient sets $a_{(j)} = (a_{j1}, \ldots, a_{jn})$, for a fixed integer p such that $p \ge n$. He was interested in the action of linear substitutions corresponding to elements $s \in G$ on the algebra $\mathbb{C}[a,x]$ generated by the coefficients a_{ji} and the variables x_{ij} of the forms under consideration.

In more detail, following Green [**166**], the action of a linear substitution corresponding to $s \in G$ on the polynomial algebra $\mathbb{C}[a]$ generated by the coefficients a_{jk} is a left action defined, as above, by the formula

$$s \circ a_{jk} = A_{jk} = \sum_\ell s_{\ell k} a_{j\ell},$$

and extended to the polynomial algebra $\mathbb{C}[a]$ in such a way that

$$s \circ f(\ldots, a_{jk}, \ldots) = f(\ldots, A_{jk}, \ldots) = f(\ldots, s \circ a_{jk}, \ldots),$$

for all polynomials $f \in \mathbb{C}[a]$. To say that it is a left action means that one has $st \circ f = s \circ (t \circ f)$ for all polynomials $f \in \mathbb{C}[a]$, and all $s, t \in G$, and of course

$1 \circ f = f$ for the identity element $1 \in G$. Similarly, there is an action by $s \in G$ on the polynomial algebra $\mathbb{C}[x]$ generated by the variables x_{ij}, given by the formula

$$g(\ldots, x_{kj}, \ldots) \to g(\ldots, X_{kj}, \ldots)$$

for each $s \in G$ and all polynomials $g \in \mathbb{C}[x]$. This is a consequence of a right action of G on $\mathbb{C}[x]$, with $x_{kj} \circ s = \sum_{\ell} s_{k\ell} x_{\ell j}$ and $g(\ldots, x_{kj}, \ldots) \circ s = g(\ldots, x_{kj} \circ s, \ldots)$. These imply that $g \circ st = (g \circ s) \circ t$, for all polynomials $g \in \mathbb{C}[x]$ and elements $s, t \in G$. It turns out that the relation between x_{kj} and X_{kj} is given by $x_{kj} \circ s^{-1} = X_{kj}$, so that

$$g(\ldots, x_{kj}, \ldots) \circ s^{-1} = g(\ldots, X_{kj}, \ldots),$$

for all polynomials $g \in \mathbb{C}[x]$ and $s \in G$. Deruyts controlled all of this information by his use of the transformed coefficients and variables A_{ij} and $X_{k\ell}$; the left and right G-actions on the polynomial algebras $\mathbb{C}[a]$ and $\mathbb{C}[x]$ are implicit in his approach.

Combining the G-actions on $\mathbb{C}[a]$ and $\mathbb{C}[x]$, one obtains a G-action on the polynomial algebra $\mathbb{C}[a, x]$. The algebra $\mathbb{C}[a, x]$ is isomorphic to the tensor product algebra $\mathbb{C}[a] \otimes \mathbb{C}[x]$, with generators, as a vector space over $\mathbb{C}$, of the form $f \otimes g$, for $f \in \mathbb{C}[a]$ and $g \in \mathbb{C}[x]$. The action of $s \in G$ on a generator $f \otimes g$ is defined by

$$s \circ (f \otimes g) = (s \circ f) \otimes (g \circ s^{-1}),$$

using the G-actions on $\mathbb{C}[a]$ and $\mathbb{C}[x]$ introduced in the preceding paragraph. The result is a left G-action on the polynomial algebra $\mathbb{C}[a, x]$, given by

$$\varphi \to s \circ \varphi \circ s^{-1},$$

for each $s \in G$ and $\varphi \in \mathbb{C}[a, x]$. From Deruyts's viewpoint, this action carries a polynomial $\varphi(\ldots, a_{ij}, \ldots, x_{k\ell}, \ldots)$ to $\varphi(\ldots, A_{ij}, \ldots, X_{k\ell}, \ldots)$, where A_{ij} and $X_{k\ell}$ are the transformed coefficients and variables resulting from the action of a linear substitution corresponding to the element $s \in G$.

In this context, an element $f \in \mathbb{C}[a]$ is an *invariant* of weight w, for an integer w, if $s \circ f = |s|^w f$ for all $s \in G$. In a subtle extension of the notion of invariant, Deruyts defined a polynomial $f \in \mathbb{C}[a]$ to be a *semi-invariant* of weight $\pi = (\pi_1, \ldots, \pi_n)$, with $\pi_j \in \mathbb{Z}$ for each j, if $s \circ f = d_1^{\pi_1} \cdots d_n^{\pi_n} f$ for each element s belonging to the group B of upper triangular matrices[8] in G with diagonal entries $d_1, \ldots, d_n$. A polynomial $\varphi \in \mathbb{C}[a, x]$ is called a *covariant* of weight w if $s \circ \varphi \circ s^{-1} = |s|^w \varphi$ for all $s \in G$. In particular, φ is a covariant of weight zero if and only if $s \circ \varphi \circ s^{-1} = \varphi$ for all $s \in G$.

The connection of all this to polynomial representations of G comes about in the following way. Each element $J \in \mathbb{C}[a, x]$ can be written (not necessarily uniquely) in the form

$$J = f_1 g_1 + \cdots + f_r g_r,$$

with $f_i \in \mathbb{C}[a]$ and $g_j \in \mathbb{C}[x]$ for each i and j. Deruyts proved that, if the number of summands r is minimal among all such expansions of the fixed polynomial J, then the set of linear combinations with coefficients in $\mathbb{C}$ of the polynomials $f_i, 1 \le i \le r$, is a uniquely determined subspace of $\mathbb{C}[a]$, of dimension r. Similarly, the set of $\mathbb{C}$-linear combinations of the polynomials $g_1, \ldots, g_r$, again assuming that r is minimal, is a uniquely determined subspace of $\mathbb{C}[x]$, of dimension r. These subspaces, associated with each covariant J of weight zero, will be denoted by $L(J)$ and $R(J)$, respectively. Moreover, he proved that, if J is a covariant of weight zero, then the subspaces $L(J)$ and $R(J)$ are stable under the left and right actions

---

[8]Today, B is called a *Borel subgroup* of G.

of G on the polynomial algebras $\mathbb{C}[a]$ and $\mathbb{C}[x]$, respectively. Now let S be the matrix of the linear transformation defined by the left action of $s \in G$ on the vector space $L(J)$, with respect to some basis of $L(J)$. Then it follows that $s \to S$ is a polynomial representation of G, which will be called a *representation afforded by* $L(J)$. Similarly, there is a polynomial representation of G afforded by $R(J)$, for each covariant J of weight zero.

The question of which of the polynomial representations afforded by the spaces $L(J)$ are irreducible required, in Deruyts's approach, a further analysis of the covariants of weight zero. For this purpose, he defined a map σ from the polynomial algebra $\mathbb{C}[a]$ to $\mathbb{C}[a,x]$, as follows. One first defines

$$\sigma(a_{jk}) = \sum_{\ell} a_{j\ell} x_{\ell k} \in \mathbb{C}[a,x],$$

and then extends the action of σ to polynomials $f \in \mathbb{C}[a]$, by setting

$$\sigma(f(\ldots, a_{jk}, \ldots)) = f(\ldots, \sigma(a_{jk}), \ldots) \in \mathbb{C}[a,x].$$

This amounts to letting an element $s \in G$ act on a_{jk}, as defined above, so that

$$s \circ a_{jk} = \sum_{\ell} s_{\ell k} a_{j\ell},$$

and then replacing the coefficients $s_{\ell k}$ by the variables $x_{\ell k}$. The resulting element of $\mathbb{C}[a,x]$ is independent of the choice of $s \in G$, and the map σ is obtained by extending this process to polynomials $f \in \mathbb{C}[a]$. If a polynomial φ is in the range of σ, with $\sigma(f) = \varphi$ for $f \in \mathbb{C}[a]$, then Deruyts called f a *source* of φ. A crucial step in his theory was the theorem that $\sigma : \mathbb{C}[a] \to \mathbb{C}[a,x]$ is a bijection from $\mathbb{C}[a]$ to the set of covariants of weight zero in $\mathbb{C}[a,x]$ ([**105**], p. 67; [**166**], Proposition 7.5). Using these concepts, Deruyts defined a *primary covariant* to be a covariant of weight zero whose source is a nonzero semi-invariant.

Some of the main consequences of Deruyts's theory for polynomial representations of $G = GL_n(\mathbb{C})$ can be stated in the following

THEOREM 1.1. (*i*) *Let* $\varphi \in \mathbb{C}[a,x]$ *be a primary covariant; then the polynomial representation of* G *afforded by* $L(\varphi)$ *is irreducible, and every irreducible polynomial representation of* G *is obtained in this way.*

(*ii*) *Every irreducible polynomial representation of* G *is afforded by a* G-*invariant subspace of* $\mathbb{C}[a]$ *generated by polynomials of the form* $g \circ a(\pi), g \in G$, *where* $a(\pi)$ *is a semi-invariant of weight* $\pi = (\pi_1, \ldots, \pi_n)$, *with* $\pi_1 \geq \pi_2 \geq \cdots$. *The irreducible polynomial representations of* G *are parametrized by weights of the form* $\pi = (\pi_1, \ldots, \pi_n)$, *with* $\pi_1 \geq \pi_2 \geq \cdots$.

(*iii*) *Every polynomial representation of* G *is completely reducible.*

A proof of the theorem, with precise information as to the relation between each step of the proof and a corresponding result in Deruyts's memoir, is contained in Green's paper [**166**].

Another noteworthy result in Deruyts's memoir was a formula for the dimension of the weight spaces in the irreducible representations described in part (*ii*) of the preceding theorem; this can be viewed as a character formula for the irreducible representations. In more detail, let V be the vector space affording such a representation, generated by the G-translates of a semi-invariant of weight π. The

α-weight space V^α, for an n-tuple $\alpha = (\alpha_1, \ldots, \alpha_n)$ of nonnegative integers, is the subspace of V consisting of all vectors v such that

$$t \circ v = t_1^{\alpha_1} \cdots t_n^{\alpha_n} v,$$

for all diagonal matrices $t \in G$ with diagonal entries $t_1, \ldots, t_n$. Deruyts obtained a formula for

$$\sum_\alpha \dim(V^\alpha) X_1^{\alpha_1} \cdots X_n^{\alpha_n},$$

which can be viewed as a generating function for the dimensions of the weight spaces. The preceding expression, in the language of representation theory, is the formal character of the polynomial representation of G afforded by V. Deruyts proved that

$$\sum_\alpha \dim(V^\alpha) X_1^{\alpha_1} \cdots X_n^{\alpha_n} = |p_{\pi_i - i + j}|,$$

where the entries of the determinant are the complete symmetric functions[9] $p_s(X_1, \ldots, X_n)$, of degree $s = 1, 2, \ldots$, in $X_1, \ldots, X_n$. In his dissertation (see §3), Schur proved a character formula for the irreducible polynomial representations of G, which is equivalent to the preceding formula of Deruyts. An outline of Deruyts's proof of his formula, in modern language, was given in a second paper [167] by J. A. Green on the subject, published in 1996.

Green characterized Deruyts's achievement in relation to modern methods this way:

> Deruyts did not talk about representations, or anything like a *matrix* representation. But he got very close to the idea of a *KG*-module (in case $K = \mathbb{C}$, $G = GL_n(\mathbb{C})$)—see p. 262 of "Classical Invariants" [166]—and he had a technique which, in effect, allowed him to describe the submodule (of a *KG*-module W, say) generated by an element $w \in W$. Another striking thing is that Deruyts defines the irreducible polynomial modules for $GL_n(\mathbb{C})$ by "inducing up" an irreducible $\mathbb{C}B$-module, where B is a Borel subgroup of $GL_n(\mathbb{C})$. In other words, he anticipates the method now used to get irreducible rational representations of an arbitrary reductive algebraic group (char. zero). So Deruyts did anticipate all the main findings of Schur's 1901 thesis, but his methods are totally different from Schur's (the symmetric group plays no part in Deruyts's paper).
>
> Schur has two references to Deruyts in his thesis: on pages 12 and 14. But these are to two minor papers of Deruyts which appeared in 1896, and I have no evidence that Schur was aware of Deruyts's major paper of 1892. What a pity![10]

2. Polynomial Representations of GL_n: Issai Schur

Apparently unaware of Deruyts's work on the subject, Schur launched his own approach to the polynomial representations of $GL_n(\mathbb{C})$ in his dissertation ([247], 1901), entitled *Über eine Klasse von Matrizen, die sich einer gegebenen Matrix*

---

[9]The complete symmetric function p_s is the sum of all monomials of degree s, in the given set of indeterminates.

[10]From a letter to the author dated August 30, 1991. Used with permission.

zuordnen lassen, and followed it up with a second paper ([**259**], 1927) which contains what is sometimes referred to today as the "Schur-Weyl reciprocity between representations of the general linear group and representations of the symmetric group."

In his dissertation, Schur stated the problem as follows. Let $A = (a_{ij})$ and $B = (b_{ij})$ be $n \times n$ matrices over $\mathbb{C}$, whose entries are independent variables, and let

$$C = (c_{k\ell}), \ c_{ik} = a_{i1}b_{1k} + \cdots + a_{in}b_{nk}$$

be the product matrix. He called an $r \times r$ matrix $T(A)$ an *invariant form* of *degree* r provided that (i) the r^2 entries of the matrix $T(A)$ are polynomials in the n^2 variables a_{ij}; and (ii), if the variables a_{ij} are replaced by b_{ij} and c_{ij}, resulting in matrices $T(B)$ and $T(C)$, then

$$T(C) = T(AB) = T(A)T(B).$$

He called $T(A)$ *homogeneous*, of *order* m, if the entries of $T(A)$ are homogeneous polynomials of degree m. He gave the determinant $|A|$ as an example of an invariant form of degree 1 and order n.

He was well aware that the concept of an invariant form was connected with representations of groups. If the entries of A are taken as arbitrary complex numbers, subject only to the condition that the determinant $|A| \neq 0$,[11] then an invariant form $A \to T(A)$ defines a representation of the infinite group $GL_n(\mathbb{C})$. In what follows, we shall concentrate on this interpretation of invariant forms, allowing the coefficients a_{ij} to vary arbitrarily in $\mathbb{C}$, and call the representations of $GL_n(\mathbb{C})$ obtained from them *polynomial representations*.

Schur defined equivalence of two invariant forms, or polynomial representations, as in the case of representations of finite groups. A polynomial representation $A \to T(A)$ was called irreducible (or primitive) if it is not equivalent to a polynomial representation of the form

$$\begin{pmatrix} T_1(A) & 0 \\ 0 & T_2(A) \end{pmatrix}$$

with T_1 and T_2 polynomial representations of lower degree.

His stated objectives were to prove that two polynomial representations are equivalent if and only if they have the same trace, and that the number of inequivalent irreducible polynomial representations of order m is equal to the number of partitions of m. In addition, he wanted to calculate the trace functions, or characters, of the irreducible polynomial representations. He planned to accomplish these aims by setting up a correspondence between homogeneous polynomial representations of order m and representations of the symmetric group S_m, and then to apply Frobenius's work on the characters of symmetric groups ([**137**], Chapter II, §5), published the preceding year. Although he emphasized group representations from the beginning, in contrast to Deruyts, this did not make things any simpler for him, as practically nothing was known at the time about representations of infinite groups such as GL_n.

Before plunging into the mathematics Schur developed to carry out his plan, we include a few words from two contributors to representation theory later in the 20th century on the influence of these works of Schur, and their own impressions

---

[11]The condition $|A| \neq 0$ is automatically satisfied in case the coefficients a_{ij} are algebraically independent over $\mathbb{Q}$, as in the definition of invariant form.

of them.[12] In the preface of his book *The Classical Groups* ([**287**], 1939), which he had dedicated to Schur, Hermann Weyl wrote:

> Ever since the year 1925, when I succeeded in determining the characters of the semi-simple continuous groups by a combination of E. Cartan's infinitesimal methods and I. Schur's integral procedures, I have looked toward the goal of deriving the decisive results for the most important of these groups by direct algebraic construction, in particular for the full group of all non-singular transformations and for the orthogonal group.[13]

In the case of the full linear group, the algebraic constructions to which he referred were those in Schur's dissertation and his second paper. Weyl continued with what can be taken as the philosophy of this sort of mathematics:

> As one sees from the above description, the subject of this book is rather special. Important though the general concepts and propositions may be with which the modern industrious passion for axiomatizing and generalizing has presented us, in algebra perhaps more than anywhere else, nevertheless I am convinced that the special problems in all their complexity constitute the stock and core of mathematics; and to master their difficulties requires on the whole the greater labor. The border line is of course vague and fluctuating. But quite intentionally scarcely more than two pages are devoted to the general theory of group representations, while the applications of this theory to the particular groups that come under consideration occupies at least fifty times as much space.[14]

While Weyl included many results from Schur's dissertation in his book, he gave the most attention to Schur's second paper [**259**].

In the introduction to his lecture notes volume *Polynomial Representations of GL_n* ([**165**], 1980), J. A. Green commented on the situation as follows:

> This pioneering achievement of Schur was one of the main inspirations for Hermann Weyl's monumental researches on the representation theory of semi-simple Lie groups [**284**]. Of course Weyl's methods, based on the representation theory of the Lie algebra of the Lie group Γ, and the possibility of integrating over a compact form of Γ, were very different from the purely algebraic methods of Schur's dissertation; in particular, Weyl's general theory had nothing to correspond to the symmetric group S_m. In 1927 Schur published another paper [**259**] on $GL_n(\mathbb{C})$, which has deservedly become a classic. In this he exploited the "dual" actions of $GL_n(\mathbb{C})$ and S_m on the mth tensor space $E^{\otimes m}$ to rederive all the results of his 1901 dissertation in a new and very economical way. Weyl publicized the method of Schur's 1927 paper, with its attractive use of the "double centralizer property" in his influential book *The Classical Groups* [**287**]. In fact the exposition in Chapters 3(B) and 4 of that book has become the

---

[12]See also the report by P. Slodowy [**264**] on the contributions of Hurwitz, Schur, and Weyl to the early development of the representation theory of semisimple Lie groups.

[13]Weyl. H., *The Classical Groups*, Copyright©1946 by Princeton University Press. Reprinted by permission of Princeton University Press.

[14]Ibid.

standard treatment of polynomial representations of $GL_n(\mathbb{C})$ (and, incidentally, of Alfred Young's representation theory of the symmetric group S_m), and perhaps this explains the comparative neglect of Schur's work of 1901. I think this neglect is a pity, because the methods of this earlier work are in some ways very much in keeping with the present-day ideas on representations of algebraic groups.[15]

We proceed now to a discussion of the main steps in Schur's dissertation. All of them are important, in one way or another, in current work on representation theory, and we include interpretations of them from a modern point of view whenever possible. A full account of Schur's two papers and subsequent developments up to 1980 has been given by J. A. Green [**165**], in a slender volume of lecture notes that has been at least partly responsible for a revival of interest in the whole area, and, as W. Ledermann remarked in [**205**], would have "intrigued and delighted Schur."

We shall restrict ourselves to polynomial representations $A \to T(A)$ of $G = GL_n(\mathbb{C})$, which are homogeneous of a fixed order m, and satisfy the condition that all the matrices $T(A)$ are invertible. The reduction of an arbitrary polynomial representation to the homogeneous case is easily done, and goes as follows. For an arbitrary polynomial representation $A \to T(A)$ of degree r, let

$$T(xE_n) = x^d C_0 + \cdots + C_d,$$

where $x \in \mathbb{C}$ is a variable, E_n is the identity matrix, d is the maximum degree of the polynomials occurring as entries of $T(A)$, and $C_0, \ldots, C_d$ are constant $r \times r$ matrices, independent of x. Upon writing out the matrix relations resulting from the equation $T(xyE_n) = T(xE_n)T(yE_n)$, with y another variable, and comparing coefficients of monomials in x and y, Schur obtained the formulas

$$C_i^2 = C_i \text{ and } C_i C_j = 0 \text{ if } i \neq j,$$

for all i, j. As the matrices $T(A)$ and C_i commute, it follows from linear algebra that the given representation is equivalent to a representation of the form

$$A \to \begin{pmatrix} T_0(A) & \ldots & 0 \\ \ldots & \ldots & \ldots \\ 0 & \ldots & T_d(A) \end{pmatrix},$$

where each submatrix $T_i(A)$ is either zero for all $A \in G$, or defines a homogeneous polynomial representation of order i. As a consequence, it follows that irreducible polynomial representations are homogeneous.

Symmetric functions had been applied with brilliant success in Frobenius's paper on the characters of the symmetric group (Chapter II, §5). Schur brought them into the picture in his set-up with the following result on the trace of a homogeneous polynomial representation.

THEOREM 2.1. *Let $A \to T(A)$ be a homogeneous polynomial representation of G, of order m. Then the trace function*

$$\text{Trace}(A) = \Phi(\omega_1, \ldots, \omega_n)$$

is a homogeneous symmetric function of degree m in the eigenvalues $\omega_1, \ldots, \omega_n$ of the matrix A.

---

[15]Green, J. A., "Polynomial Representations of GL_n," *Lecture Notes in Mathematics* **830**, Springer-Verlag, Berlin, Heidelberg, 1980. Used with permission.

The eigenvalues of A are the roots of the characteristic polynomial

$$|xE_n - A| = x^n - c_1 x^{n-1} + \cdots \pm c_n.$$

Schur then argued that A is similar to the diagonal matrix

$$D = \begin{pmatrix} \omega_1 & \cdots & \cdots \\ \cdots & \cdots & \cdots \\ \cdots & \cdots & \omega_n \end{pmatrix},$$

and to the companion matrix,

$$C = \begin{pmatrix} c_1 & -c_2 & \cdots & \pm c_n \\ 1 & 0 & \cdots & 0 \\ \cdots & \cdots & \cdots & \cdots \\ 0 & \cdots & 1 & 0 \end{pmatrix}$$

of its characteristic polynomial. At first this seems strange, as it is not true unless A is diagonalizable. Then one recalls that in his definition of invariant form, Schur considered a matrix A whose entries are algebraically independent over $\mathbb{Q}$. Such a matrix has n distinct eigenvalues, and is diagonalizable. In that case, A is similar to D and to C. Then $P^{-1}AP = D$ and $Q^{-1}AQ = C$, for invertible matrices P and Q. Consequently, $T(P)^{-1}T(A)T(P) = T(D)$ and $T(Q)^{-1}T(A)T(Q) = T(C)$, so the trace of $T(A)$ is equal to the trace of $T(D)$ and to the trace of $T(C)$. The first statement shows that the trace of $T(A)$ is a homogeneous polynomial of degree m in the eigenvalues ω_i, and the second implies that it is a symmetric function of the eigenvalues, as the coefficients c_j of the characteristic polynomial have this property. This completes the proof for matrices having distinct eigenvalues. A further argument (not given here) shows that the result holds for arbitrary matrices A (see [**165**], p. 43).

The homogeneous symmetric function $\Phi = \Phi(\omega_1, \ldots, \omega_n)$, defined in the preceding theorem, was called the *characteristic* of the homogeneous polynomial representation $A \rightarrow T(A)$. The calculation of the characteristics of the irreducible polynomial representations was one of his main objectives.

Schur obtained a connection between homogeneous polynomial representations of G of order m and representations of the symmetric group S_m by relating both to representations of a certain algebra, today called the *Schur algebra*. These connections were made in the following way. Let $A \rightarrow T(A)$ be a homogeneous polynomial representation of degree r and order m; then $T(A)$ can be expressed in the form:

(17) $$T(A) = \sum \begin{bmatrix} i_1 & \cdots & i_m \\ j_1 & \cdots & j_m \end{bmatrix} a_{i_1 j_1} \cdots a_{i_m j_m},$$

where the terms a_{ij} in the formula are the coefficients of the matrix A, for $A \in G$, while the expressions

$$\begin{bmatrix} i_1 & \cdots & i_m \\ j_1 & \cdots & j_m \end{bmatrix},$$

in Schur's notation, are constant $r \times r$ matrices, independent of the matrices A, with the property that

$$\begin{bmatrix} i_1 & \cdots & i_m \\ j_1 & \cdots & j_m \end{bmatrix} = \begin{bmatrix} i'_1 & \cdots & i'_m \\ j'_1 & \cdots & j'_m \end{bmatrix}$$

whenever the products $a_{i_1 j_1} \cdots a_{i_m j_m}$ and $a_{i'_1 j'_1} \cdots a_{i'_m j'_m}$ are obtained from one another by a rearrangement of the factors. The sum is taken over all different products $a_{i_1 j_1} \cdots a_{i_m j_m}$ of the n^2 variables a_{ij}. In what follows, the matrices

$$\begin{bmatrix} i_1 & \cdots & i_m \\ j_1 & \cdots & j_m \end{bmatrix},$$

will be denoted by $S_{\mathbf{i},\mathbf{j}}$, with $\mathbf{i} = (i_1, \ldots, i_m)$, etc.

It was clear the polynomial representation $A \to T(A)$ can be recovered from a knowledge of the matrices $S_{\mathbf{i},\mathbf{j}}$. A decisive observation in Schur's program was that the formula (17) defines an equivalence of categories, in modern terminology, from the category of polynomial representations of order m to the category of representations of the Schur algebra. This required a knowledge of the multiplicative properties of the matrices $S_{\mathbf{i},\mathbf{j}}$, which define the structure of the Schur algebra.

At this point it is useful to introduce some notation. Let $\mathcal{I}_m$ denote the set of m-tuples $\mathbf{i} = (i_1, \ldots, i_m)$ with entries from the set of integers $i, 1 \leq i \leq n$, and $\mathbf{i} \sim \mathbf{i}'$ for the equivalence relation given by place permutations of the m-tuples $\mathbf{i}$, that is, $\mathbf{i} \sim \mathbf{i}'$ whenever $\mathbf{i}' = (i_{\pi(1)}, \ldots, i_{\pi(m)})$, for $\mathbf{i} = (i_1, \ldots, i_m)$ and π an element of the symmetric group S_m. The symmetric group S_m also acts on the product set $\mathcal{I}_m \times \mathcal{I}_m$, and the resulting equivalence relation is denoted by $(\mathbf{i},\mathbf{j}) \sim (\mathbf{i}',\mathbf{j}')$.

THEOREM 2.2. *The matrices $S_{\mathbf{i},\mathbf{j}}$ appearing in (17) satisfy the following relations:*

(i) $S_{\mathbf{i},\mathbf{j}} = S_{\mathbf{i}',\mathbf{j}'}$ whenever $(\mathbf{i},\mathbf{j}) \sim (\mathbf{i}',\mathbf{j}')$; and

(ii) $S_{\mathbf{i},\mathbf{j}} S_{\mathbf{k},\ell} = \sum_{\mathbf{p},\mathbf{q}} h_{\mathbf{p},\mathbf{q},\mathbf{i},\mathbf{j},\mathbf{k},\ell} S_{\mathbf{p},\mathbf{q}}$, where the sum is over representatives of the equivalence classes in $\mathcal{I}_m \times \mathcal{I}_m$, and the coefficient $h_{\mathbf{p},\mathbf{q},\mathbf{i},\mathbf{j},\mathbf{k},\ell}$ is the cardinal number of the set of elements $\mathbf{r} \in \mathcal{I}_m$ such that $(\mathbf{i},\mathbf{j}) \sim (\mathbf{p},\mathbf{r})$ and $(\mathbf{k},\ell) \sim (\mathbf{r},\mathbf{q})$.

Schur derived the properties of the matrices $S_{\mathbf{i},\mathbf{j}}$, expressed in a different way, from equation (17) and the fact that $A \to T(A)$ is a representation ([**247**], §8, 9). The statement above is due to Green ([**165**]), and can be proved in the same way.

The fact that the relations expressed in the preceding theorem define the structure of an algebra was not stated explicitly by Schur; this step was taken later. The *Schur algebra* Σ_m is defined as the associative algebra with basis elements $\sigma_{\mathbf{i},\mathbf{j}}$ indexed by pairs $(\mathbf{i},\mathbf{j}) \in \mathcal{I}_m \times \mathcal{I}_m$. It is assumed that the basis elements satisfy the equality relation,

$$\sigma_{\mathbf{i},\mathbf{j}} = \sigma_{\mathbf{i}',\mathbf{j}'} \text{ if and only if } (\mathbf{i},\mathbf{j}) \sim (\mathbf{i}',\mathbf{j}'),$$

with the formula for multiplication of two basis elements as in part *(ii)* of the preceding theorem.

THEOREM 2.3. *There is a one-to-one correspondence between homogeneous polynomial representations $A \to T(A)$ of order m and representations $\sigma_{\mathbf{i},\mathbf{j}} \to S_{\mathbf{i},\mathbf{j}}$ of the Schur algebra Σ_m, given by (17). The correspondence preserves the relation of equivalence of two representations, irreducibility, and direct sum decompositions. Otherwise expressed, the formula (17) defines an equivalence of categories from the category of homogeneous polynomial representations of G of order m to the category of finite dimensional left modules for the Schur algebra Σ_m.*

Schur's version of the theorem was proved using again (17) and the multiplicative properties of the matrices $S_{\mathbf{i},\mathbf{j}}$; for a modern proof, see ([**165**], p. 25).

The connection with representations of the symmetric group was proved under the assumption that $m \leq n$; at the end of the paper (§32) he showed how to reduce the case $m > n$ to the first one.

In the discussion given here, we consider only the case $m \leq n$; in that situation, the m-tuple $\mathbf{i}_0 = (1, 2, \ldots, m) \in \mathcal{I}_m$. For a homogeneous polynomial representation $A \to T(A)$ of order m and degree r, Schur defined a corresponding representation of S_m by the entirely natural, in his notation, assignment of $\pi \in S_m$ to the $r \times r$-matrix $A(\pi)$ given by

$$\pi = \begin{pmatrix} 1 & 2 & \cdots & m \\ i_1 & i_2 & \cdots & i_m \end{pmatrix} \to A(\pi) = \begin{bmatrix} 1 & 2 & \cdots & m \\ i_1 & i_2 & \cdots & i_m \end{bmatrix} = S_{\mathbf{i}_0, \mathbf{i}},$$

for an arbitrary permutation $i_1, i_2, \ldots$ of $1, 2, \ldots, m$. It follows easily, using Theorem 2.2, that

$$A(\pi)A(\rho) = A(\pi\rho),$$

for all $\pi, \rho \in S_m$. There is a subtlety, however, in that the matrix $A(1) = S_{\mathbf{i}_0, \mathbf{i}_0}$ is not necessarily the $r \times r$ identity matrix; in fact it is not even clear that it is different from zero, so $\pi \to A(\pi)$ is not necessarily a representation. Schur proved that $S_{\mathbf{i}_0, \mathbf{i}_0} \neq 0$ by noting that one has, by Theorem 2.2,

$$S_{\mathbf{i}, \mathbf{j}} = S_{\mathbf{i}, \mathbf{i}_0} S_{\mathbf{i}_0, \mathbf{i}_0} S_{\mathbf{i}_0, \mathbf{j}}$$

for all $\mathbf{i}, \mathbf{j} \in \mathcal{I}_m$, so that if $S_{\mathbf{i}_0, \mathbf{i}_0} = 0$ then $S_{\mathbf{i}, \mathbf{j}} = 0$ for all $\mathbf{i}$ and $\mathbf{j}$. This would imply that $T(A) = 0$ for all $A \in G$, by (17), contrary to assumption. Moreover, a further application of Theorem 2.2 shows that

$$S_{\mathbf{i}_0, \mathbf{i}_0}^2 = S_{\mathbf{i}_0, \mathbf{i}_0}.$$

Then $S_{\mathbf{i}_0, \mathbf{i}_0}$ is similar to a matrix of the form

$$\begin{pmatrix} E_d & 0 \\ 0 & 0 \end{pmatrix}$$

for a nonzero $d \times d$ identity matrix E_d. The same similarity transformation carries

$$A(\pi) = S_{\mathbf{i}_0, \mathbf{i}}$$

to a matrix of the form

$$\begin{pmatrix} A'(\pi) & 0 \\ 0 & 0 \end{pmatrix}$$

for a $d \times d$ submatrix $A'(\pi)$. It follows from all this that $\pi \to A'(\pi)$ is a representation of S_m of degree $d \neq 0$, which Schur called the *representation of the symmetric group corresponding to the invariant operation* $T(A)$ ([**247**], §10).

With this preparation, Schur's goals were now in sight. He proved (§§16, 17, Theorems VII and IX):

THEOREM 2.4. (*i*) *Each homogeneous polynomial representation of order $m \leq n$ is completely reducible. It is irreducible if and only if the corresponding representation of S_m is irreducible.*

(*ii*) *Two homogeneous polynomial representations of order $m \leq n$ are equivalent if and only if the corresponding representations of S_m are equivalent.*

A modern proof of the theorem will be given shortly. We note, first, as Schur did, a consequence of the theorem.

COROLLARY 2.1. *The number of inequivalent irreducible homogeneous polynomial representations of order m is equal to the number of partitions of m.*

The corollary follows from parts (i) and (ii) of the theorem, together with Frobenius's theorem classifying the irreducible characters of S_m (Chapter II, §5).

Schur gave three proofs of the first statement of the theorem, one in his dissertation (Theorem IX), and two more, one algebraic, and one analytic, in his second paper [259]. In the analytic proof, he carried over his proof of complete reducibility for representations of finite groups ([249]; see Chapter IV §2, Theorem 2.3) to polynomial representations, replacing sums over a finite group by integration over the compact subgroup of a linear group consisting of unitary matrices. This procedure was later called the "unitarian trick" by Weyl ([287], Chapter VIII, §11). It had been applied to the representation theory of general semisimple Lie groups by Weyl in his series of papers on the subject [284]. We shall give here an updated version of the algebraic proof in the second paper, following Green ([165], §2.6); this contains the now famous reciprocity between the tensor representation of G and representations of the symmetric group referred to earlier in this section (in another form, this reciprocity appears again, in §3, Lemma 3.1).

By Theorem 2.3, it is sufficient to prove that the Schur algebra Σ_m is a semisimple algebra.[16] The proof is based on an analysis of the tensor representation,

$$A \to A \otimes A \otimes \cdots \otimes A \ (m \text{ factors})$$

of G, first defined by Schur in his dissertation, as an example of a homogeneous polynomial representation of G of order m. In modern terminology, G is viewed as the group of invertible linear transformations on an n-dimensional vector space E, with basis $e_1, \ldots, e_n$. Then G acts on the tensor space

$$E^{\otimes m} = E \otimes \cdots \otimes E \ (m \text{ factors}),$$

and affords the tensor representation $A \to A \otimes \cdots \otimes A$ considered by Schur. A basis of the vector space $E^{\otimes m}$ consists of the tensors

$$e_{\mathbf{i}} = e_{i_1} \otimes \cdots \otimes e_{i_m},$$

with $\mathbf{i} = (i_1, \ldots, i_m) \in \mathcal{I}_m$. There is a right action of the symmetric group S_m, and by the group algebra $\mathbb{C}S_m$, on the tensor space $E^{\otimes m}$ given by place permutations:

$$(e_{\mathbf{i}})\pi = e_{i_{\pi(1)}} \otimes \cdots \otimes e_{i_{\pi(m)}},$$

for $\mathbf{i} \in \mathcal{I}_m$, which is easily shown to commute with the action of the transformations $A \otimes \cdots \otimes A$, for $A \in G$.

LEMMA 2.1. *The Schur algebra Σ_m acts faithfully on the tensor space $E^{\otimes m}$, and is isomorphic to the centralizer algebra of the S_m-action on $E^{\otimes m}$:*

$$\Sigma_m \cong \operatorname{End}_{\mathbb{C}S_m} E^{\otimes m}.$$

The action of the Schur algebra Σ_m on $E^{\otimes m}$ is another consequence of the formula (17), applied to the homogeneous polynomial representation $A \to A \otimes \cdots \otimes A$, so that

$$A \otimes \cdots \otimes A = \sum S_{\mathbf{i},\mathbf{j}} a_{i_1 j_1} \cdots a_{i_m j_m},$$

---

[16]Schur proved this fact in his second paper, §4, using the discriminant criterion for semisimplicity (see Chapter II, §3). He attributed the discriminant criterion to Molien [215], and remarked that a simpler proof of it had been obtained in the dissertation of his doctoral student M. Herzberger.

Hermann Weyl (1885-1955)

for $A = (a_{ij})$. By Theorem 2.2, the action of the Schur algebra can be expressed in terms of the elementary matrices $T_{\mathbf{i},\mathbf{j}}$ associated with the basis elements $e_{\mathbf{i}}$ by the formulas

$$S_{\mathbf{i},\mathbf{j}} = \sum_{(\mathbf{i}',\mathbf{j}') \in \mathcal{O}} T_{\mathbf{i}',\mathbf{j}'},$$

where $(\mathbf{i},\mathbf{j})$ is a representative of an orbit $\mathcal{O}$ of the action of S_m on $\mathcal{I}_m \times \mathcal{I}_m$. An element

$$T = \sum \alpha_{\mathbf{i},\mathbf{j}} T_{\mathbf{i},\mathbf{j}} \in \text{End}_{\mathbb{C}}(E^{\otimes m})$$

commutes with the action of S_m if and only if $\alpha_{\mathbf{i},\mathbf{j}} = \alpha_{\mathbf{i}\pi,\mathbf{j}\pi}$ for all $\pi \in S_m$. Thus it is clear that the transformations $S_{\mathbf{i},\mathbf{j}}$ form a basis for the centralizer algebra $\text{End}_{\mathbb{C}S_m} E^{\otimes m}$, and that the map $\sigma_{\mathbf{i},\mathbf{j}} \to S_{\mathbf{i},\mathbf{j}}$ from Σ_m into $\text{End}_{\mathbb{C}}(E^{\otimes m})$ is an isomorphism. This completes the proof of the lemma.

We can now apply the lemma to prove that the Schur algebra Σ_m is semisimple. As S_m is a finite group, its representations are completely reducible, and the right $\mathbb{C}S_m$-module $E^{\otimes m}$ is completely reducible. By the lemma, $\Sigma_m \cong \text{End}_{\mathbb{C}S_m} E^{\otimes m}$, so the Schur algebra is isomorphic to the centralizer algebra of a completely reducible module, and is therefore a semisimple algebra.

The rest of the proof of the theorem is an application of the *Schur functor*, from representations of the Schur algebra Σ_m to representations of the symmetric group S_m ([**247**], §11 − 17). From a more general standpoint, this connection can be viewed today as a functor from left A-modules V, for a semisimple algebra A, to left eAe-modules eV, for a fixed nonzero idempotent e in A ([**165**], §6.2). In Schur's situation, A is the Schur algebra Σ_m, which is semisimple by the first part of the proof. As the hypothesis $m \le n$ remains in force, $\mathbf{i}_0 = (1, 2, \ldots, m) \in \mathcal{I}_m$. By Theorem 2.2, $\sigma_{\mathbf{i}_0,\mathbf{i}_0}$ is an idempotent in Σ_m, and it has been shown that $S_{\mathbf{i}_0,\mathbf{i}_0} \ne 0$ in every representation $\sigma_{\mathbf{i},\mathbf{j}} \to S_{\mathbf{i},\mathbf{j}}$ of the Schur algebra. This means that, for all left A-modules V, one has $eV \ne 0$, for the idempotent $e = \sigma_{\mathbf{i}_0,\mathbf{i}_0}$. Moreover the homogeneous polynomial representations of G of order m correspond to left modules V for $A = \Sigma_m$, such that $eV \ne 0$, as we have just shown.

We next turn our attention to the algebra $eAe = \sigma_{\mathbf{i}_0,\mathbf{i}_0} \Sigma_m \sigma_{\mathbf{i}_0,\mathbf{i}_0}$, and prove that it is isomorphic to the group algebra of the symmetric group $\mathbb{C}S_m$, so that left eAe-modules can be identified with representations of the symmetric group S_m. To obtain the isomorphism, one verifies first, using Theorem 2.2, that the algebra $\sigma_{\mathbf{i}_0,\mathbf{i}_0} \Sigma_m \sigma_{\mathbf{i}_0,\mathbf{i}_0}$ has a basis consisting of the elements $\sigma_{\mathbf{i}_0,\mathbf{i}}$ for permutations $\mathbf{i} = (i_1, \ldots, i_m)$ of the set $\{1, \ldots, m\}$. It is then not difficult to prove, using Theorem 2.2 again, that

$$\begin{pmatrix} 1 & \cdots & m \\ i_1 & \cdots & i_m \end{pmatrix} \to \sigma_{\mathbf{i}_0,\mathbf{i}}$$

is an isomorphism from the symmetric group S_m to a set of basis elements of the algebra eAe, and the isomorphism $eAe \cong \mathbb{C}S_m$ follows.

The next statement of the theorem, that a polynomial representation of order m is irreducible if and only if the corresponding representation of S_m is irreducible, follows from the well-known property of the functor $V \to eV$, that a left A-module V such that $eV \ne 0$ is irreducible if and only if eV is an irreducible eAe-module ([**165**], §6.2, or [**99**], §11D). The last part of the theorem follows from the properties of the functor $V \to eV$, and the fact that both algebras A and eAe are semisimple, so their modules are completely reducible.

John von Neumann (1903-1956)

In retrospect, the whole story so far, of the Schur algebra and the Schur functor, has unfolded from the layers of meaning which Schur uncovered, one by one, in the formula (17).

In Schur's second paper [**259**] on polynomial representations of $GL_n(\mathbb{C})$, and a sequel to it, published a year later [**260**], questions were also raised about analytic properties of representations of $GL_n(\mathbb{C})$ and other linear groups. These were important first steps towards understanding the connections between analytic and algebraic properties of representations of Lie groups, and arose from Schur's exchange of ideas with Hermann Weyl and John von Neumann.[17] In the introduction to [**259**], Schur wrote:

> Recently H. Weyl, in the course of his investigations on semisimple continuous groups [**284**], has arrived at new and elegant proofs of Theorems I-III [from Schur's dissertation]. His methods of proof, in which a modification of Hurwitz's integration process plays an essential role, gave at the same time an important extension of the results of my dissertation. In particular, Weyl showed that for the subgroup of $GL_n(\mathbb{C})$ consisting of all substitutions of determinant 1, there exist no representations by linear transformations other than those I have determined. As J. v. Neumann shows in the following article [**277**], the assumptions made concerning the [polynomial] dependence of the coefficients $c_{k\ell}$ [of $T(A)$] on the a_{ij} can be substantially weakened. It suffices, in particular, to take the $c_{k\ell}$ as continuous functions of the a_{ij}.[18]

In Schur's paper on the subject [**260**], representations $A \to T(A) = (c_{k\ell})$ were considered with the property that the $c_{k\ell}$ were continuous functions of the a_{ij}, for the groups, in modern notation, $GL_n(\mathbb{C})$, $GL_n(\mathbb{R})$, and their subgroups consisting of matrices of determinant one, $SL_n(\mathbb{C})$ and $SL_n(\mathbb{R})$. In each case, using algebraic methods as much as possible, he proved that the representations were completely reducible, and were either polynomial representations or generalizations of polynomial representations.

3. Schur's Character Formula

Schur's second paper on polynomial representations [**259**], published about 25 years after his dissertation, contained new and streamlined proofs, using tensor representations, of the main theorems of his dissertation. These included his formula for the characteristics of the irreducible polynomial representations, and the results, which turn out to be equivalent, that polynomial representations are determined up to equivalence by their characteristics, and are completely reducible (see Theorem 2.4 and the accompanying discussion in §2). Schur's second proof of the character formula is taken up in this section. Both of Schur's proofs of the character formula relied on the theory of symmetric functions, and are parallel, in a sense that will become clear, to Frobenius's application of symmetric functions to obtain the characters of the symmetric group (Chapter II, §5).

---

[17]J. von Neumann was a Privatdozent at the University of Berlin for three years (1927-1929).

[18]Schur, I., "Über die rationalen Darstellungen der allgemeinen linearen Gruppe," S'ber Akad. Wiss. Berlin (1927), pp. 58–75; Ges. Abh. III, pp. 68–85. Used with permission of Springer-Verlag.

As we know (see Theorem 2.1), the characteristics of all homogeneous polynomial representations of $G = GL_n(\mathbb{C})$ are themselves symmetric functions. Those attached to irreducible polynomial representations are known today as S-functions, or *Schur functions*, and play an important role in the modern theory of symmetric functions (see [210], Chapter 1, §3).

We begin with Schur's new proof, based on the Frobenius-Schur Theorem, of one of the results mentioned above.

THEOREM 3.1. *Two polynomial representations of the same degree are equivalent if and only if they have equal characteristics.*

From the preliminary discussion in §2, each polynomial representation is a direct sum of homogeneous ones, and these are completely reducible, by Theorem 2.4. The characteristic Φ of a polynomial representation $A \to T(A)$ is its trace function $\Phi(A) = \text{Trace}(T(A))$ (see the definition following Theorem 2.1). Now let $A \to T(A)$ and $A \to T'(A)$ be two polynomial representations of the same degree having equal characteristics. Then both are completely reducible, and can be expressed as direct sums of irreducible polynomial representations:

$$T = T_1 + \cdots + T_p, \ T' = T_1' + \cdots + T_q'.$$

The hypothesis implies that $\text{Trace}(T(A)) = \text{Trace}(T'(A))$ for all matrices $A \in G$. By the Frobenius-Schur Theorem (Chapter IV, Theorem 3.3), it follows that $p = q$, and that the irreducible components of T and T' are, up to equivalence, rearrangements of each other. It follows that T and T' are equivalent, completing the proof.

His first proof of the preceding result, in his dissertation, came at the end of a long journey. It required the connection between homogeneous polynomial representations and representations of the symmetric group (the Schur functor discussed in §2), and, in particular, a formula expressing the characteristic of a homogeneous polynomial representation in terms of the character of the corresponding representation of the symmetric group ([247], Theorem VIII, §16). Of course, that was before the Frobenius-Schur Theorem (Chapter IV, Theorem 3.3). He now realized that, with the intervention of the Frobenius-Schur Theorem, the preceding theorem and the complete reducibility of polynomial representations were equivalent statements: either one follows from the other ([259], p. 60).

The formula for the characteristics of the irreducible polynomial representations came in two parts. The first gave the characteristic of an arbitrary homogeneous representation of order m in terms of a corresponding character of the symmetric group:

THEOREM 3.2. *Let Φ be the characteristic of a homogeneous polynomial representation $A \to T(A)$ of order m. Then there exists a character χ of the symmetric group such that*

$$\Phi = \Phi(A) = \sum_j \frac{\chi_j}{\alpha_1! \cdots \alpha_m!} \left(\frac{s_1}{1}\right)^{\alpha_1} \cdots \left(\frac{s_m}{m}\right)^{\alpha_m},$$

where χ_j denotes the value of the character χ at an element of a conjugacy class C_j of S_m whose cycle decomposition has α_1 1-cycles, α_2 2-cycles, ..., and

$$s_i = \omega_1^i + \cdots + \omega_n^i = \text{Trace}(A^i),$$

for $i = 1, 2, \ldots$ and a matrix A with eigenvalues $\omega_1, \ldots, \omega_n$.

We shall follow Schur's proof closely. We continue to use, however, the modern notation introduced at the end of §2 for the tensor representation $A \to A^{(m)} = A \otimes \cdots \otimes A$ of G on the tensor space $E^{\otimes m}$, and the right action of $\pi \in S_m$ by place permutations on the basis elements $e_{i_1} \otimes \cdots \otimes e_{i_m}$ of the tensor space $E^{\otimes m}$. These define an action of $(A, \pi) \in G \times S_m$ on $E^{\otimes m}$, with $A \in G$ acting on the left by the tensor representation $A \to A^{(m)}$, and a commuting right action by $\pi \in S_m$.

The next result, from [259], §2, was of critical importance, and variations of it have been applied over and over again ever since to analyze the representations and characters of two commuting group actions on a vector space.

LEMMA 3.1. *(i) Let $\pi \to S^{(\lambda)}(\pi)$ be a complete set of inequivalent irreducible representations of S_m, indexed by partitions (λ) of m (see Chapter II, §5), and let f_λ be the degree of the representation $S^{(\lambda)}$. Then there exists a corresponding set of homogeneous polynomial representations of G, $A \to T^{(\lambda)}(A)$, of order m and degree g_λ, indexed by the same set of partitions, such that the action of $(A, \pi) \in G \times S_m$ on $E^{\otimes m}$ is given by*

$$(A, \pi) \to \sum_{(\lambda)} T^{(\lambda)}(A) \otimes S^{(\lambda)}(\pi).$$

(ii) The representations $A \to T^{(\lambda)}(A)$ are irreducible, inequivalent, and every irreducible polynomial representation of G of order m is equivalent to one of them.

(iii) Let $\Phi_m(A, \pi)$ be the trace of (A, π) on $E^{\otimes m}$, let $\Phi^{(\lambda)}$ be the characteristic of the representation $A \to T^{(\lambda)}(A)$ of G, and let $\chi^{(\lambda)}$ be the character of the representation $S^{(\lambda)}$ of S_m, for each (λ). Then

$$\Phi_m(A, \pi) = \sum_{(\lambda)} \Phi^{(\lambda)}(A)\chi^{(\lambda)}(\pi).$$

The first result was another application of Schur's Lemma. Schur had already used the idea in the proof of Theorem XI in his paper [249] (see the outline of the proof of Theorem 2.5 in Chapter IV, §2), and we recall it here. Assume the irreducible representation $S^{(\lambda)}$, of degree f_λ, occurs with multiplicity g_λ in the representation of S_m on $E^{\otimes m}$; then the representation of S_m, for a suitable basis of $E^{\otimes m}$, can be expressed as a direct sum:

$$\pi \to \sum_{(\lambda)} E_{g_\lambda} \otimes S^{(\lambda)}(\pi),$$

for $\pi \in S_m$. In other words, the matrix corresponding to π consists of blocks of the form $S^{(\lambda)}(\pi)$, repeated g_λ times, along the diagonal, and with zeros outside the diagonal blocks. Putting the matrices of the commuting tensor representation $A \to A^{(m)}$ of G in corresponding block form, it follows easily from Schur's Lemma that the representation of G on $E^{\otimes m}$ can be expressed as a direct sum

$$A \to \sum_{(\lambda)} T^{(\lambda)}(A) \otimes E_{f_\lambda},$$

where for each partition (λ), $A \to T^{(\lambda)}(A)$ is a homogeneous polynomial representation, of order m and degree g_λ, of G. Then the action of $(A, \pi) \in G \times S_m$ is given by

$$(A, \pi) \to \sum_{(\lambda)} (T^{(\lambda)}(A) \otimes E_{f_\lambda})(E_{g_\lambda} \otimes S^{(\lambda)}(\pi)) = \sum_{(\lambda)} T^{(\lambda)}(A) \otimes S^{(\lambda)}(\pi),$$

completing the proof of part (i). In case $n < m$, not all irreducible representations of the symmetric group appear as components of the tensor representation, so the preceding discussion has to be modified accordingly.

Part (ii) is, of course, an important result by itself. Schur's proof of it was as follows. From part (i), the character ζ of the representation of S_m on $E^{\otimes m}$ is given by

$$\zeta(\pi) = \sum_\lambda g_\lambda \chi^{(\lambda)}(\pi),$$

for $\pi \in S_m$. Using the orthogonality relations for the characters of S_m, he obtained

$$\frac{1}{m!} \sum_{\pi \in S_m} \zeta^2(\pi) = \sum g_\lambda^2.$$

(Here the fact that the characters of S_m are real valued comes into play, so that, for example, $\zeta(\pi) = \zeta(\pi^{-1})$.) He then proved, by computing the character values $\zeta(\pi)$, that the integer appearing in the preceding formula is equal to the number of distinct monomials of degree m in n^2 variables a_{ij}. Assume this result for a moment. The right hand side of the formula above is $\sum g_\lambda^2$, and is the sum of squares of the degrees of the representations $A \to T^{(\lambda)}(A)$ of G. These representations are homogeneous of order m, so their matrix coefficient functions are homogeneous polynomials of degree m in n^2 variables a_{ij}. It follows that the set of matrix coefficient functions of all the representations $A \to T^{(\lambda)}(A)$ are linearly independent. This implies that the representations $A \to T^{(\lambda)}(A)$ are irreducible, inequivalent, and form a complete set of irreducible homogeneous representations of G of order m, completing the proof of part (ii) once we fill in the missing step.

The missing step is to prove that the left side of the formula above,

$$\frac{1}{m!} \sum_{\pi \in S_m} \zeta^2(\pi),$$

is equal to the dimension of the space of homogeneous polynomials of degree m in n^2 variables a_{ij}. Instead of repeating Schur's proof by computing $\zeta^2(\pi)$ for all $\pi \in S_m$, we give a modern demonstration. By the orthogonality relations, the expression in question is equal to the multiplicity of the principal representation of S_m in the representation $V \otimes V^*$, where $V = E^{\otimes m}$ affords the representation of S_m by place permutations, and V^* affords the contragredient representation. This multiplicity is equal to the dimension of the vector space of S_m-invariant bilinear forms on the vector spaces V and V^*, and, by arguments we have used before (see the proof of Theorem 4.4 in Chapter III), is equal to the dimension of the centralizer algebra of the action of S_m on V. By the proof of Lemma 2.1, the dimension of this centralizer algebra is the number of distinct monomials of degree m in n^2 variables a_{ij}, and the proof is finished.

The proof of part (iii) is immediate from part (i), as the trace of a tensor product of matrices is the product of their traces.

One more preliminary result was needed.

LEMMA 3.2. *Let $A \in G$, and let $\pi \in S_m$ be a permutation factored as a product of disjoint cycles of orders $p, q, \ldots$. Then the trace of (A, π) on $E^{\otimes m}$ is given by the formula:*

$$\Phi_m(A \otimes \pi) = s_p s_q \cdots,$$

where $s_p = \mathrm{Trace}(A^p)$ for all p.

We first consider the case where
$$\pi = (1\,2\ldots p)(p+1\ldots p+q)\cdots.$$
Then, by computing the matrix of the linear transformation (A,π) with respect to the basis elements $e_{i_1}\otimes\cdots\otimes e_{i_m}$ of $E^{\otimes m}$, one verifies that the trace of (A,π) on $E^{\otimes m}$ is
$$\sum_{i_1,i_2,\ldots}(a_{i_1 i_2}\ldots a_{i_p i_1})(a_{i_{p+1}i_{p+2}}\ldots a_{i_{p+q}i_{p+1}})\cdots = \mathrm{Trace}(A^p)\mathrm{Trace}(A^q)\cdots = s_p s_q\cdots,$$
completing the proof in this case. The general case is easily reduced to this one by making a similarity transformation, which leaves the trace unchanged.

It was now a simple matter to finish the proof of Theorem 3.2. Schur applied the orthogonality relations for characters of S_m to part (iii) of Lemma 3.1, and obtained the formula
$$\Phi^{(\lambda)}(A) = \frac{1}{m!}\sum_{\pi\in S_m}\chi^{(\lambda)}(\pi)\Phi_m(A,\pi) = \frac{1}{m!}\sum_j |C_j|\chi_j^{(\lambda)}\Phi_m(A,\pi),$$
for all $A\in G$. By Cauchy's formula for $|C_j|$ (see Chapter II, §5) and Lemma 3.2, the formula for $\Phi^{(\lambda)}$ given in the statement of the theorem is obtained. Note that $\chi^{(\lambda)}$ is the character of the symmetric group corresponding to $\Phi^{(\lambda)}$, according to the theorem. As the $\{\Phi^{(\lambda)}\}$ form a complete set of characteristics of irreducible polynomial representations of G of order m, by part (ii) of Lemma 3.1, and the characteristic of an arbitrary homogeneous polynomial representation of order m is a sum of irreducible ones, the theorem follows.

Schur's final result is a remarkable formula for the characteristics of the irreducible polynomial representations of G. As stated earlier, it had been obtained almost a decade before by Deruyts (see §1), but apparently neither Schur nor Deruyts was aware of each other's work. We include here an outline of Schur's second proof of the formula; it will be seen to be very much in the same spirit as Frobenius's work on the characters of the symmetric group (see Chapter II, §5), in that the candidates for the irreducible characteristics are defined first, as symmetric functions, and proved to coincide with the actual characteristics by proving that a certain set of class functions on the symmetric group satisfy the orthogonality relations, using Cauchy's determinant formula. For a modern proof, see Green [165], Chapter 3, §3.5.

Schur began with the observation, due to Hurwitz (in *Zur Invariantentheorie* [185]), that for each positive integer k, there exists a homogeneous polynomial representation of G of order k, denoted by $A\to P_k(A)$, whose characteristic is the complete symmetric function p_k. In modern terminology, the representation $A\to P_k(A)$ is realized on the space of symmetric tensors in $E^{\otimes k}$. These are the tensors in $E^{\otimes k}$ which are left fixed by all elements $\pi\in S_m$, acting by place permutations on the basic tensors. As this action of S_m on $E^{\otimes k}$ commutes with the tensor representation $A\to A^{\otimes k}$ of G on $E^{\otimes k}$, the symmetric tensors afford a subrepresentation $A\to P_k(A)$ of the tensor representation, whose characteristic is easily shown to be p_k, the sum of all the monomials of degree k in the variables $\omega_1,\ldots,\omega_n$.

As the characteristics are trace functions, and the trace of a tensor product is the product of the traces, it follows that the representation
$$A\to P_{k_1}(A)\otimes\cdots\otimes P_{k_r}(A)$$

is a homogeneous polynomial representation of G of order $k_1 + \cdots + k_r$, whose characteristic is $p_1 \cdots p_r$.

Now let m be a fixed positive integer, and assume for the time being that $m \le n$. For each partition $(\mu) = (\mu_1, \ldots, \mu_n)$ of m into at most n nonnegative summands, with

$$\mu_1 + \mu_2 + \cdots = m, \ \mu_1 \ge \mu_2 \ge \cdots \ge 0,$$

Schur defined a symmetric function $\Psi^{(\mu)}$ given by the determinant

$$\Psi^{(\mu)} = |p_{\mu_i - i + j}|, \ 1 \le i, j \le n,$$

where it is understood that $p_0 = 1$ and $p_{-1} = p_{-2} = \cdots = 0$. The function $\Psi^{(\mu)}$ is called today the S-function, or *Schur function*[19] associated with (μ). Upon expanding the determinant $\Psi^{(\mu)}$, one has

$$\Psi^{(\mu)} = \sum \pm p_{k_1} \cdots p_{k_r},$$

where in each term, the sum of the indices is m. It has been shown above that products of the complete symmetric functions p_i appearing in the sum are characteristics of homogeneous polynomial representations of G of order m, and can therefore be expressed as linear combinations, with integer coefficients, of the characteristics $\Phi^{(\lambda)}$ of the irreducible polynomial representations of order m, by part (*ii*) of Lemma 3.1. Upon substituting this information in the formula for $\Psi^{(\mu)}$, one has

$$\Psi^{(\mu)} = \sum_{(\lambda)} r_{(\lambda)}^{(\mu)} \Phi^{(\lambda)},$$

where the sum is taken over all partitions of m, and the coefficients are integers.

By the assumption that $m \le n$, the partitions of m into at most n summands include all partitions of m. In this case, Schur's character formula can be stated as follows:

THEOREM 3.3. *Assume that $m \le n$. Then the symmetric functions $\Psi^{(\mu)}$, indexed by the set of partitions of m, are a complete set of distinct characteristics of the irreducible polynomial representations of G of order m.*

The matrix of coefficients $(r_{(\lambda)}^{(\mu)})$ in the formula preceding the statement of the theorem is a square matrix (as (λ), and also (μ), because of the hypothesis that $m \le n$, range over the set of all partitions of m). The proof of the theorem amounts to showing that it is a permutation matrix.

The character of S_m corresponding to the irreducible characteristic $\Phi^{(\lambda)}$, in the sense of Theorem 3.2, is the irreducible character $\chi^{(\lambda)}$. Schur's plan was to

---

[19]Schur proved that $\Psi^{(\mu)}$ can also be expressed as the quotient of two alternating expressions

$$\Psi^{(\mu)} = \frac{a_{(\mu)+(\delta)}}{a_{(\delta)}},$$

where $(\delta) = (n - 1, n - 2, \ldots, 0)$, and for an arbitrary n-tuple $(\alpha) = (\alpha_1, \ldots, \alpha_n)$ of nonnegative integers, $a_{(\alpha)}$ is the alternating expression given by

$$a_{(\alpha)} = \sum_{\sigma \in S_n} \varepsilon(\sigma) \omega_{\sigma(1)}^{\alpha_1} \cdots \omega_{\sigma(n)}^{\alpha_n}.$$

In the definition of $a_{(\alpha)}, \varepsilon(\sigma)$ denotes the sign of the permutation $\sigma \in S_n$. For a modern proof of the formula, see [**210**], Chapter 1, §3.

introduce, for each partition (μ), the generalized character $\eta^{(\mu)}$ defined by

$$\eta^{(\mu)} = \sum_{(\lambda)} r^{(\mu)}_{(\lambda)} \chi^{(\lambda)},$$

and to prove that the $\{\eta^{(\mu)}\}$ satisfy the orthogonality relations.

By Theorem 3.2, he obtained

$$\Psi^{(\mu)} = \sum_j \frac{\eta_j^{(\mu)}}{\alpha_1! \cdots \alpha_m!} \left(\frac{s_1}{1} \right)^{\alpha_1} \cdots \left(\frac{s_m}{m} \right)^{\alpha_m},$$

where, as before, $\eta_j^{(\mu)}$ denotes the value of the class function $\eta^{(\mu)}$ at an element of a conjugacy class C_j of S_m whose cycle decomposition has α_1 1-cycles, α_2 2-cycles, ... Using this formula and Cauchy's determinant identity (Chapter II, §5), he obtained:

$$\sum_{(\mu)} |p_{\mu_i - i + j}| |p'_{\mu_i - i + j}| = \frac{1}{\alpha_1! \cdots \alpha_m!} \left(\frac{s_1 s'_1}{1} \right)^{\alpha_1} \cdots \left(\frac{s_m s'_m}{m} \right)^{\alpha_m},$$

where p'_i and s'_j are the symmetric functions p_i and s_j in a new set of variables $\omega'_1, \ldots, \omega'_n$. It was then possible, as Frobenius had done in a similar situation in his determination of the characters of the symmetric group (Chapter II, §5), to compare coefficients of monomials in $s_1, \ldots, s_m, s'_1, \ldots, s'_m$ on both sides of the preceding equation to obtain the orthogonality relations for the class functions $\eta^{(\mu)}$:

$$\sum_j |C_j| \eta_j^{(\mu)} \eta_j^{(\mu)} = m!,$$

and

$$\sum_j |C_j| \eta_j^{(\mu)} \eta_j^{(\mu')} = 0 \text{ if } (\mu) \neq (\mu').$$

Combining these formulas with the orthogonality relations for the irreducible characters $\chi^{(\lambda)}$ of S_m, he obtained

$$\sum_{(\lambda)} (r^{(\mu)}_{(\lambda)})^2 = 1.$$

As the coefficients $r^{(\mu)}_{(\lambda)}$ are integers, it follows that for each partition (μ), there is exactly one partition (λ) such that $r^{(\mu)}_{(\lambda)} = \pm 1$, and for all other partitions (λ'), one has $r^{(\mu)}_{(\lambda')} = 0$. Therefore, $\Psi^{(\mu)} = \pm \Phi^{(\lambda)}$. It remains to prove that one always has the plus sign in these formulas. For this step, omitted here, it was sufficient to show that $\Psi^{(\mu)}(E) > 0$.

Schur also obtained a version of his formula for the case $m > n$. He remarked that Weyl had obtained another proof of the character formula, again based on Cauchy's determinant formula, in ([**284**] (1925), p. 299).

CHAPTER VI

Richard Brauer and Emmy Noether: 1926-1933

The scope of representation theory broadened in the decade following the first world war to include representations of rings and algebras (Richard Brauer and Emmy Noether), representations of Lie groups and Lie algebras (Élie Cartan and Hermann Weyl), and applications of representation theory to physics (Eugene Wigner and Hermann Weyl).

Brauer, in Königsberg, and Noether, in Göttingen, were leaders in the surge of activity in the structure and representation theory of algebras following the absorption of the Wedderburn theorems into the mainstream of research in algebra. This chapter contains biographical sketches of Brauer and Noether, a report on Noether's innovations in the foundations of representation theory, and a survey of Brauer's research on the theory of simple algebras, including his joint work with Noether and Helmut Hasse. Noether's approach to representation theory, and a joint work by Brauer and Schur on indecomposable representations, published in 1930, were important as background for Brauer's theory of modular representations of finite groups (Chapter VII).

1. Richard Brauer in Berlin, Königsberg, and North America

Richard Dagobert Brauer (1901-1977) was born in Berlin-Charlottenburg, the youngest of three children of Lilly Caroline and Max Brauer. His father was a well-to-do businessman in the wholesale leather trade, and was able to provide comfortably for his family. Richard Brauer attended the Kaiser-Friedrich-Schule in Charlottenburg, graduating in 1918. He was promptly drafted for civilian service in Berlin, but the war ended shortly afterwards, and he was able to resume his education. His brother Alfred, who preceded him in the pursuit of a career in mathematics, was less fortunate; he served four years in the German army, and was seriously wounded during the war.[1]

In the *Preface* [**46**] to his *Collected Papers*, written in 1977, Richard Brauer gave a brief, but absorbing, personal intellectual history. In it, he stated that he found his teachers in high school "not very competent," with one exception, a man who taught him calculus, who had taken a Ph.D. degree with Frobenius. As a youth, he continued, he dreamed of becoming an inventor, and had begun his higher education at the Technische Hochschule Berlin-Charlottenburg in February, 1919. He soon recognized that his interests "were more theoretical than practical" and transferred to the University of Berlin after one term.

---

[1] The biographical information, and part of the commentary on Brauer's mathematical work, is based to a great extent on the *Preface* [**46**] to Brauer's *Collected Papers*, and the obituary articles by J. A. Green [**164**], and. W. Feit [**117**]. These contain much more material than is included in this section, and, as the reader who turns to them will find, are hard to improve on.

Brauer's mathematical education began with the lectures of Erhard Schmidt. As Brauer explained,

> It is not easy to describe their fascination. When Schmidt stood in front of a blackboard, he never used notes and was hardly ever well prepared. He gave the impression of developing the theory right there and then. In fact, he would stop in the middle of a proof and say, "Let me start again. I see a better way of doing this." ... Occasionally Schmidt stopped in the middle of a proof and asked the class, "Do you see what will come next. You should." He then waited patiently for an answer. We were thus forced to try to work out the details of the proof mentally in the classroom.[2]

After spending one term at the University of Freiburg, following the custom of German students at the time to spend at least part of a year in a different part of the country, Brauer returned to the University of Berlin, and remained there until he completed his doctoral work. On his return, he was ready to take advanced courses in algebra and number theory with Schur, whose teaching style, as he recalled in [46], was very different from Schmidt's:

> [Schur] was very well prepared for his classes, and he lectured very fast. If one did not pay the utmost attention to his words, one was quickly lost. There was hardly any time to take notes in class; one had to write them up at home. At the end of each chapter of his course, he would give a survey of related developments and advised, "When you have the time, read Hilbert's Zahlbericht, Hilbert's work on invariants, Takagi on class field theory, ..." He conducted weekly problem hours, and almost every time he proposed a difficult problem. Some of the problems had already been used by his teacher Frobenius, and others originated with Schur. Occasionally he mentioned a problem he could not solve himself.[3]

In his dissertation, Brauer calculated the characters of the irreducible representations of the real orthogonal group. Schur had obtained partial results on the problem, using the method of integration introduced by Hurwitz and himself (see [258]) in the theory of invariants. Brauer gave a more algebraic approach, based on inductive arguments, involving, in the words of Green [164], "extensive manipulation of delicate determinantal identities, which meant that the price paid, in order to avoid the analytical element in Schur's integral method, was quite heavy." Another solution of the problem was obtained shortly afterward, and independently, by Weyl [284] in the more general setting of representations of semisimple Lie groups.

In 1920, Brauer met Ilse Karger, who was a fellow student in Schur's course on number theory. She took her Ph.D. in experimental physics in 1924. The following year, she and Brauer were married. By then, she had become more interested in mathematics, and took additional courses to prepare herself for teaching. Later, after the Brauers moved to North America, she held teaching positions at the University of Toronto, Brandeis University, and Boston University.

Brauer was awarded the Ph.D. degree *summa cum laude* in 1926. His first academic position was at the University of Königsberg, where he started out as an

---

[2]From Richard Brauer, Warren J. Wong (ed.), and Paul Fong (ed.), *Richard Brauer: Collected Papers, Volume 1* (Cambridge, MA: The MIT Press, 1980), p. xv. Used with permission.
  [3]Ibid., p. xvi.

Assistant, and was promoted to Privatdozent in 1927, a position that gave him the *venia legendi*, the right to give lectures. It was a small department, and Brauer, along with the four other senior faculty members, had to cover all of mathematics in their teaching. This was not a problem for Brauer, who by then had wide interests, and was becoming a superb teacher. The Brauers were happily settled in Königsberg, and started a family. Their two sons, George Ulrich and Fred Günther, were both born there, and were later to become productive mathematicians themselves.

Brauer remained at Königsberg until 1933. Years later, in his obituary article [**44**] on E. Artin, Brauer reminisced about that time with the words: "The intellectual atmosphere of German universities of that period is remembered with nostalgia by all who knew it." His accomplishments while there, mainly on the theory of simple algebras, were deep and influential, and established him as one of the promising mathematicians of his generation. During this period, he profited from contacts at mathematical meetings with other German mathematicians with common interests, notably Emmy Noether, and Helmut Hasse, each of whom had a significant impact on his career. He began his research at Königsberg with an investigation of the Schur index of an irreducible representation of a finite group (Chapter IV, §4) in terms of the theory of algebras. He and Noether (in Göttingen) had arrived independently at the concept of a splitting field of an irreducible representation of an algebra. In 1927, they combined their ideas, in a joint paper characterizing a splitting field as a maximal subfield of the division algebra associated with the given irreducible representation. Brauer continued his study of algebras with a determination of the structure of a simple algebra over an arbitrary field using what are called today *Brauer factor sets*. This led to his introduction of the structure of an abelian group on isomorphism classes of central simple algebras over a field, which was promptly named the *Brauer group* by Noether and Hasse, much to his embarrassment, it was reported. The Brauer group was instrumental in the solution in 1931 by Brauer, Hasse, and Noether of one of the main outstanding problems at the time in the theory of algebras. Their theorem, which had been conjectured by Dickson, stated that each central division algebra over an algebraic number field is cyclic (see §3 for further discussion).

Concerning the events of 1933 and those leading up to them, Brauer wrote (in [**46**]):

> During one of my visits to Berlin, Schur surprised me by suggesting that we should write a book jointly on all aspects of the representation theory of groups, many of which were still unexplored at that time. ... A year or so later, Schur told me that, with all his other duties and interests, he simply did not have the time to work on a book. The project would now have to include chapters on the work of E. Wigner, which had appeared in the meantime, on the application of representation theory to quantum mechanics. He suggested that now I should write a book on group representations with a young physicist whom he knew. Since Hitler seized power not long afterward, and I had to leave Germany, the project had to be dropped. I lost my position in Königsberg in the spring of 1933 after Hitler became Reichskanzler of Germany.[4]

---

[4]Ibid., p. xviii.

Richard Brauer (1901-1977)

With his invention of the Brauer group and his participation in the proof of Dickson's conjecture, Richard Brauer ordinarily would have had a bright future awaiting him in Germany. But with Hitler's assumption of dictatorial powers, and the implementation of the antisemitic policies that resulted in his dismissal from the University of Königsberg, Brauer was forced to look for a position outside Germany, as he said in his autobiographical remarks quoted above. Through the efforts of the Emergency Committee for the Aid of Displaced German Scholars,[5] based in New York, and the support of the Jewish community in Lexington, Kentucky, an offer of a visiting professorship for the academic year 1933-1934 at the University of Kentucky was arranged. Brauer arrived to take up his new post in November, 1933, speaking very little English. His wife and the two children followed him a

---

[5]The Emergency Committee of the Institute of International Education, financed jointly by the Carnegie and Rockefeller Foundations, with Edward R. Murrow as assistant director of the Institute and secretary of the Committee, was instrumental in bringing almost 300 German scholars to safety (see [195]).

few months later, and they soon adapted themselves to their new life. His brother Alfred was temporarily exempted from dismissal in 1933 because of his war service, but two years later he too was dismissed. He came to the United States in 1939. Their sister Alice remained in Germany, and died in an extermination camp during the second world war.

In 1933, the Institute for Advanced Study in Princeton was established, with Hermann Weyl as one of the original faculty members. The following year, Brauer was appointed to the Institute as Weyl's assistant. In [46], he said, "I had hoped since the days of my Ph.D. thesis to get in contact with him some day; this dream was now fulfilled." The previous year, Weyl gave a course of lectures at the Institute, on the structure of Lie groups and Lie algebras, and continued with another course, in 1934-1935, on representations of Lie groups and Lie algebras. Both courses developed the themes introduced in Weyl's great series of papers [284]. Mimeographed notes were prepared the first year by Nathan Jacobson, and the second year by Richard Brauer. Both sets of notes were marvelous expositions of Weyl's lectures, with substantial additions by the authors of the notes, and are now classics. A generation of mathematicians, including myself, learned the structure and representation theory of Lie algebras and Lie groups from them. Besides his own contributions to the notes, Brauer collaborated on several projects with Weyl, including a famous joint paper on the theory of spinors, which was used a few years later by Bruria Kaufman and Lars Onsager in a reformulation of Onsager's Nobel Prize-winning work on the Ising model of ferromagnetism (for references and further discussion, see [117]). During his time at Princeton, Brauer also gave a strikingly original approach to the structure of centralizer algebras of tensor representations of orthogonal groups, parallel to Weyl's analysis of tensor representations of the general linear group using their centralizer algebras (see Chapter V, §2). The resulting algebras, now called *Brauer algebras* [25], are very much alive today, and have been influential in current work on new combinatorial methods for investigating the structure of algebras.

After his year at the Institute for Advanced Study, Brauer needed a job. Emmy Noether had visited the University of Toronto, and it is not surprising that her report on him led to his appointment there, as an assistant professor, in the Fall of 1935. Concerning the appointment, Feit said, in [117],

> Fortunately times have changed. Today it is hard to conceive of someone, already 34, with mathematical achievements on the level of Brauer's, being offered, and accepting, an assistant professorship.

Brauer remained at the University of Toronto for 13 years, leaving in 1948 as a full professor, and a fellow of the Canadian Royal Society. J. L. Alperin, in a conversation with the author in August, 1992, said,

> The years he spent at Toronto were his most productive years. He achieved 5 or 6 great results during that time, any one of which would have established a person as a first-rank mathematician for the rest of their life. But, from interviews with Ilse Brauer, those years had their high points, but also contained fallow periods, when there was the day-to-day grind of raising a family in modest circumstances.[6]

---

[6]Used with permission.

The results Alperin referred to were in the theory of modular representations and blocks, character theory, and applications to number theory (see Chapter VII).

During his stay in Toronto, Brauer also made his mark as a teacher, with well-received courses and seminars for undergraduates and graduate students in mathematics and physics. Alperin, in the conversation mentioned earlier, said about his teaching style, "Brauer worked hard at his teaching; he never used a textbook, and always wrote a set of carefully prepared lecture notes."[7] Green, in [164], wrote, "Brauer's lectures were carefully prepared and undramatic; he was very concerned to give proofs in complete detail (in contrast to the prevailing fashion), and would sometimes go back and rephrase an argument two or three times in order to make things clearer. Some students found this tedious, but there were others who came to realize that Brauer had few equals as an expositor, both of mathematical ideas and techniques."[8] C. J. Nesbitt was Brauer's first Ph.D. student at Toronto, and was followed by R. H. Bruck, S. A. Jennings, M. S. Mendelsohn, R. G. Stanton, and R. Steinberg.

In the summer of 1948, Brauer gave the Colloquium Lectures of the American Mathematical Society, at the Summer Meeting in Madison, Wisconsin. As a second year graduate student, I attended all four lectures, even though I found the later ones tough going. Brauer had developed his own approach to representations of nonsemisimple algebras, which included modular representation theory of finite groups, and introduced the subject in a beautiful way in the lectures. I assumed at the time that the lectures would soon appear in published form as a book, as had been the case with other colloquium lectures. But the book never appeared, although he had a completed manuscript (see [117]). Brauer considered publishing a book on representation theory on at least one other occasion. In the introduction to his book with Y. Tsushima ([222], p. *xvii*), published in 1989, H. Nagao said:

> Around 1957, one of the authors, Nagao, had planned to write a book on representation theory with T. Nakayama. However Nakayama told him in his letter of November 4, 1959, that he had decided to suspend the plan because R. Brauer, who visited Japan in the spring of that year, had told Nakayama that he was planning to write a book on representation theory himself.[9]

Brauer moved from Toronto to the University of Michigan in Ann Arbor in 1948. This was a vibrant time in American mathematics, with graduate students and young faculty members back from military service and eager to make up for lost time, and mathematical meetings enlivened by the participation of refugees from Europe who had settled in the United States and Canada. Brauer brought with him the momentum from his mathematical achievements at Toronto, and Ann Arbor became a center for research in algebra. Brauer's Ph.D. students there included K. A. Fowler, W. Jenner, and D. J. Lewis. Others who studied with him were W. P. Brown and J. P. Jans, who completed their theses in ring theory, and J. Walter and W. Feit, whose interests in character theory and group theory were strongly influenced by courses they had taken with him.

---

[7]Used with permission.

[8]Green, J. A., "Richard Dagobert Brauer," *Bull. London Math. Soc.* **10** (1978), pp. 317–342. Used with permission.

[9]Nagao, H. and Tsushima, Y., *Representations of Finite Groups*, Academic Press, San Diego, 1989. Used with permission.

Snapshots from Richard and Ilse Brauer's visit to Japan, 1959

Snapshots from Richard and Ilse Brauer's visit to Japan, 1959

In 1952, Brauer was appointed to a professorship at Harvard University, and he remained there until his retirement. At Harvard, he continued to develop his theory of blocks of characters of finite groups along with its applications to finite group theory, and the ramifications of the Brauer Induction Theorem. He shared his interests enthusiastically with his students and other mathematicians he met at conferences or who visited him at Harvard. A sampling of the many joint papers resulting from these interactions appears in the bibliography. A partial list of his Ph.D. students at Harvard is as follows: D. M. Bloom, P. Fong, M. E. Harris, I. M. Isaacs, H. S. Leonard, J. H. Lindsey, D. S. Passman, W. F. Reynolds, L. Solomon, D. B. Wales, H. N. Ward, and W. Wong.

His joint paper with Fowler in 1955 on groups of even order opened a way towards the classification of finite simple groups, with their result that there are only a finite number of finite simple groups containing an involution whose centralizer is a given finite group (see Chapter VII, §3). The method of classifying simple groups in terms of the structure of the centralizer of an involution was identified with Brauer, and played a critical role in the solution of the classification problem. An antecedent of the method was discovered in the work of Burnside by W. Feit ([**117**]), who gave the following analysis of the idea:

> A few years ago, as I was preparing a lecture on the history of group theory, I came across a paper of Burnside in which he characterized the groups $SL_2(2^a)$ for $a > 1$ as the only groups of even order in which the order of every element is either 2 or odd [**75**]. In this paper, Burnside used some of the basic properties of involutions in a way quite similar to the way Brauer used them fifty years later. However Burnside did not realize the importance of this approach. His paper had an innocuous title and appeared in a journal that is not readily

available in mathematical libraries. I don't believe he referred to the paper in his book and I don't know of any mathematical paper by Burnside or anyone else that refers to this paper. When I told Brauer about this paper, he was as surprised to hear of it as I had been when I first found it. It is a tribute to Brauer's insight that he realized that these extremely elementary arguments concerning involutions are of fundamental importance. The fact that a mathematician of the caliber of Burnside had overlooked the importance of such arguments, even after proving the result in [**75**], shows that this insight was far from obvious.

2. Emmy Noether and her Impact on Representation Theory

Amalie Emmy Noether (1882-1935) was born in Erlangen, the first child of Amalie and Max Noether. Her father was a well-known mathematician at the University of Erlangen, and she received much of her early training in mathematics from her father and his colleague, Paul Gordan. She was awarded the Ph.D. degree from the University of Erlangen in 1907, as a student of Gordan, with a dissertation in Gordan's field of research, computational invariant theory.

Gordan had spent his career working on the theory of invariants of algebraic forms. One of the problems identified with him was to prove that for algebraic forms in a given number of variables, there was a finite basis for the set of invariants, that is, a finite set of invariants in terms of which all the others could be expressed. In each case, he was interested in the explicit computation of a basis for the invariants. When Hilbert, as a young man, proved a general existence theorem for a finite basis of invariants in all cases, Gordan responded with his famous statement: "That is not mathematics. That is theology."[10] In his proof, Hilbert used for the first time the result known as the *Hilbert Basis Theorem*, on the existence of finite bases for ideals in polynomial rings.

Emmy Noether absorbed all these developments, and published several papers of her own on Gordan's program in the years immediately following the completion of her dissertation. In a paper entitled *Körper und Systeme rationaler Funktionen* [**228**], published in 1915, she gave a thorough analysis of questions about the still relatively unexplored area of transcendence bases and other finiteness problems in fields of rational functions. She applied some of the ideas she had developed to give what became the standard proof used today of the construction of a basis for the invariants of the action of a finite group on a polynomial ring [**229**], and, with the help of the Hilbert Irreducibility Theorem, made an important contribution to the *inverse problem of Galois theory*, in which she gave a set of sufficient conditions to obtain a parametrization of the set of polynomial equations with coefficients in a given field, having a prescribed Galois group over the field [**230**]. By this time, Gordan had died, her father had retired, and it was time for a new chapter in her life. She moved to Göttingen in 1916, at the invitation of Klein and Hilbert.

Emmy Noether's unique contribution to mathematics, described by Hermann Weyl in his obituary article [**286**] as "a new and epoch-making style of thinking in algebra," was developed in a series of publications beginning with *Idealtheorie in Ringbereichen* [**231**], published in 1921. The paper began with the general definition of a commutative ring, followed by the definition, in modern terminology,

---

[10]Reid, C., *Hilbert*, Chapter V, Springer-Verlag, New York, 1970. Used with permission.

of a *Noetherian ring*, as a commutative ring satisfying the ascending chain condition for ideals, or the equivalent condition that each ideal is finitely generated. The main results in the paper were the theorem that each ideal in such a ring is the intersection of a finite number of primary ideals, and the uniqueness theorems associated with this decomposition. While some of the theorems were known for polynomial rings from the work of E. Lasker [**204**] and F. Macaulay [**209**], the axiomatic approach, and the directness of the proofs in the abstract setting, were fundamentally new. In the *Introduction* to her *Collected Papers*, N. Jacobson wrote:

> ... we shall begin with the truly monumental work *Idealtheorie in Ringbereichen*. This belongs to one of the mainstreams of modern algebra, commutative ring theory, and it may be regarded as the first paper in this vast subject—as distinguished from its precursor: ideal theory in rings of polynomials in several variables with complex coefficients.[11]

He continued:

> Both Lasker and Macaulay based their proofs on elimination theory and the geometry of algebraic sets defined by polynomial equations. On the other hand, Noether's proofs rested entirely on elementary consequences of the ascending chain condition and were (and remain) startling in their simplicity. They were entirely devoid of geometrical considerations. She was certainly aware of the geometry in the special case of polynomial rings and under her influence this aspect of the subject was pursued by van der Waerden and others.[12]

Her second paper on commutative algebra [**232**] contained an axiomatic approach to Dedekind's theorem that each ideal in the ring of algebraic integers of an algebraic number field can be factored uniquely as a product of prime ideals. It was known that the uniqueness of factorization of ideals also held for certain rings of algebraic functions in one variable. Noether gave a set of simple axioms for a commutative ring which are equivalent to the unique factorization of ideals as products of prime ideals. Rings satisfying these axioms are called today *Dedekind rings*.

The introduction of the abstract, or axiomatic, method in algebra was not without controversy. In his article [**286**], Weyl said, quoting from a lecture he gave at a conference on abstract algebra and topology in 1931:

> Nevertheless I should not pass over in silence the fact that today the feeling among mathematicians is beginning to spread that the fertility of these abstracting methods is approaching exhaustion. The case is this: that all these nice general notions do not fall into our laps by themselves. But definite concrete problems were first conquered in their undivided complexity, singlehanded by brute force, so to speak. Only afterwards the axiomaticians came along and stated: Instead of breaking the door with all your might and bruising your hands, you should have constructed such and such a key of skill, and by it you would have been able to open the door quite smoothly. But they can construct the key only because they are able, after the breaking in was

---

[11]Noether, E., *Gesammelte Abhandlungen*, Introduction by N. Jacobson, Springer-Verlag, Berlin, Heidelberg, 1983. Used with permission.

[12]Ibid.

Emmy Noether (1882-1935)

successful, to study the lock from within and without. Before you can generalize, formalize and axiomatize, there must be a mathematical substance.[13]

From her own work, Noether had a sure grasp of the mathematical substance underlying commutative algebra, and her influence was a lasting one. Her approach to the additive ideal theory of commutative Noetherian rings, and the multiplicative ideal theory of Dedekind rings, was included in the second volume of van der Waerden's treatise *Moderne Algebra*, and has been a standard item in graduate courses on algebra ever since.

Her impact on noncommutative algebra was equally powerful, and came mainly from two papers, *Hyperkomplexe Grössen und Darstellungstheorie* ([**233**], 1929), and *Nichtkommutative Algebra* ([**234**], 1933). The first one, which will be the main subject of the rest of this section, contained her approach to representations of algebras based on the Wedderburn theorems. It was an edited version of a set of lecture notes taken by van der Waerden from her lectures at Göttingen during the Winter semester 1927/28; she had already begun to lecture on versions of the material in 1924. In ([**16**], *Note Historique*) it was characterized as a "fundamental work," which, "for the importance of the ideas introduced there and the lucidity of the exposition, merits a place alongside the memoir of Steinitz on commutative fields, as one of the pillars of modern linear algebra."

Before taking up its contents, we should correct any impression that Noether's influence was based solely on her published work; she also communicated her results, ideas, and conjectures through the force of her personality. Hilbert was 54 years old when she arrived in Göttingen in 1916, and at the time he was absorbed in research on Einstein's theory of relativity. Through his influence, Göttingen had maintained its long and distinguished tradition as a center for research in algebra. With Hilbert's interests occupied elsewhere, the mantle now fell upon Noether's shoulders to continue the tradition. After overcoming, with the strong support of Hilbert, objections to her appointment as Privatdozent based on her sex, and her appointment to a salaried position a few years later,[14] Noether drew an ever increasing number of students and young mathematicians, including Artin, Alexandroff, and van der Waerden, to her lectures and seminars. There were also frequent social gatherings at her home, and walks in the countryside with the participants in her seminar, filled with enthusiastic mathematical discussions. Artin's wife Natascha recalled[15] that her husband had a "method" for talking about mathematics with Noether.

> They would go for walks, and he would ask her a question, and she would talk *very*, very fast. He knew he couldn't keep up with her, so he would let her talk for about half an hour, and then say, "Emmy, but I didn't understand a word; could you please tell me again." And

---

[13]Weyl, H., "Emmy Noether," *Scripta Mathematica* **3** (1935), pp. 201–220; *Gesammelte Abhandlungen*, Springer-Verlag, New York, 1968, vol. 3, pp. 425–444. Used with permission.

[14]In his address, *In memory of Emmy Noether* [**3**], P. S. Alexandrov said, "Eventually she received the appointment as Privatdozent, and later as honorary Professor; as a result of Courant's efforts she received a so-called Lehrauftrag, i.e. a small salary (200-400 marks per month) for her lectures, which required reconfirmation every year by the Ministry. It was in this position, without even a guaranteed salary, that she lived until the moment she was dismissed from the university and forced to leave Germany."

[15]See Kimberling [**195**] for his notes on the conversation with Mrs. Artin.

she would start again. But in the meantime they would walk very fast, and she would get a little slower and go more slowly through it again. The second time he would say, "Emmy, I haven't understood it yet." On the third rendition he would understand what she was talking about. By that time, you see, she was so tired that her speed would slow down. She was so amazingly lively![16]

In 1933, as a result of the decree from the newly installed Nazi government to dismiss Jewish university faculty members, Noether lost her position at the University of Göttingen, at the height of her mathematical creativity. She described the circumstances of her dismissal, and her thoughts about the future, in a letter to Brauer from Göttingen, dated September 13, 1933:

> since paragraph 3 comes into operation immediately,—I have today received the information that my permission to teach has been withdrawn in accordance with this paragraph—I should like to ask you whether you have any prospects for this or next year. Courant told me that the contract regarding the publication of Schur's lectures has been finalized; but this is hardly an annual income.
>
> I have had an inquiry from Oxford about lectures for a term, which I arranged for the period from Christmas to Easter. In addition, later on, I received an offer of a research professorship at Bryn Mawr for 1933/34. I have asked that this should be postponed until 1934/35, since I had already accepted Oxford. I have not yet had an answer; but I think it will be possible.
>
> I do not know whether consequently money will become available or whether the agreed stipend will simply be postponed for a year. Now, at the beginning of October, Weyl is going on a two months lecture tour to U. S. A., Princeton etc., and perhaps he could discuss the matter. Von Neumann, who was here over Whitsun and was asked to make a report to Veblen, has written down also your name as one of the first; whether with success—I do not know. It is always a matter of grants from the Rockefeller Foundation jointly with the American universities. According to a letter from Ore there is at the moment no money available at Yale; he would have liked to have kept Deuring there for another year with a kind of research professorship; but it could not be done. He had tried this already in January; prior to that he had something similar in mind for you. What he obtained was a stipend so drastically reduced that Deuring, who has to support his mother, preferred to come back. Incidentally, he was most enthusiastic about life at Yale and the contact with the people there.
>
> Bryn Mawr is a women's college; but there are Mitchell and other professors. Also Veblen wrote to me saying that it is very near Princeton and that it is hoped I shall make frequent visits

[16]Kimberling, C., "Emmy Noether and her influence," in *Emmy Noether: A Tribute to her Life and Work*, Marcel Dekker, New York, 1981, pp. 3–61. Reprinted by courtesy of Marcel Dekker, Inc.

there. If, in fact, funds become available through my postpone-
ment, then something might turn up elsewhere. Incidentally, in
Oxford I am also staying at a women's college, but my lectures are
available to the whole university, which is composed of the different
colleges. ...[17]

An appointment for the academic year 1934-1935 was indeed offered to her at
Bryn Mawr College, in Bryn Mawr, Pennsylvania, with partial financial support
from the Emergency Committee to Aid Displaced German Scholars, which had
also assisted Richard Brauer in finding a position at the University of Kentucky.
In 1934-1935, Noether combined her teaching at Bryn Mawr with a weekly lecture
on class field theory at the Institute for Advanced Study in Princeton, where her
contribution to the mathematical life at Princeton was received enthusiastically by
Oswald Veblen, Hermann Weyl, Richard Brauer, Nathan Jacobson, and others in
the local mathematical community. In April, 1935, she took what was to have been
a brief recess from her course to undergo surgery. She was unable to recover from
the operation, and died a few days later.

Noether's paper *Hyperkomplexe Grössen und Darstellungstheorie* [**233**] began
with the unequivocal statement, "The most important general theorems about algebras go back to Molien [**215**]." Molien had been the first to investigate representations of algebras, and, in particular, representations of finite groups, based on
a structure theory for algebras. There was also an independent approach, she recalled, starting with Dedekind's concept of the group determinant, and Frobenius's
result that the irreducible factors of the group determinant corresponded to the
equivalence classes of irreducible representations. But, she continued, "These con-
ceptually simple and transparent results have been obtained by Frobenius through
hard calculations." The main theme of her article [**233**], she explained, was that
the basic results about representations of finite groups and algebras over fields were
all special cases of a general theory of noncommutative rings satisfying certain
finiteness conditions. The foundations of representation theory now became *purely
arithmetic* in that the classification of irreducible modules, and the proof of complete reducibility, were achieved through the study of one-sided ideals in semisimple
rings (or rings without radicals). This was the arithmetical equivalent of Frobenius's results that the irreducible factors of the group determinant exhausted the
totality of classes of irreducible representations. The properties of one-sided ideals
in semisimple rings, in turn, were consequences of the Wedderburn theorems.

The rest of this section contains an abbreviated version of the contents of
the paper, following Noether's organization and way of presenting the material as
closely as possible. The paper began with the homomorphism theorems and proofs
of the Jordan-Hölder Theorem,[18] and theorems about complete reducibility, for
groups with operators satisfying the ascending and descending chain conditions for

---

[17]In a letter accompanying his translation of Noether's letter, dated December 28, 1998,
Walter Ledermann wrote, "The letter reflects the deep anxiety which had gripped so many victims
of Nazi persecution. I was among them (I was still in Berlin at the time and made desperate
efforts to find shelter and work in another country). I can well understand how much Noether was
preoccupied with her own future and that of her colleagues. Her letter also highlights the vital
role played by the U. S. A. in the rescue and rehabilitation of so many displaced persons."

[18]The Jordan-Hölder Theorem for finite groups is stated in Chapter III, §2. For a statement
and proof of the Jordan-Hölder Theorem for groups with operators, see van der Waerden [**276**],
Second Edition, §46. For a statement and proof of the Krull-Schmidt-Azumaya Theorem for
modules, see Curtis and Reiner [**99**], §6.

subgroups. Groups with operators were first introduced by W. Krull [**200**] and O. Schmidt [**246**] in order to prove Remak's theorem [**239**] about the uniqueness properties of decompositions of finite groups as direct products of indecomposable subgroups in the general setting of groups with operators. Noether mentioned the Krull-Schmidt theorem in a footnote, but placed the main emphasis on the results she needed for the representation theory of semisimple rings and algebras.

Modules over noncommutative rings were presented as examples of groups with operators, and were to play a central role in her approach to representation theory. A *left module*, for example, over a ring $\mathfrak{o}$ is an additive abelian group $\mathfrak{M}$, with an operation of composition which assigns to each pair $(r, m) \in \mathfrak{o} \times \mathfrak{M}$ a unique element $rm \in \mathfrak{M}$ satisfying the conditions:

$$r(m + m') = rm + rm'; \ (r + r')m = rm + r'm; \ \text{and} \ (rr')m = r(r'm),$$

for all elements $r, r' \in \mathfrak{o}$ and $m, m' \in \mathfrak{M}$. In particular, left ideals $\mathfrak{l}$ in $\mathfrak{o}$ are examples of left $\mathfrak{o}$-modules, with the module composition defined by the operation of multiplication in $\mathfrak{o}$. Similarly, right ideals in $\mathfrak{o}$ are examples of right $\mathfrak{o}$-modules, while two-sided ideals in $\mathfrak{o}$ are examples of $(\mathfrak{o}, \mathfrak{o}')$-bimodules, consisting of an abelian group with a pair of commuting module compositions for two rings $\mathfrak{o}$ and $\mathfrak{o}'$, one from the left and one from the right.

An *algebra* $\mathfrak{o}$ *over a field* K [hyperkomplexes System in Bezug auf K] was defined as a ring $\mathfrak{o}$, with the additional structure of a finite dimensional (left) vector space over K, satisfying the condition

$$\lambda(ab) = (\lambda a)b = a(\lambda b),$$

for all $\lambda \in K$ and $a, b \in \mathfrak{o}$. It was also assumed that the identity element of K acts as the identity operator on $\mathfrak{o}$. The group algebra of a finite group over a field K was mentioned as an example of an algebra.

Noether's plan was to develop *noncommutative ideal theory* in a ring $\mathfrak{o}$ satisfying the ascending and descending chain conditions for left ideals,[19] and then to apply it to the representation theory of rings and algebras. The chain conditions had already proved themselves to be useful for organizing ideal theory in commutative rings, and Noether was about to demonstrate that the same was true for noncommutative rings. In the case of an algebra $\mathfrak{o}$ over a field K, it was understood that left, right, or two-sided ideals were ideals in the ring $\mathfrak{o}$ which were also subspaces of the vector space $\mathfrak{o}$ over K. Therefore finite dimensional algebras over fields were examples of rings satisfying the ascending and descending chain conditions for left ideals. In 1927, Artin published a proof of a generalization of Wedderburn's theorem (see Theorem 2.1 below) for rings satisfying the chain conditions [**8**], and had pointed out that the generalization from algebras over fields to rings with chain conditions was not an empty one. Algebras over fields suffered from the disadvantage that, for example, the quotient $\mathfrak{o}/\mathfrak{m}$ of a commutative ring $\mathfrak{o}$ by a maximal ideal $\mathfrak{m}$ was an algebra over the field $\mathfrak{o}/\mathfrak{m}$, but $\mathfrak{o}/\mathfrak{m}^k$ is not, if $k > 1$.

---

[19]A ring satisfies the *ascending chain condition* for left ideals if every ascending chain $\mathfrak{l}_1 \subseteq \mathfrak{l}_2 \subseteq \ldots$ of left ideals has the property that all the left ideals $\mathfrak{l}_i$ from some point on are equal; the *descending chain condition* is defined similarly, with $\subseteq$ replaced by $\supseteq$. It will be convenient to use the fact that the ascending chain condition for left ideals is equivalent to the *maximum condition*, which states that a nonempty family of left ideals with a certain property contains a maximal one with this property. Similarly, the descending chain condition for left ideals is equivalent to the *minimum condition* for left ideals.

Noether began by proving that a decomposition of a ring $\mathfrak{o}$ with identity element 1 as a direct sum

$$\mathfrak{o} = \mathfrak{l}_1 \oplus \cdots \oplus \mathfrak{l}_n$$

of left ideals $\mathfrak{l}_i$ was equivalent to a decomposition of the identity element 1 as a sum of orthogonal idempotents,[20]

$$1 = e_1 + \cdots + e_n,$$

with

$$e_i^2 = e_i, \ e_i e_j = 0 \text{ if } i \neq j, \text{ and } \mathfrak{l}_i = \mathfrak{o} e_i,$$

for all i and j. She continued with a discussion of decompositions of $\mathfrak{o}$ as direct sums of two-sided ideals, and proved results such as the fact the summands $\mathfrak{a}_i$ in a decomposition

$$\mathfrak{o} = \mathfrak{a}_1 \oplus \cdots \oplus \mathfrak{a}_m$$

of $\mathfrak{o}$ as a direct sum of indecomposable two-sided ideals, are uniquely determined. These facts were by then well-known for ideals in algebras over fields, from Wedderburn's paper [279] and the book *Algebras and their Arithmetics* [111] by L. E. Dickson.

The first inkling of how these results, and others leading up to the Wedderburn theorem, might be related to representation theory was the observation that in a decomposition of $\mathfrak{o}$ as a direct sum of two-sided ideals $\mathfrak{a}_i$, a nonzero left ideal $\mathfrak{l}$ contained in one summand $\mathfrak{a}_i$ was never isomorphic, as a left $\mathfrak{o}$-module, with a left ideal $\mathfrak{l}'$ contained in a different summand $\mathfrak{a}_j$ with $i \neq j$. For if $\theta : \mathfrak{l} \to \mathfrak{l}'$ is an $\mathfrak{o}$-isomorphism, then for all $a \in \mathfrak{a}_i$ and $\ell \in \mathfrak{l}$, one has $\theta(a\ell) = a\theta(\ell) = 0$ because $\mathfrak{a}_i \mathfrak{l}' \subseteq \mathfrak{a}_i \cap \mathfrak{a}_j = 0$. But $\mathfrak{a}_i \mathfrak{l} = \mathfrak{l}$ because $\mathfrak{o}$ has an identity element, and we are led to the impossible conclusion $\mathfrak{l}' = \theta(\mathfrak{l}) = 0$. This completes the proof.

The next part of Noether's account of noncommutative ideal theory was the theory of nilpotent ideals and the radical of a ring $\mathfrak{o}$ satisfying the maximum and minimum conditions for left ideals. An ideal $\mathfrak{c}$ in a ring $\mathfrak{o}$ is called *nilpotent* if $\mathfrak{c}^t = 0$ for some integer $t > 0$. It is easily shown that the sum of two nilpotent left ideals is nilpotent. By the maximum condition for left ideals, it follows that there exists a maximal nilpotent left ideal $\mathfrak{r}$. Then $\mathfrak{r}$ contains all other nilpotent left ideals. Otherwise, if $\mathfrak{l}$ is a nilpotent left ideal not contained in $\mathfrak{r}$, then $\mathfrak{r} + \mathfrak{l}$ is a nilpotent left ideal not contained in $\mathfrak{r}$, contrary to the assumption that $\mathfrak{r}$ is maximal. Now consider $\mathfrak{r}\mathfrak{o}$; this is a left ideal, and is nilpotent, because

$$(\mathfrak{r}\mathfrak{o})^2 \subseteq \mathfrak{r}^2 \mathfrak{o}, \ (\mathfrak{r}\mathfrak{o})^3 \subseteq \mathfrak{r}^3 \mathfrak{o}, \ \ldots.$$

It follows that $\mathfrak{r}\mathfrak{o} \subseteq \mathfrak{r}$, so that $\mathfrak{r}$ is a two-sided ideal, and one has:

PROPOSITION 2.1. *In a ring $\mathfrak{o}$, satisfying the ascending chain condition for left ideals, there is a unique maximal nilpotent left ideal $\mathfrak{r}$. Moreover, $\mathfrak{r}$ is a two-sided ideal, and contains all nilpotent right ideals.*

The ideal $\mathfrak{r}$ defined in the preceding result is called the *radical* of the ring $\mathfrak{o}$. A ring $\mathfrak{o}$ satisfying the ascending and descending chain conditions for left ideals is called *completely reducible* if it is a completely reducible left $\mathfrak{o}$-module, that is, $\mathfrak{o}$ is a direct sum of a finite number of simple, or minimal, left ideals.

---

[20]Idempotents $e_i, i = 1, 2, \ldots$, are called *orthogonal* if $e_i e_j = 0$ whenever $i \neq j$.

PROPOSITION 2.2. *A completely reducible ring $\mathfrak{o}$ with identity element, satisfying the ascending and descending chain conditions, contains no nonzero nilpotent left ideals. The radical of $\mathfrak{o}$ is the zero ideal.*

By one of the conditions equivalent to complete reducibility, each nonzero left ideal $\mathfrak{l}$ has a complement: $\mathfrak{o} = \mathfrak{l} \oplus \mathfrak{l}'$ for some left ideal $\mathfrak{l}'$. Then $\mathfrak{l} = \mathfrak{o}e$ for some idempotent $e \in \mathfrak{l}$, by the previous discussion. If $\mathfrak{l}$ is a nilpotent ideal, then $e = e^2 = e^3 = \ldots 0$, and $\mathfrak{l} = 0$, which is impossible. This completes the proof.

The converse is a deeper result.

PROPOSITION 2.3. *Let $\mathfrak{o}$ be a ring with identity element, satisfying the ascending and descending chain conditions for left ideals. If the radical of $\mathfrak{o}$ is zero, then $\mathfrak{o}$ is a completely reducible ring.*

The proof depends on the following:

LEMMA 2.1. *Let $\mathfrak{o}$ be a ring satisfying the hypotheses of the Proposition. Then each nonzero left ideal $\mathfrak{l}$ in $\mathfrak{o}$ contains a nonzero idempotent element.*

The lemma goes back to Wedderburn (**[279]**, 1908), who proved it in the form:

Every potent [nonnilpotent] algebra A contains an idempotent element.

His proof may well have contained the first application of the descending chain condition; he used it to deduce that for a nonnilpotent element $x \in A$, one has $Ax^{2n+1} = Ax^n$ for some positive integer n, and by a further argument, omitted here, proved that A contains a nonzero idempotent element. Noether's proof is also interesting, and goes as follows. By the minimum condition, there exists a minimal left ideal $\mathfrak{l} \neq 0$. By the hypothesis of the lemma, $\mathfrak{l}^2 \neq 0$, and hence $\mathfrak{l}^2 = \mathfrak{l}$. Then $\mathfrak{l}a \neq 0$ for some $a \in \mathfrak{l}$, and $\mathfrak{l}a = \mathfrak{l}$, as $\mathfrak{l}a$ is a nonzero left ideal contained in $\mathfrak{l}$. The set of all elements $b \in \mathfrak{l}$ such that $ba = 0$ is a left ideal contained in $\mathfrak{l}$, and is properly contained in $\mathfrak{l}$, so it must be zero. As $\mathfrak{l}a = \mathfrak{l}$, it must be possible to express a in the form ca, for some element $c \in \mathfrak{l}$. Then $a = ca$, $c \neq 0$, and one has

$$ca = c^2a, \ (c - c^2)a = 0, \ \text{and} \ c - c^2 = 0,$$

by what has been shown. Then c is a nonzero idempotent element in $\mathfrak{l}$, completing the proof of the lemma.

The proposition follows easily from the lemma. Let $\mathfrak{l}$ be a minimal nonzero left ideal. Then $\mathfrak{l}$ contains an idempotent element $e \neq 0$, by the Lemma, and one has $\mathfrak{o}e = \mathfrak{l}$. The next step is to prove that $\mathfrak{l}$ is a direct summand of $\mathfrak{o}$, viewed as a left $\mathfrak{o}$-module. To prove this, Noether used the one-sided *Peirce decomposition*[21] to express every element $a \in \mathfrak{o}$ in the form

$$a = ae + (a - ae).$$

The elements $a - ae$, for $a \in \mathfrak{o}$, form another left ideal $\mathfrak{l}'$ such that $\mathfrak{l}'e = 0$, because e is an idempotent element. It follows that $\mathfrak{o}$ is the direct sum of $\mathfrak{l}$ and $\mathfrak{l}'$. If $\mathfrak{l}'$ is not a minimal ideal, the process can be repeated, and shows that $\mathfrak{l}'$ is the direct

---

[21]B. Peirce, in his paper *Linear associative algebra*, first published privately in 1870, and later in the *American Journal of Mathematics* (**[237]**, 1881), was the first to investigate the structure of algebras over fields using the concepts of nilpotent and idempotent elements, and to apply these ideas to classify algebras having small dimensions over the field of complex numbers (see **[16]**, *Note historique*, for further discussion).

sum of a minimal ideal and some other left ideal. The descending chain condition implies that, after a finite number of steps, one obtains

$$\mathfrak{o} = \mathfrak{l}_1 \oplus \cdots \oplus \mathfrak{l}_m,$$

for minimal left ideals $\mathfrak{l}_i$, completing the proof.

A ring $\mathfrak{o}$ with minimum condition satisfying either of the equivalent conditions stated in the two preceding propositions is called today a *semisimple* ring. The next result gave a two-sided decomposition for a semisimple ring.

PROPOSITION 2.4. *Let $\mathfrak{o}$ be a semisimple ring. Then*

$$\mathfrak{o} = \mathfrak{a}_1 \oplus \cdots \oplus \mathfrak{a}_m,$$

where the $\mathfrak{a}_i$ are minimal two-sided ideals. The two-sided ideals $\mathfrak{a}_i$ are themselves simple rings, that is, having no nontrivial two-sided ideals, and are uniquely determined.

Let $\mathfrak{a}$ be a minimal two-sided ideal in $\mathfrak{o}$. Then $\mathfrak{a}$ is a direct summand of $\mathfrak{o}$, viewed as a left $\mathfrak{o}$-module:

$$\mathfrak{o} = \mathfrak{a} \oplus \mathfrak{l},$$

for some left ideal $\mathfrak{l}$. As noted earlier, such a decomposition implies that there exist orthogonal idempotents e_1 and e_2 such that $1 = e_1 + e_2$, $\mathfrak{a} = \mathfrak{o}e_1$, and $\mathfrak{l} = \mathfrak{o}e_2$. It follows that one also has a decomposition of $\mathfrak{o}$ as a direct sum of right ideals: $\mathfrak{o} = e_1\mathfrak{o} \oplus e_2\mathfrak{o}$. It was then not difficult to show (it is a nice exercise for the reader) that $\mathfrak{l} = e_2\mathfrak{o}$, so that $\mathfrak{l}$ is a two-sided ideal. The rest of the proof of the first statement of the theorem follows as in the proof of Proposition 2.3. The second statement follows from the fact that the minimal two-sided ideals $\mathfrak{a}_i$ satisfy the condition $\mathfrak{a}_i\mathfrak{a}_j \subseteq \mathfrak{a}_i \cap \mathfrak{a}_j = 0$; the uniqueness is left as an exercise.

The structure of a semisimple ring is completely determined by the preceding result, and the following:

THEOREM 2.1 (Wedderburn's Theorem). *A simple ring $\mathfrak{o}$ with minimum condition, which is completely reducible as a left $\mathfrak{o}$-module, is isomorphic to the ring of $n \times n$ matrices with entries in a division ring D. The integer n and the division ring D are uniquely determined.*

We include an outline of Noether's proof of this fundamental theorem; there are several other proofs available today, which the reader may wish to consult for further insight and comparison with this one. Noether's proof consisted of two main steps: first, the description of the division ring D; and second, the construction of a set of matrix units c_{ij} in order to exhibit $\mathfrak{o}$ as a matrix ring over D. It is really not all that different from the proof Wedderburn himself gave in [**279**], and has much in common with Artin's proof of a more general result ([**8**], 1928).

Step (i). The division ring D. As a completely reducible left $\mathfrak{o}$-module, $\mathfrak{o}$ is a direct sum of minimal left ideals,

$$\mathfrak{o} = \mathfrak{l}_1 \oplus \cdots \oplus \mathfrak{l}_n,$$

for a uniquely determined integer n (the uniqueness of n follows from the Jordan-Hölder Theorem). The hypothesis that $\mathfrak{o}$ is a simple ring implies that the minimal left ideals $\mathfrak{l}_i$ are all isomorphic, as left $\mathfrak{o}$-modules. To prove this, take a decomposition of 1 as a sum of orthogonal idempotents e_i, with $\mathfrak{l}_i = \mathfrak{o}e_i$, and observe that, for each i, $\mathfrak{o}e_i\mathfrak{o}$ is a nonzero two-sided ideal in $\mathfrak{o}$, so $\mathfrak{o}e_i\mathfrak{o} = \mathfrak{o}$, because $\mathfrak{o}$ is a simple ring. Then $\mathfrak{o}e_i\mathfrak{o}e_j = \mathfrak{o}e_j \neq 0$, and it follows that $\mathfrak{o}e_j = \mathfrak{o}e_i a$ for some element

$a \in e_i \mathfrak{o} e_j$. This proves that $\mathfrak{l}_i$ and $\mathfrak{l}_j$ are isomorphic as left $\mathfrak{o}$-modules, for all i, j. It follows that the rings of $\mathfrak{o}$-endomorphisms (or the centralizer rings) of the left ideals $\mathfrak{l}_i$ are all isomorphic to each other, and are division rings, by the modern version of Schur's Lemma:

LEMMA 2.2 (Schur's Lemma: modern version). *The ring of $\mathfrak{o}$-endomorphisms of a simple left $\mathfrak{o}$-module is a division ring.*

As Noether observed, this result is an immediate consequence of the homomorphism theorems for groups with operators.

Step (ii). The matrix units. Keep the notation from Step (i). Let γ_{11} be the identity automorphism of $\mathfrak{l}_1 = \mathfrak{o} e_1$, and let γ_{i1} be a fixed $\mathfrak{o}$-isomorphism from $\mathfrak{l}_1$ to $\mathfrak{l}_i$, for each i. Put $\gamma_{ij} = \gamma_{i1} \gamma_{j1}^{-1}$; then γ_{ij} is an isomorphism of left $\mathfrak{o}$-modules from $\mathfrak{l}_j$ to $\mathfrak{l}_i$, and one has

$$\gamma_{ij} \gamma_{jk} = \gamma_{ik},$$

for all i, j, k. By Step (i), γ_{ij} is realized by an element $c_{ji} \in e_j \mathfrak{o} e_i$, so that $\gamma_{ij}(u) = u c_{ji}$ for all $u \in \mathfrak{l}_j$. It follows that

$$c_{ji} c_{ik} = c_{jk} \text{ and } c_{ji} c_{\ell k} = 0 \text{ if } i \neq \ell,$$

for all i, j, k, ℓ, so the elements c_{ij} are a set of matrix units.

By Step (i), $e_1 \mathfrak{o} e_1$ is a division ring, anti-isomorphic to the ring of $\mathfrak{o}$-endomorphisms of the simple left $\mathfrak{o}$-module $\mathfrak{l}_1$. For each element $a_{11} \in e_1 \mathfrak{o} e_1$, put $a_{ii} = c_{i1} a_{11} c_{1i}$; then $a_{ii} \in e_i \mathfrak{o} e_i$, and the correspondence $a_{11} \to a_{ii}$ is an isomorphism of rings. Now let D be the set of all elements

$$\alpha = a_{11} + \cdots + a_{nn} = a_{11} + c_{21} a_{11} c_{12} + \cdots,$$

with $a_{11} \in e_1 \mathfrak{o} e_1$. Then D is a division ring isomorphic to $e_1 \mathfrak{o} e_1$. It is easily checked that, for all i, j, one has

$$\alpha c_{ij} = c_{ij} \alpha, \text{ for } \alpha \in D,$$

and that

$$\mathfrak{o} = \sum_{i,j} D c_{ij}.$$

From these facts, it follows that $\mathfrak{o}$ is isomorphic to the ring of $n \times n$ matrices with entries in D, completing the proof of the first statement of the theorem. The proof of the uniqueness statement is left as an exercise for the reader.

Noether's approach to representations of algebras was based on the idea that representations of an algebra $\mathfrak{o}$ correspond to left $\mathfrak{o}$-modules, in such a way that two representations are equivalent if and only if the modules associated with them are isomorphic. Thus the main problem of representation theory, for an algebra $\mathfrak{o}$, becomes the construction and classification of the left $\mathfrak{o}$-modules. This problem made sense for algebras over arbitrary fields, semisimple or not. For semisimple algebras the problem was solved, as we shall see, by the preceding results on the structure and ideal theory of semisimple rings with minimum condition. Noether also included a few results that pointed the way towards the representation theory of nonsemisimple algebras.

The correspondence between representations and modules was defined as follows. A *representation of degree d* of an algebra $\mathfrak{o}$ over a field K was defined as a homomorphism of algebras T from $\mathfrak{o}$ to the algebra of linear transformations $\mathfrak{L}(\mathfrak{M})$ of a d-dimensional vector space $\mathfrak{M}$ over K. Two representations $T : \mathfrak{o} \to \mathfrak{L}(\mathfrak{M})$ and

$T' : \mathfrak{o} \to \mathfrak{L}(\mathfrak{M}')$ are *equivalent* if and only if there exists an isomorphism of vector spaces $S : \mathfrak{M} \to \mathfrak{M}'$ such that

$$T'(a)Su = ST(a)u,$$

for all $a \in \mathfrak{o}$ and $u \in \mathfrak{M}$. The left $\mathfrak{o}$-module corresponding to a representation $T : \mathfrak{o} \to \mathfrak{L}(\mathfrak{M})$ is the vector space $\mathfrak{M}$, with the module composition defined by

$$au = T(a)u,$$

for all elements $a \in \mathfrak{o}$ and $u \in \mathfrak{M}$. The defining properties of a module are easily shown to be satisfied. The process is clearly reversible: each left $\mathfrak{o}$-module defines a representation, and two left $\mathfrak{o}$-modules are isomorphic if and only if the corresponding representations are equivalent. The preceding definitions are independent of the choice of a basis in the underlying vector space of a representation $T : \mathfrak{o} \to \mathfrak{L}(\mathfrak{M})$ of degree d. If a basis is chosen, the map that assigns to $a \in \mathfrak{o}$ the matrix of $T(a)$ with respect to the given basis is a homomorphism of algebras, called a *matrix representation*, from $\mathfrak{o}$ to the algebra of $d \times d$ matrices with entries in K. Two matrix representations are equivalent whenever one is carried to the other by a similarity transformation.[22]

   In the correspondence between representations and modules, irreducible representations of $\mathfrak{o}$ correspond to simple left $\mathfrak{o}$-modules, that is, modules $\mathfrak{M} \neq 0$ having no nontrivial submodules. For example, minimal left ideals in a semisimple ring are simple left $\mathfrak{o}$-modules. Completely reducible representations correspond to completely reducible modules, that is, modules which are direct sums of simple modules. The general facts about representations of semisimple algebras are summarized in the following theorem.

   THEOREM 2.2. *Let $\mathfrak{o}$ be a semisimple algebra over the field K. Then:*
   (i) *Each finitely generated left $\mathfrak{o}$-module is completely reducible;*
   (ii) *Each simple left $\mathfrak{o}$-module is isomorphic to a minimal left ideal in $\mathfrak{o}$; and,*
   (iii) *The number of isomorphism classes of simple left $\mathfrak{o}$-modules is equal to the number of simple rings in a two-sided decomposition of $\mathfrak{o}$ (see Proposition 2.4). In particular, each simple algebra has exactly one isomorphism class of simple modules.*

   The proof is straightforward. Let

$$\mathfrak{o} = \mathfrak{l}_1 \oplus \cdots \oplus \mathfrak{l}_m$$

be a decomposition of $\mathfrak{o}$ as a direct sum of minimal left ideals, and let $\mathfrak{M}$ be a finitely generated left $\mathfrak{o}$-module, with a set of generators $u_1, u_2, \ldots$, so that

$$\mathfrak{M} = \mathfrak{o}u_1 + \cdots + \mathfrak{o}u_k.$$

Then

$$\mathfrak{M} = \mathfrak{o}\mathfrak{M} = \sum_{i,j} \mathfrak{l}_i u_j.$$

For each pair i and j, the map $a \to au_j$ is a homomorphism of $\mathfrak{o}$-modules from $\mathfrak{l}_i$ to $\mathfrak{l}_i u_j$. As $\mathfrak{l}_i$ is a minimal left ideal, it follows from the homomorphism theorems for groups with operators that $\mathfrak{l}_i u_j$ is either zero or a simple left $\mathfrak{o}$-module. Therefore

---

[22]In [233], Noether gave first the definition of matrix representations and modules associated with them. In a note added in proof, which she attributed to van der Waerden, she noted that matrices stood for linear transformations of vector spaces, and gave the above definition of representations of algebras by linear transformations, which appeared in van der Waerden's book [276], and is the definition used today.

$\mathfrak{M}$ is a sum of simple left $\mathfrak{o}$-modules, hence a direct sum, and $\mathfrak{M}$ is completely reducible.

For the proof of the second statement, let $\mathfrak{M}$ be a simple left $\mathfrak{o}$-module. Then $\mathfrak{M} = \mathfrak{o}\mathfrak{M} \neq 0$, so $\mathfrak{l}_i\mathfrak{M} \neq 0$ for one of the minimal left ideals $\mathfrak{l}_i$ appearing in the decomposition of $\mathfrak{o}$. Therefore $\mathfrak{l}_i u \neq 0$ for some element $u \in \mathfrak{M}$, and $\mathfrak{l}_i u = \mathfrak{M}$ because $\mathfrak{M}$ is irreducible. Finally, $\mathfrak{l}_i \cong \mathfrak{l}_i u = \mathfrak{M}$ by an argument used in the proof of the first part of the theorem. Part (iii) follows from part (ii), Proposition 2.4, Theorem 2.1, and the remark made earlier to the effect that no two minimal left ideals contained in different two-sided direct summands of $\mathfrak{o}$ are isomorphic.

For a nonsemisimple algebra $\mathfrak{o}$, Noether made the observation that $\mathfrak{o}$ is a direct sum of indecomposable left ideals, and that the number of summands and their isomorphism classes, up to a reordering, are uniquely determined, by the Krull-Schmidt Theorem. She also noted that the simple modules for such an algebra are related to the simple modules of the semisimple algebra $\mathfrak{o}/\mathfrak{r}$, where $\mathfrak{r}$ is the radical of $\mathfrak{o}$:

COROLLARY 2.1. *Let $\mathfrak{r}$ be the radical of an algebra $\mathfrak{o}$. Then $\mathfrak{o}/\mathfrak{r}$ is a semisimple algebra, and each simple left $\mathfrak{o}$-module is isomorphic to a minimal left ideal in $\mathfrak{o}/\mathfrak{r}$.*

The first statement follows from the theory of the radical (see Proposition 2.1). Now let $\mathfrak{M}$ be a simple $\mathfrak{o}$-module. Then $\mathfrak{r}\mathfrak{M} = 0$, otherwise $\mathfrak{r}\mathfrak{M}$ is a proper submodule of the irreducible module $\mathfrak{M}$, because $\mathfrak{r}$ is a nilpotent ideal. It follows that $\mathfrak{M}$ is a simple $\mathfrak{o}/\mathfrak{r}$-module, and the result follows from part (ii) of the preceding theorem.

The main general theorems about the representations of a finite group G in the field of complex numbers $\mathbb{C}$ were immediate consequences of the preceding results. The first step was to observe that each representation of G extends, by linearity, to a representation of the group algebra $\mathbb{C}G$, and conversely, representations of the group algebra $\mathbb{C}G$ restrict to representations of G. By the general theory of representations of algebras given above, the construction and classification of representations of G boils down to the construction and classification of left $\mathbb{C}G$-modules. These are described as follows.

THEOREM 2.3. (i) *The group algebra $\mathbb{C}G$ of a finite group G is a semisimple algebra.*

(ii) *All left $\mathbb{C}G$-modules are completely reducible. The simple $\mathbb{C}G$-modules are isomorphic to minimal left ideals in $\mathbb{C}G$.*

(iii) *Let $n_1, n_2, \ldots$ be the dimensions (over $\mathbb{C}$) of the simple left $\mathbb{C}G$-modules. Then*

$$|G| = n_1^2 + n_2^2 + \cdots .$$

(iv) *The number of isomorphism classes of simple left $\mathbb{C}G$-modules is equal to the number of conjugacy classes in G.*

The group algebra $\mathbb{C}G$ is semisimple by Proposition 2.2 and Maschke's Theorem (Chapter IV, Theorem 2.3). Part (ii) follows from Theorem 2.2.

By Theorem 2.2, the number of isomorphism classes of simple $\mathbb{C}G$-modules is the number of simple algebras $\mathfrak{a}_i$ in the two-sided decomposition of $\mathbb{C}G$. By Wedderburn's Theorem 2.1, the simple algebra $\mathfrak{a}_i$, for each i, is isomorphic to the algebra of all $n_i \times n_i$ matrices with entries in a division ring D_i. By the proof of Wedderburn's Theorem, D_i is isomorphic to the centralizer algebra of a simple module, so it is a finite dimensional algebra over $\mathbb{C}$; we shall prove that it coincides with $\mathbb{C}$. This follows from the fact that each element of D_i satisfies a polynomial equation with coefficients in $\mathbb{C}$, and hence has a minimal polynomial, as in the theory of algebraic extensions of fields (see Chapter I, §1). As D_i has no divisors of zero, it is easily checked that the minimal polynomial of each element of D_i is an irreducible polynomial over $\mathbb{C}$. On the other hand, $\mathbb{C}$ is an algebraically closed field, so the irreducible polynomials over $\mathbb{C}$ all have degree equal to one. This proves that $D_i = \mathbb{C}$, and hence, by Wedderburn's Theorem and part (iii) of Theorem 2.2,

$$\mathfrak{a}_i \cong M_{n_i}(\mathbb{C}),$$

the algebra of $n_i \times n_i$ matrices with entries in $\mathbb{C}$, where n_i is the dimension of the unique (up to isomorphism) simple module for the simple algebra $\mathfrak{a}_i$. Then the two-sided decomposition of $\mathbb{C}G$ is:

$$\mathbb{C}G \cong M_{n_1}(\mathbb{C}) \oplus \cdots \oplus M_{n_s}(\mathbb{C}),$$

and part (iii) of the theorem follows by comparing dimensions on both sides.

The proof of part (iv) requires an examination of the center of the group algebra. It is easily verified that the center of the algebra $\mathbb{C}G$ is isomorphic to the direct sum of the centers of the matrix algebras $M_{n_i}(\mathbb{C})$ in the decomposition given above. The center of each matrix algebra consists of the scalar matrices, so the dimension (over $\mathbb{C}$) of the center of the matrix algebra is one. Therefore the dimension of the center of the group algebra is the number of simple algebras in a two-sided decomposition, and is the number of isomorphism classes of simple modules. By an observation that goes back at least to Frobenius (see Chapter II, §4), the center of the group algebra has a basis consisting of the class sums, so its dimension is also equal to the number of conjugacy classes in G. Part (iv) is proved by comparing these two ways of computing the dimension of the center.

Noether's approach to representation theory also yielded a new proof of a generalization of Burnside's theorem (Chapter III, Theorem 4.5), for irreducible representations of finite or infinite groups.

THEOREM 2.4 (Burnside's Theorem). *Let $\mathfrak{o}$ be an algebra (not assumed to be finite dimensional) over an algebraically closed field K, and let $T : \mathfrak{o} \to \mathfrak{L}(\mathfrak{M})$ be an irreducible representation of $\mathfrak{o}$, of degree d. Then there exist d^2 linearly independent transformations $T(a)$ corresponding to elements of $\mathfrak{o}$.*

For the proof, Noether observed that the set of all linear transformations $T(a), a \in \mathfrak{o}$, is a simple algebra $\mathfrak{O}$ of linear transformations, and that $\mathfrak{M}$ is a simple $\mathfrak{O}$-module, of dimension d. By Wedderburn's Theorem, and the fact that $D_i = \mathbb{C}$ by the proof of the preceding theorem (all that is required is the assumption that K is algebraically closed), it follows that

$$\mathfrak{O} \cong M_d(K),$$

and the theorem follows.

As she observed, the theorem applies to finite dimensional irreducible representations of a possibly infinite group G, with $\mathfrak{o}$ the group algebra of G consisting of all linear combinations of finitely many elements of G.

A simple extension of the preceding argument gives a new proof of the Frobenius-Schur Theorem (Chapter IV, Theorem 3.2.)

COROLLARY 2.2 (Frobenius-Schur Theorem). *The set of linear transformations $T(a), a \in \mathfrak{o}$, in a completely reducible representation $T : \mathfrak{o} \to \mathfrak{L}(\mathfrak{M})$, with* *r inequivalent irreducible summands of degrees* $d_1, d_2, \ldots$, *contains* $d_1^2 + \cdots + d_r^2$ *linearly independent linear transformations* $T(a)$.

3. Simple Algebras and the Brauer Group: 1926-1933

As a student in Berlin, Brauer was an active participant in Schur's seminar. He reported jointly with Heinz Hopf on Schur's paper *Neue Begründung der Theorie der Gruppencharaktere*, and by himself on *Arithmetische Untersuchungen über endliche Gruppen linearer Substitutionen* and *Neue Anwendung der Integralrechnung auf Probleme der Invariantentheorie* (see [**46**]). The latter led to his Ph.D. thesis, on representations of the real orthogonal group. Soon after taking up his new position at Königsberg, he turned to the theory of irreducible groups of matrices, and simple algebras associated with them, starting from Schur's paper [**250**]. In ([**253**], 1909), Schur had extended his theory of the index for representations of finite groups (Chapter IV, §4) to semigroups of matrices, which were not assumed to be finite. Brauer, and, independently, Emmy Noether, realized that the theory of the index, combined with the Wedderburn theorems, provided a link between representation theory and the theory of simple algebras. Brauer's pursuit of this idea led him to a theory of factor sets for irreducible groups of matrices, and to the concept of the Brauer group, while Noether took a direct approach to the theory of simple algebras.

Brauer started with a study of irreducible matrix groups, and their behavior under extensions of the ground field, continuing the work of Schur [**253**]. After making the connection between irreducible matrix groups and simple algebras, he was able to interpret his results as new theorems about central simple algebras. In this section, we describe in some detail the results Brauer obtained in making the transition from the representation theory of matrix groups to the theory of simple algebras. We also include some letters[23] from Noether to Brauer leading to their joint paper ([**47**], 1927) on splitting fields of simple algebras, and some discussion of the famous Brauer-Hasse-Noether paper [**49**] containing a proof of Dickson's conjecture on division algebras over algebraic number fields.

We begin with a review of some facts concerning the Schur index for semigroups of matrices; the proofs are essentially the same as for the case of representations of finite groups (see Chapter IV, §4). Following Schur and Brauer, a *matrix group* $\mathfrak{H}$ of degree f is understood to be a set, not necessarily finite, of $f \times f$ matrices with entries in $\mathbb{C}$, with the property that the product of any two matrices in $\mathfrak{H}$ belonged to $\mathfrak{H}$, in other words, a semigroup of matrices, as we would say today, or a group in the extended sense used by Frobenius and Schur (see Chapter IV, §3). The notions of irreducibility and equivalence of matrix groups were defined as in Chapter IV, §3. A matrix group $\mathfrak{H}$ is called *rational* in a field K, or simply, K-rational, if $\mathfrak{H}$ is equivalent to a matrix group whose elements all have entries in K. An irreducible matrix group $\mathfrak{H}$ consisting of matrices with entries in a field K is said to be *absolutely irreducible* if it is irreducible when viewed as an L-rational matrix group, for all extension fields L of K. The absolutely irreducible matrix groups are those which are irreducible over $\mathbb{C}$, although they may, of course, be

---

[23]I am indebted to B. Srinivasan for calling my attention to a collection of letters and postcards from Noether to Brauer, dated from 1927 until 1934, located in the Archives of the Bryn Mawr College Library. I am also grateful for assistance from the Archivist, L. Treese, who sent me copies of them, and from W. Ledermann, who translated two letters which appear later in this section and the letter quoted in the preceding section.

rational in some subfield of $\mathbb{C}$. By Burnside's Theorem 2.4, an irreducible matrix group $\mathfrak{H}$ of degree f is absolutely irreducible if and only if $\mathfrak{H}$ contains f^2 linearly independent matrices.

Let K be a given ground field, contained in $\mathbb{C}$ as usual. A field L containing K is called a *splitting field* for an absolutely irreducible matrix group $\mathfrak{H}$ if $\mathfrak{H}$ is L-rational. A splitting field L contains the field $K(\chi)$ generated by the character values $\chi(a) =$ Trace$(a), a \in \mathfrak{H}$. The *index* (or *Schur index*) of an absolutely irreducible matrix group with respect to the ground field K is defined to be the minimum degree $(L : K(\chi))$ of all possible splitting fields L of $\mathfrak{H}$. (The fact that splitting fields of finite degree over $K(\chi)$ always exist follows from the connection between irreducible matrix groups and simple algebras, and will be explained presently (see Theorem 3.4).)

THEOREM 3.1. *Let $\mathfrak{H}$ be an absolutely irreducible matrix group, of Schur index m with respect to a field K. Assume that K contains the character values $\chi(a) =$ Trace(a), for all elements $a \in \mathfrak{H}$. Then the direct sum of m copies of $\mathfrak{H}$ is a K-rational, irreducible matrix group $\mathfrak{F}$.*

For representations of finite groups, the theorem follows from Theorems 4.1 and 4.2 of Chapter IV. In the more general setting, further discussion is needed; this was provided by Schur [**253**].

The preceding theorem gave the connection between representation theory and the theory of simple algebras.

THEOREM 3.2. *Let $\mathfrak{F}$ be an irreducible, K-rational, matrix group. Assume that K and $\mathfrak{H}$ satisfy the hypothesis of the preceding theorem, for some absolutely irreducible component $\mathfrak{H}$ of $\mathfrak{F}$. Then the set of K-linear combinations of the elements of $\mathfrak{F}$ is a simple algebra over K, whose center consists of the multiples of the identity matrix by elements of K.*

This theorem was proved by Brauer ([**20**], 1929, §3) in a more general form. We sketch a modern proof. The set of K-linear combinations of the elements of $\mathfrak{F}$ is an algebra A over the field K, because the set of matrices $\mathfrak{F}$ is closed under the operation of matrix multiplication. Moreover, the algebra A has a faithful[24] irreducible representation $a \to T(a)$ by linear transformations on the underlying vector space V over K, of the matrix group $\mathfrak{F}$. From this it follows from the results in §2, first, that the radical of A is zero, so that A is semisimple, and, second, that A has, up to isomorphism, only one simple module, so that A is a simple algebra. It also follows from the results in §2 that the center of a simple algebra is a field; the subtle point is to show that it coincides with the multiples of the identity transformation 1 by elements of K.

Now let $\mathfrak{H}$ be an L-rational absolutely irreducible component of $\mathfrak{F}$, for some extension field L of K, such that K and $\mathfrak{H}$ satisfy the hypothesis of the preceding theorem. By Theorem 3.1, the representation $a \to 1 \otimes T(a)$ of A on the vector space over L, obtained by extension of the field from K to L, is a direct sum of copies of the absolutely irreducible representation defined by $\mathfrak{H}$. The centralizer of this absolutely irreducible representation consists of multiples of the identity by elements of L, by Burnside's Theorem 2.4, and Schur's Lemma. Therefore, each element c belonging

---

[24]A representation $a \to T(a)$ of an algebra $\mathfrak{o}$ by linear transformations on a vector space V is called *faithful* if it is an isomorphism of algebras from $\mathfrak{o}$ to a subalgebra of the algebra of linear transformations on V.

to the center of A is represented in each of the absolutely irreducible components of the representation $a \to 1 \otimes T(a)$ by a multiple of the identity transformation by an element of L. As the absolutely irreducible representations are all equivalent, the same multiple appears each time, and c is represented on V by $\gamma 1$, for some element $\gamma \in L$. On the other hand, $c = \sum_i \alpha_i a_i$ for some elements $\alpha_i \in K$ and $a_i \in \mathfrak{F}$. Taking traces on $L \otimes V$, one has

$$\sum \alpha_i \text{Trace}(a_i) = f\gamma,$$

where f is the dimension of V. By the hypothesis of the theorem, Trace $(a_i) \in K$, for each i, and it follows that $\gamma \in K$, completing the proof.

A simple algebra over a field K whose center consists of the multiples of the identity element by elements of K is called a *central simple algebra* (or *normal simple algebra* in the older literature). The preceding theorem showed that central simple algebras occur naturally in carrying over the theory of the Schur index to representations of algebras.

The next step was to consider the behavior of a central simple algebra over a field K with regard to extensions of the base field. For an algebra A over K, and an extension field L of K, we denote by A^L the algebra $L \otimes_K A$ over L.

THEOREM 3.3. *Let A be a central simple algebra over the field K, and let L be an extension field of K. Then $A^L = L \otimes A$ is a central simple algebra over L.*

We give an outline of a modern proof of the result. By Wedderburn's Theorem 2.1, A is isomorphic to the ring of $n \times n$ matrices with entries in a division ring D:

$$A \cong M_n(D) \cong D \otimes M_n(K).$$

It is easily verified that the center of A coincides with the scalar matrices with entries in the center of D, so that D is a central division algebra over K. Moreover

$$L \otimes A \cong L \otimes D \otimes M_n(K),$$

and the theorem will follow if it is shown that $L \otimes D$ is a central simple algebra. We first prove that it is a simple algebra. Let $B \neq 0$ be a two-sided ideal in D^L, and let

$$b = \sum_{i=1}^{s} \ell_i \otimes d_i \in B$$

be a nonzero element of B with the ℓ_i linearly independent elements of L, and $d_i \neq 0$ in D. Assume s is minimal, for nonzero elements $b \in B$. If $s = 1$, then b is clearly an invertible element of D^L, so $1 \in B$, and $B = D^L$. Now assume $s > 1$. Then $b(1 \otimes d_1^{-1}) \in B$, and, changing notation, we may assume that

$$b = \ell_1 \otimes 1 + \sum_{i=2}^{s} \ell_i \otimes d_i \in B.$$

Then, for each nonzero element $d \in D$, one has

$$b - (1 \otimes d)b(1 \otimes d^{-1}) = \sum_{i=2}^{s} \ell_i \otimes (d_i - dd_i d^{-1}) \in B.$$

As s is minimal, the preceding expression is zero, and, because the elements ℓ_i are linearly independent, we obtain $d_i = dd_i d^{-1}$ for each i. Then the elements d_i belong to the center of D. As D is a central division algebra over K, the elements d_i all belong to K, and the expression for b can be simplified, so that $s = 1$, a contradiction. This completes the proof that D^L is a simple ring. A similar argument shows that the center of D^L is $L \otimes 1$.

A simple algebra over a field K has, up to equivalence, exactly one irreducible representation, by Theorem 2.2, (iii). It follows from this fact and the preceding theorem that a central simple algebra A has exactly one equivalence class of absolutely irreducible representations. Indeed, A is contained in the algebra $A^{\mathbb{C}}$, and the latter, by Theorem 3.3, is a simple algebra over $\mathbb{C}$, with one absolutely irreducible representation, up to equivalence. The unique absolutely irreducible representation of A is rational in $\mathbb{C}$, or in any subfield L of $\mathbb{C}$ with the property that A^L is a total matrix algebra over L. Such a field L is a splitting field for the absolutely irreducible representation of A according to the definition given earlier, in connection with the theory of the Schur index.

Therefore it was natural to define a *splitting field* for a central simple algebra A over a field K to be an extension field L of K such that $A^L \cong M_d(L)$, for some positive integer d. The next problem was to prove that splitting fields L of finite degree over K exist, and to characterize them. This problem was solved by Brauer and Noether, at first working independently, Brauer in the context of representation theory, and Noether in the context of simple algebras. Noether stated her solution of the problem, and several other results we recognize today as standard theorems in the theory of simple algebras, in a letter to Brauer dated March 28, 1927.

> I am very glad that you have now also recognized the connection between representation theory and the theory of noncommutative rings or "algebras," and the connection between the Schur index and division algebras. In regard to these fundamentals our investigations are, of course, in agreement; but then it seems to me there is a divergence. What you are doing is the actual construction of the noncommutative field (division algebra) which is associated with the representations that are irreducible in K and the formation of the irreducible representations by means of factor systems, all of this with restriction to perfect fields.
>
> I have dropped this restriction and with it the transition to r conjugate fields in the case of fields of degree r; my results are valid generally, that is, also for an extension of the second kind [an inseparable extension], if this corresponds to an irreducible representation in an imperfect ground field and its transition to an absolutely irreducible representation. More precisely, I have established the following theorems:
>
> (1) *If a noncommutative field K is of finite rank relative to its center Z (which may be a perfect or an imperfect field), then this rank is more precisely equal to n^2.*
>
> (2) *Each maximal commutative subfield P of K (of the first or second kind [separable or inseparable]) is of degree n relative to Z.*
>
> (3) *When Z is extended by the adjunction of a field P^* which is isomorphic with such a commutative subfield P, then K is transformed into a matrix ring whose individual matrices are of degree n with elements from P^*. Expressed differently, an irreducible representation of the noncommutative field K over Z becomes in P^**

the direct sum of an absolutely irreducible representation n times
repeated.

(4) *Each minimal commutative extension field Σ^* of Z in which
the representation becomes the sum of absolutely irreducible representations is isomorphic with a maximal commutative subfield P of
K and is therefore of degree n over Z. Thus n is the Schur index—
its properties having now been significantly sharpened.* ...

I hope to elaborate these matters in the course of the summer.
The above theorems are not formulated very clearly, and I cannot
judge how far you understand them. But you will gather from them
that you may well elaborate independently your points about the
factor systems. I should only like to suggest to you that you treat
the connection between Loewy and Frobenius etc. rather briefly.
When the connection has been expounded *in general*, then all of
this becomes rather obvious. It seems to me that you put the
emphasis on the construction by means of the factor system.

I should like to propose that we send each other proof sheets;
my paper is due to appear in the Zeitschrift; it is going to be rather
long, since at the same time I elaborate the foundations which I
have presented in my lecture. ...

The paper to appear in the *Mathematische Zeitschrift*, to which she referred,
was *Nichtkommutative Algebren*, ([**234**], 1933). The elaboration of the theorems
stated in her letter is given in §7 and §8 of [**234**].

Soon after the first letter, Noether and Brauer began their collaboration. In a
letter from Göttingen dated October 10, 1927, Noether wrote:

I have thought about your example—that the degree of a minimal
decomposition field can be greater than its index—and I have seen
that, by means of a small modification, it can be realized directly
in the field of quaternions (to which incidentally Schur's example
refers.) Moreover, after Hasse furnished me with the necessary
number-theoretical existence proof, I was able to show that the
degree of this minimal decomposition field is in fact unbounded,
as you had already conjectured at Kissingen.[25] Now I have been
thinking of a joint publication, provided, of course, that you have
not in the meantime proved your conjecture independently of the
example. Even if you have, the example may still be valuable; only
the text of the paper will have to be altered; for I have sketched
such a paper in which I want to indicate, on the one hand, what
is known to both of us, and on the other hand, where our investigations diverge. I have in mind a publication in the Berlin
Proceedings, simultaneously with a brief note by Hasse. ...

Please let me have your wishes for changes in the paper, both
in regard to citations and intelligibility. I have tried to stay as long
as possible in the language of representation theory, in order to be

---

[25]In a note following his translation of the letter, Walter Ledermann wrote, "Kissingen is a
fashionable spa in Germany. I guess that a mathematical conference took place there some time
before Emmy Noether wrote this letter, and that she had met Richard Brauer there and discussed
mathematics with him."

intelligible to a wider circle of readers. Decomposition into a direct sum is presumably also intelligible thanks to Dickson's book. ...

I enclose Hasse's proof and your example; please return everything to me—possibly with the exception of the example. Hasse is quite ready for printing; I shall write to him still today regarding my plan for publication.

Their joint paper [**47**] appeared later in the same year. It contained a statement of the following theorem, and an example to show that the degrees of minimal splitting fields of a central simple algebra are in general unbounded.

THEOREM 3.4 (Brauer-Noether Theorem). (*i*) *Let* D *be a central division algebra over a field* K. *Then the dimension* $(D : K) = m^2$ *for some positive integer* m. *Moreover,* m *is the degree of the unique absolutely irreducible representation of* D, *and is the Schur index of this absolutely irreducible representation with respect to* K.

(*ii*) *A central division algebra* D *over* K, *of dimension* m^2 *over* K, *has splitting fields of degree* m *over* K; *these are characterized as the maximal subfields of* D, *and are the splitting fields of* D *of minimal degree over* K. *The Schur index* m *divides the degree* $(L : K)$ *of an arbitrary splitting field* L *of* D.

(*iii*) *The splitting fields of* D *are the same as the splitting fields of central simple algebras of the form* $M_r(D)$, *for a positive integer* r.

This theorem settled the question, raised earlier, of the existence of splitting fields of finite degree over K, for central simple algebras over K. Furthermore, it gave a new interpretation of the Schur index, independent of character theory. Brauer and Noether also removed the assumption that K was a subfield of $\mathbb{C}$, and assumed only that the ground field K was perfect. It is not easy to extract a proof of the theorem from their paper, however, as they referred to their separate proofs of it, which were unpublished at the time. Noether's proofs used her results on the structure and representations of algebras ([**233**], [**234**]), while Brauer proved the theorem using his theory of factor sets of central simple algebras ([**19**], [**20**]). For a modern proof of the theorem, see [**276**], 1937, §132.

Brauer's theory of factor sets was developed in his Habilitationsschrift [**19**], which he presented to the Philosophical Faculty at the University of Königsberg in the Winter semester 1926-1927. He attached a factor set to an absolutely irreducible matrix group $\mathfrak{H}$, and proved that the Schur index of the matrix group $\mathfrak{H}$, with respect to a given field K, was determined by properties of the factor set. The idea of using factor sets to investigate rationality questions for matrix groups was due to Andreas Speiser, in Zürich ([**265**], 1919). Schur, in the paper [**256**] immediately following Speiser's in the *Mathematische Zeitschrift* — remember that Schur was an editor of the journal in its early years — gave new proofs of some of Speiser's results. They were refined and extended in Brauer's first paper [**18**] after his appointment at Königsberg, and led to the results of his Habilitationsschrift.

We begin with the definition of factor sets, from [**19**]. Let K be a given field, and let $\mathfrak{H}$ be an absolutely irreducible matrix group of degree f, such that the character values Trace $(a), a \in \mathfrak{H}$, belong to K (as in Theorem 3.1), and assume that $\mathfrak{H}$ is rational in a finite extension field $K(\theta)$ of K. We shall refer to these conditions as the *basic hypothesis*. Let $\theta_1 = \theta, \theta_2, \ldots, \theta_r$ be the distinct conjugates of θ in $\mathbb{C}$. The field $K(\theta_1, \ldots, \theta_r)$ is a Galois extension of K, and there exists, for each i, an element σ_i, of the Galois group $\mathcal{G}$ of $K(\theta_1, \ldots, \theta_r)$ over K, which carries θ to θ_i.

Upon applying σ_i to the matrix coefficients of elements of $\mathfrak{H}$, one obtains another absolutely irreducible matrix group $\mathfrak{H}_i$, which is equivalent to $\mathfrak{H}$ by the Frobenius-Schur Theorem 2.2, as corresponding elements of $\mathfrak{H}$ and $\mathfrak{H}_i$ have the same character, by the basic hypothesis. It follows that for each pair $i, j, 1 \le i, j \le r$, there exists an invertible $f \times f$ matrix P_{ij} such that

$$\mathfrak{H}_i P_{ij} = P_{ij} \mathfrak{H}_j.$$

The matrices P_{ij} can be chosen so that the following conditions hold: (a) the entries of P_{ij} belong to $K(\theta_i, \theta_j)$; (b) if an element $\sigma \in \mathcal{G}$ takes $\theta_i \to \theta_k, \theta_j \to \theta_\ell$, then σ takes $P_{ij} \to P_{k\ell}$; and (c), P_{ii} is the identity matrix, for each i.

From the definition of the matrices P_{ij}, it follows that, for all i, j, k, one has

$$\mathfrak{H}_i P_{ij} P_{jk} = P_{ij} P_{jk} \mathfrak{H}_k \quad \text{and} \quad \mathfrak{H}_i P_{ik} = P_{ik} \mathfrak{H}_k.$$

As the matrix groups $\mathfrak{H}_i$ are absolutely irreducible, one can apply Schur's Lemma to the preceding equations, and obtain, for each i, j, k, a complex number c_{ijk} such that

$$P_{ij} P_{jk} = c_{ijk} P_{ik}.$$

The set of complex numbers c_{ijk} is called a *factor set* associated with the irreducible matrix group $\mathfrak{H}$. If one replaces the matrices P_{ij} with a second set of matrices $P'_{ij} = P_{ij} \alpha_{ij}$ for a set of nonzero complex numbers $\alpha_{ij} \in K(\theta_i, \theta_j)$, then a second factor set c'_{ijk} is obtained, which is related to the first one by the condition

$$c'_{ijk} = c_{ijk} \frac{\alpha_{ij} \alpha_{jk}}{\alpha_{ik}},$$

for all i, j, k. The factor sets c_{ijk} and c'_{ijk} are said to be *associated*. The factor set such that $c_{ijk} = 1$ for all i, j, k was called the *unit* factor set.

THEOREM 3.5. (i) *An absolutely irreducible matrix group $\mathfrak{H}$, satisfying the basic hypothesis, is rational over K if and only if the factor set of $\mathfrak{H}$ is associated with the unit factor set.*

(ii) *Let $\mathfrak{G}$ be a second absolutely irreducible matrix group, satisfying the basic hypothesis. Then the tensor product $\mathfrak{G} \otimes \mathfrak{H}$ is an absolutely irreducible matrix group, and also satisfies the basic hypothesis. If c_{ijk} and d_{ijk} are factor sets of $\mathfrak{G}$ and $\mathfrak{H}$, respectively, then their product $c_{ijk} d_{ijk}$ is a factor set of $\mathfrak{G} \otimes \mathfrak{H}$.*

(iii) *Let $\mathfrak{H}$ be an absolutely irreducible group, as above, and let $\mathfrak{H}'$ be the matrix group whose elements are transposes of the elements of $\mathfrak{H}$. Then $\mathfrak{H} \otimes \mathfrak{H}'$ is a K-rational, absolutely irreducible matrix group.*

Brauer's proof of part (i) appeared in [18], §1; it was a subtle, but wholly elementary, argument based on the theory of elementary divisors and calculations with matrices, in much the same spirit as Schur's proof of Theorem 4.2 in Chapter IV. Brauer noted that a version of part (i) had been obtained by Speiser [265].

The proof of the first statement of part (ii) follows easily from Burnside's Theorem. Once it has been proved that $\mathfrak{G} \otimes \mathfrak{H}$ is an absolutely irreducible group satisfying the basic hypothesis, a straightforward calculation shows that its factor set is the product of the factor sets of $\mathfrak{G}$ and $\mathfrak{H}$.

Part (iii) is also proved by a direct calculation using the definition of a factor set, namely that c_{ijk}^{-1} is a factor set of $\mathfrak{H}'$ if c_{ijk} is a factor set of $\mathfrak{H}$. Then the factor set of $\mathfrak{H} \otimes \mathfrak{H}'$ is the unit factor set, and the conclusion follows from part (i).

Using the connection between irreducible matrix groups and central simple algebras discussed earlier, Brauer carried over the theory of factor sets to central simple algebras over a perfect field K. The procedure described above associates to each central simple algebra A over K, with a splitting field $L = K(\theta)$ as above, a factor set c_{ijk}, consisting of elements in a Galois extension of K containing L, with indices $i, j, \ldots$ corresponding to conjugates $\theta = \theta_1, \theta_2, \ldots$ of θ. From the preceding theorem, one obtains:

THEOREM 3.6. (i) *Let A and B be central simple algebras over K, with a fixed splitting field $L = K(\theta)$. Then $A \otimes B$ is a central simple algebra over K, with the splitting field L. If c_{ijk} and d_{ijk} are factor sets of A and B, respectively, then $c_{ijk}d_{ijk}$ is a factor set of $A \otimes B$.*

(ii) *Let A be a central simple algebra over K, with the splitting field $L = K(\theta)$, and let A' be its opposite algebra, that is the vector space A, with the multiplication of two elements given by multiplying them in A, but in the reverse order. Then $A \otimes A' \cong M_d(K)$, a total matrix algebra over K.*

The first statement follows from part (ii) of Theorem 3.5, while the second follows from part (iii). A direct proof, without using representation theory, of the result that $A \otimes B$ is a central simple algebra can be given along the lines of the proof of Theorem 3.3.

These results prepared the way for Brauer's definition of the abelian group of classes of central simple algebras over a field K, today called the *Brauer group*,[26] and denoted by $B(K)$. Two central simple algebras A and B are said to belong to the same *class* if the division algebras associated with them by the Wedderburn Theorem 2.1 are isomorphic. The relation of belonging to the same class is an equivalence relation. Brauer defined the product of two classes of central simple algebras $\{A\}$ and $\{B\}$ by setting

$$\{A\}\{B\} = \{A \otimes B\},$$

and observed that the product is a well-defined operation on the set $B(K)$ of algebra classes, by Theorem 3.6, satisfying the commutative and associative laws. The class $\{K\}$ is the identity element in $B(K)$, and for each class $\{A\}$, there is, by Theorem 3.6(ii), an inverse class $\{A\}^{-1}$ given by $\{A'\}$, where A' is the opposite algebra of A.

The *Schur index* m of a central simple algebra A over the field K is the Schur index with respect to K of the unique absolutely irreducible representation of A; it is the minimum degree $(L : K)$ of splitting fields L of A. The Schur index of a central simple algebra A coincides with the Schur index of the division algebra associated with A, by part (iii) of the Brauer-Noether Theorem, and hence is the same for all central simple algebras belonging to the class of A. One of the first main results ([22], Satz 2) concerning the group $B(K)$ was the following:

THEOREM 3.7. *Each element $\{A\}$ of the Brauer group $B(K)$ has finite order. The order of an element $\{A\} \in B(K)$ divides the Schur index m of A.*

Let $L = K(\theta)$ be a splitting field of a given central simple algebra A, and let c_{ijk} be a factor set associated with A by the procedure described above, using an absolutely irreducible representation $\mathfrak{H}$ of A, of degree f. Then there exist invertible $f \times f$ matrices P_{ij} such that

$$P_{ij}P_{jk} = c_{ijk}P_{ik},$$

---

[26]The Brauer group $B(K)$ was defined formally for the first time in [22], but the results needed for its definition had all been established earlier, in [20].

for all i, j, k. Let δ_{ij} be the determinant $|P_{ij}|$ for each pair i, j; then one has

$$\delta_{ij}\delta_{jk} = c^f_{ijk}\delta_{ik},$$

for all i, j, k, by the preceding equation. This shows that the factor set c^f_{ijk} is associated with the unit factor set, and hence the irreducible matrix group $\mathfrak{H}^{(f)} = \mathfrak{H} \otimes \cdots \otimes \mathfrak{H}$ (f factors) is K-rational, by Theorem 3.5. It follows that $\{A\}^f = 1$ in $B(K)$, and we have proved that each element of $B(K)$ has finite order. Further reasoning, not given here (see [19], Satz III) proves that $\{A\}^m = 1$ in $B(K)$, and completes the proof of the theorem.

The order of the class $\{A\}$ of a central simple algebra A as an element of the group $B(K)$ was called the *exponent* of the class $\{A\}$; it turned out to be another important numerical invariant, besides the Schur index, for investigating central simple algebras.

Brauer's proof of the preceding theorem, showing that the exponent divides the index, was an application of the theory of factor sets to the group $B(K)$. A more precise relation between factor sets and the structure of the group $B(K)$ was obtained by combining results in his papers [19] and [20]. Let $L = K(\theta)$ be a finite extension field of K. It is easily shown that the set of classes $\{A\}$ of central simple algebras A having L as a splitting field forms a subgroup $B_L(K)$ of the group $B(K)$. On the other hand, Brauer had been able to characterize the sets of elements c_{ijk} which occurred as factor sets of central simple algebras over K, having L as a splitting field, in terms of the condition

$$c_{ijk}c_{ik\ell} = c_{ij\ell}c_{jk\ell},$$

and other properties describing the Galois group action on the elements c_{ijk} of a factor set (see the discussion preceding Theorem 3.5). He defined an equivalence relation on the set of factor sets, with two factor sets belonging to the same equivalence class if and only if they are associates. With the definition of product of two factor sets given in Theorem 3.5, the set of equivalence classes of factor sets forms an abelian group $H_L(K)$. Using the theory of factor sets and its connection with central simple algebras, Brauer proved:

THEOREM 3.8. *There is an isomorphism of abelian groups:*

$$B_L(K) \cong H_L(K).$$

This theorem, along with other results such as *Hilbert's Theorem 90*, led to the subject known today as *Galois cohomology* (see [164] for further discussion and references to the literature).

In his book *Algebras and their arithmetics* [111], Leonard Eugene Dickson (1874–1954) introduced a type of central simple algebras, called *cyclic algebras*, whose structure is described as follows. Let K be a field, and let L be a finite Galois extension field of K with a cyclic Galois group $\mathcal{G}$ of order n. A *cyclic algebra* A over the field K is a vector space over L of dimension n, generated by elements $1, u, u^2, \ldots$, satisfying the relations $u^i u^j = u^{i+j}$, and $u^n = \alpha$ for some element $\alpha \neq 0$ in K. One also assumes that multiplication of elements of L by the element u satisfies the relations

$$\ell u = u\ell^\sigma,$$

for all $\ell \in L$, where $\ell \to \ell^\sigma$ denotes the action of a generator σ of $\mathcal{G}$ on elements of the field L. In his book, Dickson proved:

THEOREM 3.9. *The algebra A described above is a central simple algebra of dimension n^2 over K. The elements in A of the form $\ell \cdot 1$, corresponding to the elements $\ell \in L$, form a maximal subfield of A, isomorphic to L. The algebra A is a division algebra if α^n is the least power of α which is the norm of an element of L.*

Dickson proved ([**111**], §31 and §32) that every central division algebra of dimension n^2 is cyclic, if $n = 2$, and Wedderburn had proved the corresponding result for $n = 3$ [**279**]. Dickson conjectured, in [**111**], that all central division algebras over algebraic number fields are cyclic.

Dickson's book was translated into German in 1927. The problem about division algebras over algebraic number fields set off concentrated efforts on both sides of the Atlantic to solve it. In 1931, H. Hasse published a paper (in English) entitled *Theory of cyclic algebras over an algebraic number field* [**170**], in the *Transactions of the American Mathematical Society*, with the following introductory paragraph:

> I present this paper for publication to an American journal and in English for the following reason: The theory of linear algebras has been greatly extended through the work of American mathematicians. Of late, German mathematicians have become active in this theory. In particular, they have succeeded in obtaining some apparently remarkable results by using the theory of algebraic numbers, ideals, and abstract algebra, highly developed in Germany in recent decades. These results do not seem to be as well known in America as they should be on account of their importance. This fact is due, perhaps, to the language difference or to the unavailability of the widely scattered sources.

One American mathematician to whom the results mentioned in the preceding paragraph were well known was A. Adrian Albert (1905-1972), at the University of Chicago. A student of L. E. Dickson, who received his Ph.D. degree in 1928, Albert was well on his way to a proof of Dickson's conjecture in 1931. Brauer was also in the thick of the activity connected with the problem; in [**20**], he extended the results of Dickson and Wedderburn to the case $n = 6$, and in [**21**], he gave an example of a central divison algebra over a field K, not an algebraic number field, which was not a cyclic algebra.

Dickson's conjecture was proved by Brauer, Hasse, and Noether in 1931 [**49**]:

THEOREM 3.10 (Brauer-Hasse-Noether Theorem). *Each central division algebra over an algebraic number field is cyclic (or of Dickson's type).*

The proof used the theory of division algebras over p-adic fields, and the theory of the norm-residue symbol, which Hasse had developed in [**170**] and in earlier publications. Meanwhile, Albert had obtained proofs of some of the main steps needed for the proof, and the main theorem itself for division algebras of dimension a power of 2. After some correspondence, Albert and Hasse published a joint paper [**2**], also in 1932, entitled *A determination of all normal division algebras over an algebraic number field*, containing a modified proof of the main theorem stated above. In the first sentence of the paper, they stated the problem. Note that they used *degree* for the square root of the *order*, or dimension, of a central simple algebra over the ground field.

The principal problem in the theory of linear algebras is that of the determination of all normal [central] division algebras (of order n^2, degree n) over a field F.

The following excerpt from a section of their paper entitled *The history of our proof* contains a chronology of the steps leading to a solution of the problem for algebras over a number field F:

At the time (April, 1931) when Hasse presented his paper [**170**] to these Transactions, he had also outlined a proof of the following existence theorem. Let A be a normal [central] simple algebra of degree n over F. Then there exists a cyclic field C of the same degree n such that A^C splits everywhere. Hasse then conjectured that it follows that A^C is a total matric algebra. As an immediate consequence, A is cyclic with a subfield equivalent to C.

Hasse had thus reduced the proof of our principal theorem to the proof of his conjecture. He attempted to prove the latter result by the same method which had been successful for a cyclic algebra A, but did not succeed in this attempt at first. Later (October, 1931) he could however use his results to prove the principal theorem for the case where A has a splitting field with regular abelian group and degree n. Albert used Hasse's communicated result to prove the principal theorem for $n = 2^e$. Albert also proved that F could be extended to K over F such that a normal division algebra A, of degree $n = p^e$ over F, p a prime, has the property that A^K is a cyclic normal division algebra over K. Albert communicated these results to Hasse. They were very close to the principal theorem.

Shortly before this time (October, 1931, presented, September, 1931) Albert published certain algebraic theorems (amounting to the latter result just mentioned) from which the proof of Hasse's conjecture follows immediately. When these theorems as well as the above mentioned communication from Albert were still unknown to Hasse and throughout Germany, while Hasse's existence theorem (even yet unpublished in complete form) was still unknown to Albert (November 11, 1931), R. Brauer, Hasse, and E. Noether succeeded in completing a proof of Hasse's conjecture and hence of the principal theorem. However they used a reduction not as simple as the one by Albert already in print. The authors of the present article feel that it is desirable to show how the proof of the main theorem is an immediate consequence of Hasse's arithmetic and Albert's algebraic results.

Brauer's theory of factor sets attaches a factor set to each central simple algebra A with an arbitrary splitting field $K(\theta)$ of finite degree over F. When $K(\theta)$ is a Galois extension of K, Noether developed a simpler version of the factor set, and a type of central simple algebra called a *crossed product* [verschränkte Produkt]. Cyclic algebras are crossed product algebras, in case the Galois group of $K(\theta)$ over K is a cyclic group. In the last two sections of the second edition of his book [**276**], van der Waerden gave a self-contained exposition of the proofs of Theorems

3.3-3.9, based on the theory of crossed product algebras. For a proof of the Brauer-Hasse-Noether Theorem, and references to the literature on simple algebras, the reader is referred to Deuring's survey volume ([**106**], 1935) and Albert's Colloquium Publication of the American Mathematical Society ([**1**], 1938).

CHAPTER VII

Modular Representation Theory

The new developments in the representation theory of finite groups and algebras by Brauer and Noether in the years before 1933 (see Chapter VI) did not include applications of character theory to the structure of finite groups comparable to those achieved by Frobenius, Burnside, and Schur in the early years of the twentieth century. This aspect of representation theory had become somewhat dormant, until Brauer published the first definitive results on representations of finite groups in fields of characteristic $p > 0$, in 1935. This new area of research was to lead, in Brauer's hands, to a revival of interest in the whole subject of representations of finite groups and its applications to finite group theory.

This chapter contains an historical account of Brauer's research, in its formative stages, on modular representation theory, the theory of blocks, the classification of finite simple groups, and algebraic number theory. The work reported on here was done in the period, roughly, from 1935-1960, while he was a faculty member at the University of Toronto, the University of Michigan, and at Harvard University. Some research by Tadasi Nakayama on nonsemisimple algebras, and by Michio Suzuki on character theory and the classification of simple groups, had important interactions, at different times, with Brauer's program, and is included in our survey.

1. Brauer and Nesbitt: Opening Moves

After Dickson's conjecture on central division algebras over number fields was proved (see Chapter VI, §3), interest in the theory of simple algebras began to subside, and Brauer decided it was time to move on. By the time he took up his new position at the University of Toronto in 1935, he had changed the direction of his research, and had begun a systematic investigation of *modular representation theory*, the term L. E. Dickson had introduced for the representation theory of finite groups in fields of characteristic $p > 0$. The subject was first explored by Dickson in three papers, published in 1902 and 1907 ([**108**], [**109**], [**110**]). In the first, he proved that Frobenius's result on the factorization of the group determinant of a finite group G (see Chapter II, §3) held for algebraically closed fields of characteristic p not dividing the group order $|G|$. But in the last two, he obtained results that clearly showed that the theory was altogether different in case the order of G was divisible by p.

It will be convenient, in the discussion to follow, to adopt some of Brauer's terminology. Let p be a prime number. Elements of a finite group of order prime to p will be called *p-regular elements*, and conjugacy classes containing p-regular elements, *p-regular classes*.

The determination of the number of irreducible representations of a finite group in an algebraically closed field of characteristic $p > 0$ was of course a fundamental problem in the new theory. In [**110**], Dickson settled the problem for finite abelian groups. Brauer's first paper on the subject [**23**], published almost 30 years later, solved the problem in the general case, and also contained a result on the connection between irreducible representations and indecomposable representations that proved to be useful later, for the proof of Theorem 1.2.

THEOREM 1.1. *Let G be a finite group, and let K be an algebraically closed field of characteristic $p > 0$. The number of equivalence classes of irreducible representations of G in the field K is the number of p-regular conjugacy classes in G.*

In the introduction to the paper, Brauer noted that, in case the characteristic p of K divides the order of G, there exist representations of G which are not completely reducible. This meant that the group algebra was no longer semisimple, so the usual methods of proof of the theorem for fields of characteristic zero led to difficulties, and a new approach had to be found.

His first proof of the theorem was indirect, and used properties of the modular characters, or trace functions, of the irreducible representations, such as the Frobenius-Schur Theorem (Chapter VI, Corollary 2.2), which implies that two irreducible representations of G in the field K are equivalent if and only if their characters coincide. Brauer returned to Theorem 1.1, and others he considered important, again and again in his later writings, reviewing them from different points of view, and giving new proofs whenever possible. Here we shall give one of Brauer's later proofs of the theorem, from his paper ([**42**], 1956, §3), in the volume, mentioned earlier, of the *Mathematische Zeitschrift* dedicated to the memory of Issai Schur. Instead of using characters, the argument is based on another way of counting irreducible representations of a nonsemisimple algebra, and applying it to the group algebra KG.

LEMMA 1.1. *Let A be a finite dimensional algebra over an algebraically closed field K of characteristic $p > 0$. Let S be the subspace of A spanned by the elements $ab - ba, a, b \in A$, and let T be the set of elements $r \in A$ with the property that $r^q \in S$ for some power q of p. Then T is a subspace of A, and the number of inequivalent irreducible representations of A is equal to the dimension of the quotient space A/T.*

Let $a, b \in A$; then write out the terms in the expansion $(a + b)^p$. The terms in the expansion besides a^p and b^p can be grouped into sets of p terms, with the terms in each set obtained from each other by cyclic permutations of the factors. It is readily shown that the elements of each set of p terms are all congruent to each other (mod S), and hence

$$(a + b)^p \equiv a^p + b^p \pmod{S}.$$

Now let $u, v \in A$, and let $w = v(uv)^{p-1}$. By the preceding result, one has

$$(uv - vu)^p \equiv (uv)^p - (vu)^p \equiv uw - wu \equiv 0 \pmod{S}.$$

It follows that the p-th power of each element of S is in S, and that T is a subspace of A, containing S.

The rest of the proof was a simple application of the results on representations of algebras, from Noether's paper [**233**] (see Chapter VI, §2). If A is a simple algebra, then it is a full matrix algebra over K, by Wedderburn's Theorem, and S is the set of matrices in A of trace zero. If u is an idempotent in A with nonzero trace, then $u \notin S$, and $u \notin T$. The dimension of A/S is equal to one, and it follows that the dimension of A/T is one, completing the proof of the lemma in this case.

In the general case, the radical N of A is contained in T, because it is a nilpotent ideal. Moreover, the number of equivalence classes of irreducible representations of A is equal to the number of classes of irreducible representations of the semisimple algebra A/N, by Corollary 2.1, Chapter VI. The semisimple algebra A/N is the direct sum of two-sided ideals A_i, which are simple algebras, and the number of simple algebras in the two-sided decomposition of A/N is the number of equivalence classes of irreducible representations of A, by Theorem 2.2, Chapter VI. Letting T_i be defined for A_i as T was defined for A, it is easily verified that the dimension of A/T is equal to the sum of the dimensions of A_i/T_i. These dimensions are all equal to one, by the preceding discussion, and the lemma follows.

Brauer proved the theorem by showing that the dimension of the space A/T, in case A is the group algebra KG, is the number of p-regular conjugacy classes in G, and then applying the preceding lemma. Each element $x \in G$ can be expressed uniquely in the form $x = st$, where s and t are both powers of x, s is a p-regular element, and the order of t is a power of p. Then the first congruence in the proof of Lemma 1.1, with $a = st, b = -s$, becomes:

$$(st - s)^p \equiv s^p t^p - s^p \pmod{S}.$$

Upon replacing p by q, the order of t, he obtained

$$(st - s)^q \equiv s^q t^q - s^q \equiv s^q - s^q \equiv 0 \pmod{S},$$

so that $st - s \in T$, and $x \equiv s \pmod{T}$. It follows that the general element of KG is congruent (mod T) to a linear combination of the p-regular elements of G. Furthermore, if x and y are conjugate elements in G, then $x \equiv y \pmod{S}$, and hence $x \equiv y \pmod{T}$. Combining these remarks, one sees that the general element of KG is congruent (mod T) to an expression of the form $\sum_r a_r r$, with the elements r a set of representatives of the p-regular classes of G, and $a_r \in K$. The final step was to show that the representatives r of the p-regular classes are linearly independent (mod T). Assume a relation of linear dependence (mod T) of the form $\sum_r a_r r \equiv 0 \pmod{T}$. It is not difficult to show that there exists a power q of p such that $r^q = r$ for each p-regular class representative r, and such that the q-th power of the preceding congruence becomes

$$\sum_r a_r^q r \equiv 0 \pmod{S}.$$

A further observation was that the set S consists of those elements of KG with the property that the sum of the coefficients of elements in a conjugacy class is zero, for each conjugacy class. It follows that the coefficients a_r in the proposed relation of linear dependence satisfy the condition $a_r^q = 0$, and hence are all equal to zero. This proves that the set of representatives of the p-regular classes give a basis of KG/T, and completes the proof of the theorem.

Three of the most important papers ([**50**], [**51**], [**52**]) in the early development of modular representation theory were published jointly by Brauer and Cecil J. Nesbitt, Brauer's first Ph.D. student at the University of Toronto. Nesbitt gave the following account of their collaboration:[1]

---

[1]The quotation to follow is taken from [**164**], p. 320. Used with permission of the London Mathematical Society.

Curiously, as thesis advisor, he did not suggest much preparatory reading or literature search. Instead, we spent many hours exploring examples of the representation theory ideas that were evolving in his mind. Eventually, I pursued a few of these ideas for thesis purposes; they received some elegant polishing by him, and later were abstracted and expanded by another great friend, Tadasi Nakayama. Professor Brauer generously ascribed joint authorship to several papers that came out of these discussions, but my part was more that of interested auditor.

In the first two joint papers, both published in 1937, Brauer and Nesbitt investigated the regular representation of a finite group G in an algebraically closed field K of characteristic $p > 0$. This amounts to studying the regular representation of the group algebra KG, which is not semisimple whenever p divides the order of the group.[2] Their first paper [50] contained three of the topics that turned out to be of central importance for the later development of the theory: a formula relating the *Cartan invariants* of the algebra KG to the *decomposition numbers*, the introduction of what are called today *Brauer characters*, and the definition and some of the basic properties of *blocks*.

Their result on Cartan invariants and decomposition numbers provided an important link between properties of the nonsemisimple algebra KG, and the character theory of G in the field of complex numbers. Our account of it begins with the Cartan invariants.[3] Let $F^{(1)}, \ldots, F^{(t)}$ be a set of representatives of the equivalence classes of irreducible representations of the algebra KG. The representations $F^{(i)}$ all occur as composition factors of the regular representation of KG, by the Jordan-Hölder Theorem, and Chapter VI, Corollary 2.1. On the other hand, the regular representation of KG is a direct sum of indecomposable representations $U^{(j)}$, which are not necessarily the same as the irreducible representations whenever the group algebra is not semisimple. By the Krull-Schmidt Theorem for groups with operators, the number of indecomposable summands of the regular representation, and their equivalence classes, are uniquely determined.[4] In [51], Brauer and Nesbitt took another important step, and announced without proof[5] that there was a natural bijective correspondence between the equivalence classes of indecomposable

---

[2]One way to show this is to observe that, in case p divides $|G|$, the sum of the basis elements

$$\sum_{x \in G} x$$

is a nonzero nilpotent element belonging to the center of KG. It follows that the radical of KG is different from zero, so KG is not semisimple.

[3]The Cartan invariants of the regular representation of a finite dimensional algebra were first defined by E. Cartan in his paper [90] on the structure of finite dimensional algebras over the field of complex numbers; see §2 for further discussion.

[4]In their only joint paper ([48], 1930), Brauer and Schur gave new proofs of the Krull-Schmidt Theorem, for matrix groups in the extended sense used earlier, that is, for semigroups of matrices. Their aim was to show that, while Krull and Schmidt had established the result using "important abstract group-theoretic principles" such as the chain conditions, it was possible, for matrix groups, to prove the uniqueness theorem in the "framework of the special theory of groups of linear substitutions." In fact, they gave two proofs, in sections entitled, "Erster Beweis des Krullschen Satzes (von I. Schur)", and "Zweiter Beweis des Krullschen Satzes (R. Brauer)," without attempting to reconcile the rather different approaches taken in their separate accounts.

[5]A proof was given in Nesbitt's Ph.D. thesis [226], and extended to rings satisfying the chain conditions by Nakayama (see §2, Theorem 2.1).

summands of the regular representation, and the classes of irreducible representations of KG. For a set of representatives $U^{(1)}, \ldots, U^{(t)}$ of the equivalence classes of indecomposable summands of the regular representation, the *Cartan invariants* KG are the multiplicities c_{ij} with which $F^{(j)}$ occurs as a composition factor of $U^{(i)}$, for $i, j = 1, \ldots, t$. Brauer and Nesbitt summarized this information with the notation:

$$U^{(i)} \leftrightarrow \sum_{j=1}^{t} c_{ij} F^{(j)}.$$

The *Cartan matrix* $C = (c_{ij})$ collects all this information, and is a square $t \times t$ matrix.

We next consider the decomposition numbers. The idea was to associate, with each irreducible representation $x \to A(x)$ of G in the field of complex numbers, a modular representation $x \to \bar{A}(x)$ of G in the algebraically closed field K of characteristic p. Nothing like this had been tried before for general finite groups, and their attempt to carry it out raised some rather subtle questions. The first step was to replace the representation A by an equivalent one which is rational in an algebraic number field E. Let $\mathfrak{o}$ be the ring of algebraic integers in E, and let $\mathfrak{p}$ be a prime ideal in $\mathfrak{o}$ dividing the rational prime p. The localization[6] of $\mathfrak{o}$ at the prime ideal $\mathfrak{p}$ is the ring $\mathfrak{o}_{\mathfrak{p}}$ consisting of elements a/b belonging to the quotient field of $\mathfrak{o}$, such that $a, b \in \mathfrak{o}, b \notin \mathfrak{p}$. The ring $\mathfrak{o}_{\mathfrak{p}}$ is a principal ideal domain, and from this it follows that the given representation $x \to A(x)$ is equivalent to a representation $x \to \mathbf{A}(x)$ by matrices with entries in $\mathfrak{o}_{\mathfrak{p}}$. The ideal $\mathfrak{p}\mathfrak{o}_{\mathfrak{p}}$, which we shall also denote by $\mathfrak{p}$, is a maximal ideal in $\mathfrak{o}_{\mathfrak{p}}$, and the quotient field $k = \mathfrak{o}_{\mathfrak{p}}/\mathfrak{p}$ is a finite field of characteristic p, which we may assume is a subfield of K. Let $a \to \bar{a}$ denote the natural homomorphism from $\mathfrak{o}_{\mathfrak{p}} \to k$ with kernel $\mathfrak{p}$. When this homomorphism is applied to the entries of the matrices $\mathbf{A}(x), x \in G$, it is easily checked that the resulting matrices $\bar{A}(x), x \in G$, define a representation of G by matrices with entries in the finite field k of characteristic p. A representation obtained in this way is called a *modular representation* associated with the given representation $x \to A(x)$.

Of course, the preceding construction can be carried out in more than one way. There may be other representations of G with matrix entries in $\mathfrak{o}_{\mathfrak{p}}$ which are equivalent to the given representation $x \to A(x)$, and, as Brauer and Nesbitt pointed out, the modular representations resulting from them need not be equivalent. They overcame this apparent impasse by proving:

LEMMA 1.2. *Any two modular representations associated with a given representation $x \to A(x)$ of G have the same absolutely irreducible composition factors.*

The lemma made it possible to define the *decomposition numbers* as follows. Let $x \to A^{(1)}(x), \ldots, x \to A^{(s)}(x)$ be representatives of the equivalence classes of irreducible representations of G in the field of complex numbers. The decomposition number d_{ij}, for $1 \leq i \leq s, 1 \leq j \leq t$, is the multiplicity of $F^{(j)}$ as an absolutely irreducible composition factor of a modular representation $x \to \bar{A}^{(i)}(x)$ obtained from the representation $x \to A^{(i)}(x)$ by the construction given above; it is independent of the choice of the modular representation by Lemma 1.2. The $s \times t$ matrix $D = (d_{ij})$ is called the *decomposition matrix* of the finite group G, for the prime p. Their theorem was as follows.

---

[6]See [**191**] for the theory of Dedekind domains and their localizations.

THEOREM 1.2 (Brauer-Nesbitt Theorem). *The Cartan invariants c_{ij}, for a finite group G, and a prime p, and the decomposition numbers d_{ij} are connected by the relations*

$$c_{ij} = \sum_{k=1}^{s} d_{ki} d_{kj},$$

for $1 \leq i, j \leq t$. In other words, the Cartan matrix C and the decomposition matrix D are related by the formula

$$C = D'D,$$

where D' denotes the transpose of the matrix D.

We include a sketch of their proof of the remarkable formula stated in the theorem; it is interesting as it was based on an equally remarkable factorization of a determinant by Frobenius. The papers of Frobenius ([**144**], [**145**]) in which the factorization was proved also contained, as Brauer, Nesbitt, and Nakayama realized, other ideas which proved to be useful in the study of nonsemisimple group algebras (see §2 for further discussion, and a proof of Frobenius's formula (Theorem 2.7)). New proofs of the Brauer-Nesbitt Theorem were given shortly afterwards, by Nakayama [**223**] and Brauer [**24**], based on the fact that the indecomposable representations U_i can be lifted to indecomposable representations $\hat{U}_i$ with entries in the completion of the $\mathfrak{p}$-adic ring $\mathfrak{o}_\mathfrak{p}$.

Let $g_1, \ldots, g_n$ be the elements of G, and let

$$x = x_1 g_1 + \cdots + x_n g_n, y = y_1 g_1 + \cdots + y_n g_n$$

be elements of the group algebra of G over the field of rational functions over the algebraic number field E introduced earlier, in indeterminates $x_1, \ldots, x_n; y_1, \ldots, y_n$. Let L and R' denote the first and second regular representations of the group algebra of G over this field,[7] and let $\tilde{L}$ and $\tilde{R}'$ be the first and second regular representations of the group algebra over the field $K(x_1, \ldots, x_n; y_1, \ldots, y_n)$, where K is an algebraically closed field of characteristic $p > 0$, as above. Frobenius's formula, from [**144**] and [**145**], states that

$$|\tilde{L}(x) + \tilde{R}'(y)| = \prod_{k,\ell} \tilde{\psi}_{k\ell}^{c_{k\ell}},$$

where $|\cdot|$ is the determinant, the $c_{k\ell}$ are the Cartan invariants of the group algebra KG, and the $\tilde{\psi}_{k\ell}$ are distinct irreducible polynomials in $K[x_1, \ldots, x_n; y_1, \ldots, y_n]$ defined as follows. Let $F^{(1)}, \ldots, F^{(t)}$ be a set of representatives of the equivalence classes of irreducible representations of the group algebra KG, as above, extended to the field $K(x_1, \ldots, x_n; y_1, \ldots, y_n)$, and let $\tilde{u}_a^{(k)}, \tilde{v}_b^{(\ell)}$ run over the eigenvalues of the matrices $F^{(k)}(x)$ and $F^{(\ell)}(y)$, respectively. Then

$$\tilde{\psi}_{k\ell} = \prod_{a,b} (\tilde{u}_a^{(k)} + \tilde{v}_b^{(\ell)}),$$

for $1 \leq k, \ell \leq t$.

The preceding formula of Frobenius can also be applied to the first and second regular representations L and S' of EG. Let $A^{(1)}, \ldots, A^{(s)}$ be a full set of inequivalent irreducible, E-rational representations of G, and let x and y be the elements of the extended group algebra of G introduced above. The matrix of the Cartan

---

[7]The representation L is given by left multiplication, while R' is the representation obtained from the right multiplication map R, by taking the transpose R'; for further discussion, see §2.

invariants of the group algebra, in this situation, can be taken to be the identity matrix, and one has

$$|L(x) + R'(y)| = \prod_{i=1}^{s} \psi_{ii},$$

with

$$\psi_{ii} = \prod_{a,b} (u_a^{(i)} + v_b^{(i)}),$$

where $u_a^{(i)}$ and $v_b^{(i)}$ run over the eigenvalues of the matrices $A^{(i)}(x)$ and $A^{(i)}(y)$, respectively. We now consider this relation (mod $\mathfrak{p}$); the eigenvalues of $A^{(i)}(x)$ and $A^{(i)}(y)$ can be shown to belong to the ring $\mathfrak{o}_{\mathfrak{p}}[x_1, \ldots, x_n; y_1, \ldots, y_n]$, so that the homomorphism from $\mathfrak{o}_{\mathfrak{p}} \to \mathfrak{o}_{\mathfrak{p}}/\mathfrak{p} \subseteq K$ can be applied to them. By the definition of the decomposition numbers, it follows that each $\tilde{u}_a^{(k)}$ occurs d_{ik} times among the images (mod $\mathfrak{p}$) of the $u_a^{(i)}$, and each $\tilde{v}_b^{(\ell)}$ occurs $d_{i\ell}$ times among the images of the $v_b^{(i)}$. It follows that the image $\bar{\psi}_{ii}$ of ψ_{ii} taken (mod $\mathfrak{p}$) is given by

$$\bar{\psi}_{ii} = \prod_{k,\ell} \tilde{\psi}_{k\ell}^{d_{ik}d_{i\ell}},$$

and hence, by the Frobenius formula,

$$\overline{|L(x) + R'(y)|} = \prod_{i=1}^{s} \prod_{k,\ell} \tilde{\psi}_{k\ell}^{d_{ik}d_{i\ell}} = \prod_{k,\ell} \tilde{\psi}_{k\ell}^{c_{k\ell}}.$$

As the $\tilde{\psi}_{k\ell}$ are distinct irreducible polynomials, one has

$$c_{k\ell} = \sum_{i=1}^{s} d_{ik}d_{i\ell}$$

for all i and j, completing the proof of the theorem.

We turn now to a discussion of two more fundamental new ideas introduced in the first Brauer-Nesbitt paper: *Brauer characters* and *blocks*. In [**117**], Feit wrote:

> Brauer once told me that one of his motives in studying modular representations was the hope of characterizing certain classical groups over finite fields. If for instance G is supposed to be isomorphic to $GL_n(q)$ then one tries to show that G has a faithful n dimensional representation over the field of q elements and so is isomorphic to a subgroup of $GL_n(q)$. This was exactly the method used in [**29**] and [**30**]. [see §3] However this method doesn't seem to work for any larger groups of Lie type. ... It was soon realized by Brauer that the importance of modular representations lies in a different direction. Namely, the theory can be used to give new information about the character table of a group, which in turn can be applied to questions about the structure of the group.

The introduction of Brauer characters, or modular characters as understood in §5 of their paper [**50**], gave a way to attach complex-valued class functions to modular representations of a finite group G, and hence a direct connection between modular representations over the algebraically closed field K of characteristic p and the ordinary character theory of G. Let $g = |G|$, and let $g = p^a g'$, with g' relatively prime to p. Let E be an algebraic number field containing the $|G|$-th roots of unity, such that all the representations $x \to A^{(i)}(x)$ are rational in E, let $\mathfrak{o}$ be the ring of algebraic integers in E, and let $\mathfrak{p}$ be a prime ideal in $\mathfrak{o}$ containing p, as above. Here is their own account of the procedure, for a given modular representation $x \to A(x)$ of G (from [50], §5):

> It is, however, advisable to change the point of view. Every value of a character Trace $(A(x))$, x of an order m prime to p, is a sum of m-th roots of unity. Set $g' = g/p^a$. The g'-th roots of unity in the algebraic number field form a cyclic group H of order g'. No two of them are congruent mod $\mathfrak{p}$, since the difference of any two distinct ones is a divisor of g'. If we replace each of these roots by its residue class $(\bmod\,\mathfrak{p})$, we obtain an isomorphic mapping of H upon the group of g'-th roots of unity in K. We replace now every modular root of unity in Trace $(A(x))$ by the corresponding complex root of unity. Then Trace $(A(x))$ becomes a *complex number*. If this number is known for every x of order prime to p, we know Trace $(A(x)^r)$, that is the sum of the r-th powers of the characteristic roots of $A(x)$. Since we are now in the field of complex numbers, we may find the characteristic roots of $A(x)$.[8]

The complex valued class function $\varphi(x)$, defined as above on the set of p-regular elements $x \in G$, with the property that $\overline{\varphi(x)} = $ Trace $(A(x))$ for all p-regular elements $x \in G$, is today called the *Brauer character* of the modular representation $x \to A(x)$. Brauer and Nesbitt proved, in [50], Theorem II, the following basic property of the Brauer characters.

THEOREM 1.3. *If two modular representations $x \to A(x)$ and $x \to B(x)$ have the same Brauer character, for elements of G whose orders are prime to p, then they have the same irreducible composition factors.*

From the properties of the Brauer characters obtained above, it follows that $A(x)$ and $B(x)$ have the same eigenvalues, for all p-regular elements of G. By a lemma proved earlier, using the Frobenius-Schur Theorem ([50], §2; see also [100], §17), it follows that the representations $x \to A(x)$ and $x \to B(x)$ have the same composition factors.

Brauer and Nesbitt gave several applications of the preceding results. These included a new proof of Theorem 1.1, based on the result that the Brauer characters of the irreducible modular representations, which are complex-valued class functions on the p-regular classes of G, are linearly independent. We include another application here, namely a formula for the Brauer character of an indecomposable direct summand of the regular representation of KG, in terms of the irreducible complex characters $\chi^{(k)}(x) = $ Trace $(A^{(k)}(x)), 1 \leq k \leq s$. It illustrates how they used Theorem 1.2 to express Brauer characters of modular representations in terms of ordinary characters.[9]

THEOREM 1.4. *The Brauer character $\eta^{(i)}$ of the indecomposable direct summand $U^{(i)}$ of the regular representation of KG is given by the formula*

$$\eta^{(i)} = \sum_{k=1}^{s} d_{ki}\chi^{(k)},$$

---

[8]From Richard Brauer, Warren J. Wong (ed.) and Paul Fong (ed.), *Richard Brauer: Collected Papers, Volume 1* (Cambridge, MA: The MIT Press, 1980), p. 345. Used with permission.

[9]Here, and in what follows, we shall use the term *ordinary character* for characters of representations of G over the complex field $\mathbb{C}$.

where it is understood that both sides are viewed as complex-valued functions on the set of p-regular elements of G. In the formula, the d_{ki} are decomposition numbers, and the $\chi^{(k)}$ are ordinary irreducible characters.

By the definition of the decomposition numbers, the Brauer character of a modular representation $x \to \bar{A}^{(k)}(x)$ is given by:

$$\chi^{(k)} = \sum_{j=1}^{t} d_{kj}\varphi^{(j)},$$

for $1 \leq k \leq s$, where $\varphi^{(j)}$ is the Brauer character of the irreducible modular representation $F^{(j)}$. As Brauer and Nesbitt emphasized, "This is now an equality, not a congruence (if we consider only elements in the first h classes [the p-regular classes])." The definition of the Cartan invariants $c_{k\ell}$ gives a companion formula:

$$\eta^{(k)} = \sum_{\ell=1}^{t} c_{k\ell}\varphi^{(\ell)},$$

for $1 \leq k \leq t$. Applying the relation $C = D'D$ from Theorem 1.2 and the formula given at the beginning of the proof, one has

$$\eta^{(i)} = \sum_{j=1}^{t} c_{ij}\varphi^{(j)} = \sum_{j=1}^{t}\sum_{k=1}^{s} d_{ki}d_{kj}\varphi^{(j)} = \sum_{k=1}^{s} d_{ki}(\sum_{j=1}^{t} d_{kj}\varphi^{(j)}) = \sum_{k=1}^{s} d_{ki}\chi^{(k)},$$

as required.

In case p does not divide the order of G, the group algebra KG is semisimple, and is a direct sum of simple algebras as in the characteristic zero theory. In this situation, the ordinary irreducible characters can be identified with the Brauer characters of the irreducible representations of G in the field K, and the decomposition matrix can be taken to be the identity matrix. When p divides the order of G, the group algebra is not semisimple, and not all the indecomposable representations $U^{(i)}$ are irreducible. In [50], Brauer and Nesbitt said, "It is this second case in which we are mainly interested." In the last section of their paper, they described partitions of the sets of ordinary irreducible representations, modular indecomposable representations, and modular irreducible representations into subsets, called *blocks*, or *p-blocks*, which all flow from the decomposition of the nonsemisimple algebra KG as a direct sum of indecomposable two-sided ideals, called *block ideals*.

They began with the modular indecomposable representations. Assume that the regular representation of KG is the direct sum of indecomposable representations $U_1, U_2, \ldots$. Two indecomposable summands U_i and U_j are said *to belong to the same block* if there is a sequence

$$U_i = U_{i_0}, U_{i_1}, \ldots, U_{i_r} = U_j$$

with the property that any two adjacent terms in the sequence have at least one composition factor in common. This is an equivalence relation, and divides the indecomposable representations U_i into subsets $B_i, i = 1, 2, \ldots$, called *blocks*. Brauer and Nesbitt proved ([50], Theorem VI):

THEOREM 1.5. *Let $\mathfrak{b}_i$ be the set of elements in the group algebra KG which are represented by the zero matrix in every indecomposable representation U_j not belonging to the block B_i. Then $\mathfrak{b}_1, \mathfrak{b}_2, \ldots$ are indecomposable two-sided ideals in KG, and KG is their direct sum.*

A proof of the theorem is given later (see Theorem 2.2). The theorem also divides the irreducible representations into blocks, as it is easily shown that no irreducible representation of KG can occur as a composition factor of the representations afforded by two different blocks.

The distribution of the ordinary irreducible representations of G into p-blocks, for a prime number p, was obtained in [50] using Frobenius's description of the representations of the center of the group algebra over $\mathbb{C}$ in terms of the characters (Chapter II, Theorem 4.2). Let E be an algebraic number field in which the irreducible representations $A^{(i)}$ are rational, and keep the notation introduced earlier for the definition of the modular representations $\bar{A}^{(i)}$, $i = 1, \ldots, s$. The center of the group algebra EG has a basis consisting of class sums c_k, and by Schur's Lemma, $A^{(i)}(c_k) = \omega_k^{(i)} E_{f_i}$, where

$$\omega_k^{(i)} = \frac{h_k \chi_k^{(i)}}{f_i},$$

h_k is the number of elements in the conjugacy class C_k, f_i is the degree of the character $\chi^{(i)}$, and E_{f_i} is the $f_i \times f_i$ identity matrix. The function $\omega^{(i)}$ is an irreducible representation of the center of EG, whose value at a class sum c_k is $\omega_k^{(i)}$. It was known that the expressions $\omega_k^{(i)}$ were algebraic integers (see Chapter III, Lemma 5.1), and hence belong to $\mathfrak{o} \subseteq \mathfrak{o}_\mathfrak{p}$. It follows that c_k, now viewed as a basis element of the center of the group algebra KG, is represented in the modular representation $\bar{A}^{(i)}$ by a diagonal matrix with the diagonal entry $\overline{\omega_k^{(i)}}$. Consequently, all the irreducible modular representations $F^{(j)}$ occurring as composition factors of $\bar{A}^{(i)}$ represent the center of the group algebra KG in the same way. This means that all the composition factors of $\bar{A}^{(i)}$ belong to the same block,[10] and we can assign the ordinary irreducible representation $A^{(i)}$ to that block. The conclusion is that two ordinary representations $A^{(i)}$ and $A^{(i')}$ belong to the same p-block if and only if

$$\omega_k^{(i)} \equiv \omega_k^{(i')} \pmod{\mathfrak{p}},$$

for each k. This criterion can be checked from the character table of G, and was another step in Brauer's search for connections between modular representations and properties of the character table of G.

2. Nonsemisimple Algebras: Brauer, Nakayama, and Nesbitt

Brauer, Nesbitt, and Nakayama realized from the beginning that progress on the modular representation theory of finite groups depended on a better understanding of the structure of nonsemisimple algebras. Their collaboration began when Tadasi Nakayama, with the support of a fellowship, spent two years, from 1937-1939, as a visiting member at the Institute for Advanced Study in Princeton. While in the U.S., he met Brauer and Nesbitt, and joined them in their investigation of modular representation theory. He shared an interest with Brauer in algebraic number theory, as well as in representation theory of finite groups and algebras, and they developed a lasting friendship. Some years later, Brauer was appointed to a visiting professorship at Nagoya University in 1959-1960, at the invitation of Nakayama, who had been a professor there from the time he returned to Japan after

---

[10]This is easily shown, as the block ideals $\mathfrak{b}_\ell$, $\ell = 1, 2, \ldots$ are generated by idempotents ε_ℓ belonging to the center of KG. An irreducible modular representation $F^{(j)}$ belongs to the block B_ℓ if and only if ε_ℓ is represented by 1 in the representation $F^{(j)}$.

his stay in Princeton. The Brauers traveled extensively in Japan, and Brauer had an opportunity to meet other Japanese mathematicians who had made, and were to make, substantial contributions to the development of modular representation theory and the theory of blocks (see §3).

The proofs of the main results of Brauer and Nesbitt in the first papers on modular representation theory [**50**] and [**51**] relied on the theory of matrix representations of group algebras. Nakayama's first contribution to modular representation theory ([**223**], 1938) was to establish the main facts in the setting of what Emmy Noether had called "noncommutative ideal theory," this time for a general nonsemisimple ring satisfying the maximum and minimum conditions,[11] just as Noether had done for semisimple rings in [**233**] (see Chapter VI, §2). The following result, from [**223**], §1, established in a general setting a bijection between isomorphism classes of indecomposable summands of the regular representation of a ring satisfying the chain conditions, and the isomorphism classes of simple modules. Brauer and Nesbitt had used this bijection for the case of group algebras (proved by Nesbitt in his Ph.D. dissertation [**226**]) in their definition of the Cartan matrix.

THEOREM 2.1. *Let A be a ring with identity element 1, satisfying the maximum and minimum condition for left ideals, and let N be the radical of A. Let $\bar{A}$ denote the semisimple ring A/N, and let $a \to \bar{a}$ be the natural homomorphism from A to $\bar{A}$. Then the following statements hold:*

(i) The ring A is a direct sum,

$$A = Ae_1 \oplus \cdots \oplus Ae_m,$$

of indecomposable left ideals Ae_i generated by orthogonal idempotents e_i, and

$$\bar{A} = \bar{A}\bar{e}_i \oplus \cdots \oplus \bar{A}\bar{e}_m$$

is a decomposition of the semisimple ring $\bar{A}$ as a direct sum of minimal left ideals $\bar{A}\bar{e}_i$, for $i = 1, \ldots, m$;

(ii) Each left ideal Ae_i has a unique maximal subideal $Ne_i = Ae_i \cap N$, and $Ae_i/Ne_i \cong \bar{A}\bar{e}_i$, for each i.

(iii) Two indecomposable left ideals Ae_i and Ae_j are isomorphic as left A-modules if and only if the minimal left ideals $\bar{A}\bar{e}_i$ and $\bar{A}\bar{e}_j$ in $\bar{A}$ are isomorphic $\bar{A}$-modules.

Part (i) was known ([**106**], Chapter II, §6). Here is an outline of a proof of the fact that the ideals $\bar{A}\bar{e}_i$ are minimal left ideals. The radical of $e_i Ae_i$ is $e_i Ne_i$, for each i. The condition that Ae_i is indecomposable is equivalent to the statement that the ring $e_i Ae_i$ contains exactly one idempotent. This condition, in turn, means that $e_i Ae_i / e_i Ne_i = \bar{e}_i \bar{A} \bar{e}_i$ is a division ring, and hence that $\bar{A}\bar{e}_i$ is a minimal left ideal in $\bar{A}$.

For part (ii), following [**223**], we have $Ne_i = Ae_i \cap N$, so

$$Ae_i/Ne_i \cong Ae_i/(Ae_i \cap N) \cong (Ae_i + N)/N = \bar{A}\bar{e}_i,$$

by one of the isomorphism theorems for groups with operators. As $\bar{A}\bar{e}_i$ is a simple left $\bar{A}$-module, it follows that Ne_i is a maximal submodule of Ae_i. If M is another maximal submodule of Ae_i, then Ae_i/M is a simple module. Then $N(Ae_i/M) = 0$,

---

[11]A year later, C. Hopkins (Ann. of Math. 40 (1939), 712-730) proved that it was enough to assume the minimum condition: a ring with identity element, which satisfies the minimum condition for left ideals, also satisfies the maximum condition.

Tadasi Nakayama (1912-1964)

and $Ne_i \subseteq M$. This implies that $Ne_i = M$, and completes the proof that Ne_i is the unique maximal submodule of Ae_i.

We shall prove one implication in part (iii), namely the statement that if $\bar{f}$: $\bar{A}\bar{e}_i \to \bar{A}\bar{e}_j$ is an isomorphism of left $\bar{A}$-modules, then there exists an isomorphism of A-modules: $f : Ae_i \to Ae_j$. Let

$$\bar{f}(\bar{e}_i) = \bar{f}(e_i + N) = u + Ne_j \in \bar{A}\bar{e}_j,$$

with $u \in Ae_j$. Then $u \notin Ne_j$. Moreover, $\bar{f}(\bar{e}_i^2) = \bar{e}_i\bar{f}(\bar{e}_i) = \bar{f}(\bar{e}_i)$, because $e_i^2 = e_i$, so $e_i u \notin Ne_j$. Replacing u by $e_i u$, we obtain an element $u \in e_i Ae_j$ with $u \notin Ne_j$. Then the map $f : Ae_i \to Ae_j$ defined by putting $f(a) = au$, for $a \in Ae_i$, is a well-defined homomorphism of A-modules; and is surjective, otherwise $Au \subseteq Ne_j$ by part (ii) of the theorem, and this cannot happen. It remains to prove that f is an isomorphism. This was done in [**223**] using the theory of primary rings. Another approach was to prove that the kernel of f, if not zero, is a direct summand of the A-module Ae_i, contradicting the assumption that Ae_i is indecomposable. This last step was taken a few years later, in a paper by Nagao and Nakayama ([**221**], 1953). They proved:

LEMMA 2.1. *The left ideals of A of the form Ae, with e an idempotent element, have the following property. Whenever a left A-module M has a submodule X such that $M/X \cong Ae$, there exists a submodule X' of M such that $M = X \oplus X'$.*

The proof is almost immediate. Let $u \in M$ be an element mapped to e by the given isomorphism, $M/X \cong Ae$. Then eu also maps onto e, because e is idempotent. Replace u by eu. Then, as in the first part of the proof of part (iii), the map $a \to au$ is a well-defined homomorphism of A-modules from $Ae \to Au$. It is then easy to verify that the submodule $X' = Au$ has the required property. This completes the proof of the lemma, and of part (iii) of the theorem.

The property of left A-modules $P = Ae$ stated in the lemma becomes the assertion, in modern language, that every short exact sequence of left A-modules,

$$0 \to Q \to M \to P \to 0,$$

is a split exact sequence. As Nagao and Nakayama said, in a note added at the end of their paper [**221**], this condition was equivalent to the assertion that P is a *projective* left A-module[12] in the sense of the book *Homological Algebra*, by Cartan and Eilenberg [**91**], which appeared shortly after their paper. One of the main results in [**221**] was the theorem that the projective left A-modules, for a finite dimensional algebra A over a field, are characterized by the property that they can be expressed as direct sums, finite or infinite, of indecomposable left A-modules of the form Ae, for an idempotent $e \in A$.

Nakayama applied Theorem 2.1 to give an interpretation of blocks in the setting of a general nonsemisimple ring A satisfying the chain conditions, and, in particular, a new proof of Theorem 1.5. Let

$$A = Ae_1 \oplus \cdots \oplus Ae_m,$$

where the Ae_i are indecomposable left ideals, as in part (i) of the theorem. Two left ideals Ae_i and Ae_j are said to belong to the same block if and only if there is a sequence of idempotents starting with e_i and ending with e_j, such that the left ideals

---

[12]For a modern proof of part (iii) of the theorem, using projective modules and the result known today as *Nakayama's Lemma*, see [**99**], Proposition 6.6.

generated by any two adjacent ones have a composition factor in common. This is the same definition given earlier for the case of group algebras, and divides the indecomposable left ideals Ae_i into equivalence classes, called blocks, and denoted by $B_1, B_2, \ldots$.

LEMMA 2.2. *The left A-module Ae_j, has a composition factor isomorphic to the simple module Ae_i/Ne_i, for $1 \leq i, j \leq m$, if and only if $e_i Ae_j \neq 0$.*

A proof is easily supplied, using the fact that a simple left A-module $\mathfrak{m}$ is isomorphic to Ae_i/Ne_i if and only if $e_i\mathfrak{m} \neq 0$.

THEOREM 2.2. *For each block B_i, $i = 1, 2, \ldots$, let $\mathfrak{b}_i$ be the sum of the left ideals Ae_j belonging to the block B_i. Then*

$$A = \mathfrak{b}_1 \oplus \mathfrak{b}_2 \oplus \cdots,$$

and the block ideals $\mathfrak{b}_i$ are uniquely determined, indecomposable, two-sided ideals.

We include a sketch of Nakayama's proof. The first step was to prove that the left ideals $\mathfrak{b}_i$ are, in fact, two-sided ideals. For this, it was sufficient to prove that $Ae_i Ae_j = 0$ whenever Ae_i and Ae_j belong to different blocks. Assume, to the contrary, that $Ae_i Ae_j \neq 0$. Then one has $e_i Ae_j \neq 0$, and this means that Ae_i and Ae_j belong to the same block, by Lemma 2.2. This completes the proof that the block ideals are two-sided ideals.

The chain conditions imply that A is a direct sum of indecomposable two-sided ideals $\mathfrak{b}_1', \mathfrak{b}_2', \ldots$. Nakayama's idea was to prove that the ideals $\mathfrak{b}_i'$ coincide with the block ideals. For each left ideal $Ae_i, 1 \leq i \leq m$, one has

$$Ae_i = \mathfrak{b}_1' e_i \oplus \mathfrak{b}_2' e_i \oplus \cdots.$$

As Ae_i is indecomposable, it follows that $Ae_i = \mathfrak{b}_j' e_i$ for some j, and hence Ae_i is contained in one of the ideals $\mathfrak{b}_r'$, for $r = 1, 2, \ldots$. The proof is completed by showing that if Ae_j belongs to the same block as Ae_i, then both are contained in the same two-sided ideal $\mathfrak{b}_r'$. Let Ae_u, Ae_v be two adjacent ideals in the sequence from Ae_i to Ae_j. Then they have a common composition factor Ae_h/Ne_h for some h. By Lemma 2.2, one has $e_h Ae_u \neq 0$ and $e_h Ae_v \neq 0$. Then $e_h Ae_u \subseteq Ae_h A, Ae_u A$ and $e_h Ae_v \subseteq Ae_h A, Ae_v A$. Therefore Ae_u and Ae_v belong to the same two-sided ideal $\mathfrak{b}_r'$. As this is true for each adjacent pair in the sequence, it follows that Ae_i and Ae_j are both contained in $\mathfrak{b}_r'$, completing the proof.

An important step towards understanding the structure of nonsemisimple group algebras was Brauer's, Nesbitt's, and Nakayama's grasp of the ramifications of an idea introduced by Frobenius in two papers on algebras over the field of complex numbers ([**144**], [**145**], 1903). These papers also contained the formula they used in their proof, given above, of the relation $C = D'D$.

Frobenius's idea arose from a continuation of his analysis of the structure constants of an associative algebra in the work leading up to his factorization of the group determinant (see Chapter II, §3). Let A be an associative algebra with a basis $a_1, \ldots, a_n$ over an arbitrary field F, and let the equations

$$a_j a_k = \sum_{i=1}^{n} \alpha_{ijk} a_i$$

define the multiplication of the basis elements.[13] The associative law implies that the structure constants satisfy the relations

$$\sum_i \alpha_{ijk}\alpha_{hi\ell} = \sum_i \alpha_{ik\ell}\alpha_{hji}.$$

Frobenius defined matrices $L(a_j), R(a_\ell)$, and P_h in terms of the structure constants α_{ijk} by setting

$$L(a_j)_{ik} = \alpha_{ijk}, \ R(a_\ell)_{ik} = \alpha_{ik\ell}, \text{ and } (P_h)_{ij} = \alpha_{hij},$$

where, for example, $L(a_j)_{ik} = \alpha_{ijk}$ means that the entry in the i-th row and k-th column of the $n \times n$ matrix $L(a_j)$ is α_{ijk}. Then $a_i \rightarrow L(a_i)$, extended by linearity, is the (first) regular representation of A, given by the left multiplications by elements of A; while the linear map that takes a_ℓ to the transpose $R(a_\ell)'$ of the matrix $R(a_\ell)$ of right multiplication by a_ℓ is another representation of A, called the *second regular representation*. As Frobenius said, this follows from the fact that the matrices of left multiplication $L(a)$ and right multiplication $R(a)$ satisfy the equations

$$L(ab) = L(a)L(b) \text{ and } R(ab) = R(b)R(a),$$

for all elements $a, b \in A$.

The new observation was that the matrices of the form $P(\xi) = \sum_h \xi_h P_h$, with $\xi_h \in F$ for $h = 1, \ldots, n$, which Frobenius called *parastrophic matrices*, have the property that

$$P(\xi)L(a) = R'(a)P(\xi)$$

for all elements $a \in A$. In other words, the parastrophic matrices $P(\xi)$ intertwine the (first) regular representation and the second regular representation.

THEOREM 2.3. (i) *The matrices* $\{L(a), a \in A\}$ *and* $\{R(b), b \in A\}$ *commute:*

$$L(a)R(b) = R(b)L(a),$$

for all $a, b \in A$, *and each set is the centralizer of the other.*

(ii) *The first and second regular representations* $a \rightarrow L(a)$ *and* $a \rightarrow R'(a)$ *are equivalent if and only if there exists an invertible parastrophic matrix* $P(\xi)$.

The proof of the theorem is immediate from the preceding discussion, and is based on the identities for the structure constants α_{ijk} implied by the associative law.

Algebras with the property that the first and second regular representations are equivalent were called *Frobenius algebras*[14] by Brauer, Nesbitt, and Nakayama ([51], [226], [224]). In [51], Brauer and Nesbitt stated a criterion for an algebra to be a Frobenius algebra which was independent of the choice of a basis, and did not involve the parastrophic matrices:

THEOREM 2.4. *A finite dimensional algebra* A *over an arbitrary field* F *is a Frobenius algebra if and only if there exists a hyperplane in* A *containing no nonzero right ideal.*

---

[13]By this time (1903), Frobenius was fully aware of the extent of Molien's work on the structure of algebras, and acknowledged, in the introduction to [144], Molien's "fundamental article" [215] on the subject.

[14]In [226], Nesbitt wrote, "The author, in collaborating with T. Nakayama, adopted the term Frobeniusean algebra, but now, quailing before our critics, we return to simply Frobenius algebra."

The criterion means that for some nonzero linear function μ, one has: $\mu(aA) = 0$ implies $a = 0$, for all elements $a \in A$. Semisimple algebras, and group algebras of finite groups, even in case they are not semisimple, are easily shown to be Frobenius algebras. For example, let FG be a group algebra of a finite group G over an arbitrary field F. Let μ be the linear function on FG defined by:

$$\mu(\sum_{x \in G} \alpha_x x) = \alpha_1,$$

where 1 is the identity element of G. Then the set of zeros of the linear function μ is a hyperplane in FG containing no nonzero right ideal, so FG is a Frobenius algebra, by the preceding theorem.

It was now a short step to the modern approach to Frobenius algebras, which appeared for the first time in Nakayama's paper published two years later [**224**]. The (first) regular representation of an algebra A over a field F is the matrix representation afforded by the left A-module A, with the module operation given by left multiplication in the ring A. Now consider a finitely generated right A-module $\mathfrak{M}$. The dual vector space $\mathfrak{M}^*$, consisting of the linear functions $\lambda : \mathfrak{M} \to F$, is a finite dimensional vector space over F, and becomes a left A-module with the module operation $(a, \lambda) \to a\lambda \in \mathfrak{M}^*$ defined by

$$(a\lambda)(m) = \lambda(ma),$$

for $a \in A, \lambda \in \mathfrak{M}^*, m \in \mathfrak{M}$. When this construction is applied to the right A-module A, with the module operation given by right multiplication, the resulting left A-module A^* affords the second regular representation, so that A is a Frobenius algebra if and only if there is an isomorphism of left A-modules $A \cong A^*$.

THEOREM 2.5. *Let A be a finite dimensional algebra over a field F. There exists an isomorphism of left A-modules $A \cong A^*$ if and only if there exists a nondegenerate bilinear form $f : A \times A \to F$ which is associative, in the sense that*

$$f(ab, c) = f(a, bc),$$

for all $a, b, c \in A$.

The theorem is easily proved using the theory of dual vector spaces, and implies the two preceding theorems.

Let A be a finite dimensional algebra, and let

$$A = Ae_1 \oplus \cdots \oplus Ae_m = e_1 A \oplus \cdots \oplus e_m A,$$

for a set of orthogonal idempotents e_i such that the left A-modules Ae_i and $(e_j A)^*$ are indecomposable, for all i and j. The duals $(e_i A)^*$ are the indecomposable direct summands of the dual module A^*, so that, if A is a Frobenius algebra, there exists a permutation π of $1, 2, \ldots, m$ such that $(e_i A)^* \cong Ae_{\pi(i)}$ for each i, by Theorem 2.5. In [**51**], Brauer and Nesbitt raised the question, under what circumstances is it true that there is an isomorphism of left A-modules $(e_i A)^* \cong Ae_i$ for each i? A family of algebras with this property were the *symmetric algebras*, which Brauer and Nesbitt defined as algebras A, with a hyperplane containing all commutators $ab - ba$, but no nonzero right ideal.[15] Symmetric algebras are, of course, Frobenius algebras. The hyperplane used above to prove that group algebras of finite groups

---

[15]Alternatively, an algebra A is symmetric if and only if there exists a nondegenerate, associative bilinear form $f : A \times A \to F$, as in Theorem 2.5, which is symmetric: $f(a, b) = f(b, a)$, for all $a, b \in A$.

are Frobenius algebras can easily be shown to contain all commutators, so that, in fact, all group algebras are symmetric algebras. Therefore the following theorem, proved by Brauer and Nesbitt for symmetric algebras, applied to group algebras of finite groups, and provided additional information for the theory of modular representations.

THEOREM 2.6. *Let A be a symmetric algebra. Then each indecomposable left ideal Ae generated by an idempotent element e is isomorphic to the dual module $(eA)^*$. Moreover, the indecomposable module Ae contains a unique maximal submodule Ne, and a unique minimal submodule, which is isomorphic to Ae/Ne. In particular, the first and last composition factors of Ae are isomorphic.*

In their paper [221], Nagao and Nakayama also gave a characterization of the dual modules $(eA)^*$ for an arbitrary algebra A. They proved that a left A-module M has the property that every short exact sequence

$$0 \to M \to N \to Q \to 0$$

is a split exact sequence if and only if M is a finite or infinite direct sum of modules of the form $(eA)^*$, where eA is an indecomposable right ideal generated by an idempotent element e. The splitting condition for a left A-module M was equivalent, as they pointed out, to the condition that M is an *injective* module, in the sense of Cartan and Eilenberg's book [91].

We turn now to the formula of Frobenius [144] which Brauer and Nesbitt used in their proof of the formula $C = D'D$.

THEOREM 2.7. *Let A be an algebra over an algebraically closed field F, and let L and R' denote the first and second regular representations of A. Let $x = x_1 a_1 + \cdots + x_n a_n$ and $y = y_1 a_1 + \cdots + y_n a_n$, for a basis $a_1, a_2, \ldots$ of A, and indeterminates $x_1, \ldots, x_n; y_1, \ldots, y_n$. Then*

$$|L(x) + R'(y)| = \prod_{k,\ell} \psi_{k\ell}^{c_{k\ell}},$$

where $|\cdot|$ is the determinant, the $c_{k\ell}$ are the Cartan invariants of the algebra A, and the $\psi_{k\ell}$ are distinct irreducible polynomials in $K[x_1, \ldots, x_n; y_1, \ldots, y_n]$ defined as follows. Let $F^{(1)}, \ldots, F^{(t)}$ be a set of representatives of the equivalence classes of irreducible representations of the algebra A, extended to the field $K(x_1, \ldots, x_n; y_1, \ldots, y_n)$, and let $u_a^{(k)}, v_b^{(\ell)}$ run over the eigenvalues of the matrices $F^{(k)}(x)$ and $F^{(\ell)}(y)$, respectively. Then

$$\psi_{k\ell} = \prod_{a,b} (u_a^{(k)} + v_b^{(\ell)}),$$

for $1 \le k, \ell \le t$.

The matrices $L(x)$ and $R(y)$ commute, by the associative law, and the eigenvalues of the matrix $R'(y)$ are the same as those of $R(y)$. The commuting matrices $L(x)$ and $R(y)$ can simultaneously be put in triangular form, so the eigenvalues of $L(x) + R'(y)$ can be expressed as sums $\alpha + \beta$, where α runs through the eigenvalues of $L(x)$, and for each α, β is a corresponding eigenvalue of $R(y)$. Frobenius proved, in ([144], §12), that the product of these eigenvalues $|L(x) + R'(y)|$ has a factorization as stated in the theorem by analyzing the linear factors of the determinant $|uL(x) + vR(y) + wE|$, where u, v, and w are additional variables, and E is the identity matrix. For this purpose, he referred to §10 of his paper [**132**], where he used his analysis of the factors of the group determinant to obtain the linear factors of the polynomial

$$|ux_{pq^{-1}} + vy_{q^{-1}p} + w\varepsilon_{pq^{-1}}|,$$

where p, q are elements of a finite group G, $(x_{pq^{-1}})$ is the group determinant, and $(\varepsilon_{pq^{-1}})$ is the identity matrix.

A conceptual proof of the factorization of the determinant $|L(x) + R'(y)|$ can be obtained as follows. The first step is to prove that the composition factors of the algebra A, viewed as a two-sided A-module, are isomorphic to bimodules of the form $\mathfrak{m} \otimes \mathfrak{n}$, where $\mathfrak{m}$ is a simple left A-module, and $\mathfrak{n}$ is a simple right A-module. For this, one shows that a minimal two-sided submodule $\mathfrak{a}$ of A can be expressed as

$$\mathfrak{a} = \mathfrak{l}A = A\mathfrak{r}$$

for some minimal left ideal $\mathfrak{l}$ and minimal right ideal $\mathfrak{r}$, so that $\mathfrak{a} \cong \mathfrak{l} \otimes \mathfrak{r}$. Then the determinant of the linear transformation $L(x) \otimes 1 + 1 \otimes R(y)$, restricted to $\mathfrak{m} \otimes \mathfrak{n}$, is an irreducible polynomial[16] ψ in the variables x_i and y_j, and has the form:

$$\psi = \prod_{a,b}(u_a + v_b),$$

where the elements u_a are the eigenvalues of the restriction of $L(x)$ to $\mathfrak{m}$, and v_b are the eigenvalues of the restriction of $R(y)$ to $\mathfrak{n}$.

It remains to prove that the multiplicities with which the irreducible factors $\psi_{k\ell}$ appear in $|L(x) + R'(y)|$ are the Cartan invariants. We give a proof[17] from Nesbitt's thesis [**226**], based on the following result of Brauer ([**23**], §2).

THEOREM 2.8. *Let A and B be two algebras of $n \times n$ matrices with entries in an algebraically closed field K, such that each is the centralizer of the other. Let $\{F_i, i = 1, 2, \dots\}$ be the composition factors of A, and let $\{U_j, j = 1, 2, \dots\}$ be the indecomposable direct summands of B. Then there is a bijection $F_i \leftrightarrow U_i$ from representatives of equivalence classes of the irreducible representations F_i of A to the equivalence classes of the indecomposable representations U_i of B. For a suitable ordering, the multiplicity of F_i as a composition factor of A is equal to the degree u_i of U_i, and the multiplicity of U_i as a summand of B is equal to the degree f_i of F_i.*

Nesbitt began with the observation, which goes at least as far back as Frobenius's paper [**144**] (see Theorem 2.3), that the algebras of left multiplications $\{L(a), a \in A\}$ and right multiplications $\{R(b), b \in A\}$ are centralizers of each other. Upon pairing the irreducible composition factors F_i of the first regular representation of A with the indecomposable summands U_i of the second, it follows from Theorem 2.8 that one has

$$|L(x) + R'(y)| = \prod_k |E_{f_k} \otimes U_k(y) + F_k(x) \otimes E_{u_k}|,$$

where f_k is the degree of F_k, u_k is the degree of U_k, and the E_{f_k} etc. are identity matrices. Now choose elements $x, y \in A$ such that $F_k(x) = \alpha_k E_{f_k}$ and $F_\ell(y) =$

---

[16]See the proof of Corollary 2.2 in Chapter IV.

[17]In [**144**] and [**145**], Frobenius proved the result by comparison of the irreducible factors of the polynomials $|L(x)|$, for A, and for the semisimple algebra A/N, where N is the radical of A.

$\beta_\ell E_{f_\ell}$, for elements $\alpha_k, \beta_\ell \in K$. Using the fact that the Cartan invariant $c_{k\ell}$ is the multiplicity of F_ℓ as a composition factor of U_k, for each pair k, ℓ, it follows that

$$|E_{f_k} \otimes U_k(y) + F_k(x) \otimes E_{u_k}| = \prod_\ell (\alpha_k + \beta_\ell)^{c_{k\ell} f_k f_\ell}.$$

On the other hand, for the same choice of x and y, one obtains

$$\psi_{k\ell} = (\alpha_k + \beta_\ell)^{f_k f_\ell}.$$

Therefore, for this choice of x and y, it follows that

$$|L(x) + R'(y)| = \prod_{k,\ell} (\alpha_k + \beta_\ell)^{c_{k\ell} f_k f_\ell} = \prod_{k,\ell} (\alpha_k + \beta_\ell)^{e_{k\ell} f_k f_\ell},$$

where $e_{k\ell}$ is the multiplicity of $\psi_{k\ell}$ in the factorization of $|L(x) + R'(y)|$ with indeterminate coefficients x_i and y_j. As the preceding formula holds for all choices of α_k and β_ℓ, it follows that $c_{k\ell} = e_{k\ell}$ for all k and ℓ, completing the proof.

3. Blocks and the Classification of Finite Simple Groups

The theory of blocks, and its application to finite group theory, especially to the problem of classifying finite simple groups, became central themes in Brauer's research soon after blocks were first introduced, in [**50**]. His idea was to use the theory of blocks to obtain information about the values of the ordinary irreducible characters of a finite group G satisfying some general hypotheses, such as having Sylow subgroups of a given structure, and then using the character-theoretic information to narrow the list of possibilities for G itself, at least in case G is a simple group. His first results in this direction were enormously successful, and we begin with a report on some of them. The early work led to general results on blocks, and problems and conjectures concerning them, that occupied him for the rest of his career. As he said in his *Colloquium Lectures*, given at the Summer Meeting of the American Mathematical Society in Madison, Wisconsin, 1948, "The blocks can be made the subject of an elaborate theory."

The partition of the set of ordinary irreducible characters of a finite group G into p-blocks, for a prime p, was explained at the end of §1. In the last of the joint papers with Nesbitt [**52**], the authors defined a numerical invariant, called the *type*, for each p-block of ordinary characters. A p-block of ordinary characters was said to have type α if the degrees of all the characters in the block are divisible by p^α, but the degree of at least one character in the block is not divisible by $p^{\alpha+1}$. This concept was understood with reference to Frobenius's theorem that the degree of each irreducible character divides the order of the group (Chapter III, Theorem 3.5). Let $|G| = p^a h$, with h not divisible by p. Then the possible types of p-blocks of G are the integers $0, 1, \ldots, a$, by Frobenius's theorem. The main results in their paper [**52**] were two quite remarkable theorems about the blocks of highest and lowest type. We state only one here, for blocks of highest type.

THEOREM 3.1. *Let G be a finite group of order $p^a h$, with h not divisible by p. Each ordinary irreducible character χ, of degree divisible by p^a, forms a p-block by itself. If $x \to A(x)$ is an ordinary irreducible representation affording χ, then a modular representation $x \to \bar{A}(x)$ obtained by reduction $(\bmod\, \mathfrak{p})$ is irreducible, and so is equivalent to one of the irreducible modular representations F_i. Then the corresponding indecomposable representation U_i is irreducible, $U_i = F_i$. Finally, the character χ vanishes on elements of order divisible by p.*

Finite groups G of order divisible by a prime p to the first power, but not divisible by p^2, have the property that all p-blocks are either of the highest or lowest type. Brauer carried out a profound investigation of the ordinary character theory of finite groups G satisfying this hypothesis, by obtaining precise information about the character values of ordinary irreducible characters belonging to a p-block of lowest type. He announced the results in [26], (1939), along with some applications to problems concerning finite simple groups; the proofs of the theorems appeared in a series of papers published later ([29], [30], [31]), and in a joint paper on simple groups of finite order, with Hsio-Fu Tuan ([53]). In [29], he wrote,

> To be sure, this assumption [the hypothesis that the order of G contains a prime to the first power] is of a very restrictive nature. It may, however, be mentioned that Dickson's list ([107], pp. 309, 310) of 78 simple groups of an order smaller than 10^9 contains only one group for which there is no prime p such that [the hypothesis] holds.[18]

Brauer found a nice way to explain his results, at least in outline form, in terms of the character table. Recall from Chapter II, §3, that the character table Z of a finite group G is a square matrix of complex numbers, whose rows are indexed by the irreducible characters, and the columns by the conjugacy classes. An entry of the character table matrix Z is the value of an irreducible character at an element of a conjugacy class. For a finite group G satisfying the hypothesis above, let P be a Sylow p-subgroup, and let N denote the normalizer of P; then P is cyclic of order p. The degrees of the irreducible characters divide the order of the group, and consequently are either relatively prime to p, or are divisible by p to the first power only. The rows and columns of the character table Z are arranged so that the characters of degree relatively prime to p are taken before those of degree divisible by p, and the conjugacy classes are ordered so that the p-regular classes are taken first. Then Z has the form

$$Z = \begin{pmatrix} Z_1 & Z_2 \\ Z_3 & Z_4 \end{pmatrix}.$$

The characters of degree divisible by p belong to blocks of highest type, so that $Z_4 = 0$, by Theorem 3.1. The main result of [29] states that the matrix Z_2 is determined by the structure of the normalizer N of the Sylow p-subgroup P, with only some $\pm$ signs missing. More precisely, if G is replaced by N, the matrix Z_2 remains essentially the same, except that the signs of some rows may have to be changed. In the case of the group N, the second row (Z_3, Z_4) of the character table matrix, arranged as above, is missing. This implies that the number of characters of G whose degrees are not divisible by p is equal to the number of conjugacy classes of N.

The paper contained more explicit versions of the preceding statements; in particular, a surprisingly detailed description of the characters of degree prime to p was obtained, by partitioning them in blocks of lowest type, and carrying out a painstaking analysis of the characters in each block. Several ideas which proved to be important for subsequent research on blocks of characters in more general situations appeared here for the first time. These included the identification of a certain family of p-conjugate characters belonging to a block of lowest type,

---

[18]R. Brauer, On groups whose order contains a prime to the first power I. *American Journal of Mathematics* **64** (1942), pp. 401–420. Used with permission of The Johns Hopkins University Press.

called *exceptional* characters ([**29**], Theorem 2), and a graph associated with the characters belonging to a block of lowest type, today known as the *Brauer tree* ([**29**], Theorem 3). For modern versions of Brauer's results on characters of groups whose orders contain a prime to the first power, with new proofs, additional results, and applications, see Feit [**118**], Chapters VII and VIII.

In [**117**], Feit wrote, "At the time he discovered these results, he tried to check them on known character tables. In one case the results did not check and, as he told me many years later, he was rather unhappy for a while until he had shown that the error was in the character table and not in his results."

From the results stated above, some information can be obtained concerning the submatrices Z_1 and Z_3 of the character table Z, for example by applying the orthogonality relations. As an illustration, Brauer mentioned two cases where he had been able to calculate the full character table by these methods, in connection with the question as to whether there was more than one simple group of a given order.[19] He wrote, in [**29**],

> Heretofore it has not been known whether more than one simple group exists for the orders $|G| = 5616$ and $|G| = 6048$. Assuming that G is a simple group of one of these orders, we may find by elementary methods the structure of N, where $p = 13$ in the first case and $p = 7$ in the second case. It turns out that the results obtained are sufficient to construct the complete table of group characters in either case. From this table we can derive the modular group characters [Brauer characters] of G for the prime $p = 3$ (here the theorem given above does not hold since 3 divides the order of the group to a higher power than the first). The values of the modular characters show that a simple group G of order 5616 must have an absolutely irreducible representation of degree 3 in the Galois field $GF(3)$. Hence it is a subgroup of $PSL_3(3)$, and since G and $PSL_3(3)$ have the same order, we find $G \cong PSL_3(3)$. Similarly, it can be shown that a simple group of order 6048 is isomorphic to $U_3(3)$ since it must have a unitary representation of degree 3 in $GF(9)$.[20]

Three theorems announced in [**24**] were of a more general nature, and were the first applications of modular representation theory to the structure of finite groups.

THEOREM 3.2. *Let G be an irreducible group of linear transformations of degree n which has no normal subgroup of order p. If the order of G is divisible by the prime p to the first power only, then $p \leq 2n + 1$. If $p = 2n + 1$, then G, viewed as a collineation group, is isomorphic to $PSL_2(p)$.*

This theorem improved, for groups of order not divisible by p^2, a theorem of H. F. Blichfeldt,[21] who proved that $p \leq (2n+1)(n-1)$. It also established, for groups of order not divisible by p^2, a conjecture of G. de B. Robinson, a colleague of Brauer's at Toronto, namely that a finite group with a faithful complex representation of

---

[19]In his list of simple groups of order less than 10^9, Dickson (loc. cit.) had noted the existence of two nonisomorphic simple groups of order 20160, and the fact that there is an infinite sequence of numbers with this property.

[20]R. Brauer, On groups whose order contains a prime to the first power I. *American Journal of Mathematics* **64** (1942), pp. 401–420. Used with permission of The Johns Hopkins University Press.

[21]Trans. Amer. Math. Soc. **4** (1903), 387–397.

degree $d < \frac{1}{2}(p-1)$ has a normal Sylow p-subgroup. The conjecture was proved twenty years later by Walter Feit and John Thompson [121], who reduced the general question to the case which had been settled by Brauer.

By Burnside's $p^a q^b$-theorem (Chapter III, §5), the order of a finite simple group of composite order contains at least 3 distinct prime factors. Brauer's second theorem began the classification of simple groups whose orders contain 3 distinct prime factors.

THEOREM 3.3. *Let G be a simple group of order pqr^m, for distinct primes p, q, r, and a positive integer m. Then $|G| = 60$ or $|G| = 168$.*

The proof of the theorem appeared later, in a joint paper ([53], 1945) with Tuan.[22] In their paper, Brauer and Tuan obtained the more general result that a simple group G of order $pq^b t$, where p and q are two primes, and b and t are positive integers with $t < p - 1$, is isomorphic either to $PSL_2(p)$, with $p = 2^m \pm 1, p > 3$, or to $SL_2(2^m)$ with $p = 2^m + 1, p > 3$. The proof once again used the full strength of the results in [29], together with a theorem of Tuan [273] on the classification of finite simple groups whose order contain a prime to the first power in terms of a condition on an irreducible character belonging to the principal p-block.[23]

The classification of finite simple groups G whose orders are divisible by 3 primes was not achieved by character-theoretic methods alone; a new method, called *local group-theoretic analysis*, was required.[24] Local analysis was at the heart of the proofs of the odd-order theorem [122] by Feit and Thompson, and Thompson's results on the classification of simple N-groups, that is, finite simple groups all of whose local subgroups are solvable [272]. From Thompson's N-group paper, it follows that if the order of a finite simple group G has just 3 distinct prime factors, then two of the factors are 2 and 3, and the remaining one is $5, 7, 13$, or 17. From this information, the possibilities for the group G can be determined.

The order of a transitive permutation group of prime degree p has the form $qp(1 + np)$, where q divides $p - 1$, and is divisible by p to the first power only, so that Brauer was able to apply his powerful new results on blocks to this family of groups, which had previously been considered by Frobenius, Burnside, Schur, and others (see Chapter III, Theorem 3.8, and the subsequent discussion). Brauer announced a classification of a wide class of simple groups belonging to this family:

THEOREM 3.4. *Let G be a simple group of order $qp(1 + np)$, where q divides $p - 1$, and assume that elements of order p commute only with their powers. If $n \leq (2p + 7)/3$, then either (i) G is cyclic, or (ii) $G \cong PSL_2(p)$, or (iii) p is a prime of the form $2^h \pm 1$ and $G \cong SL_2(2^h)$.*

The hypothesis of the theorem is satisfied not only by simple permutation groups of prime degree, but also by other families of simple groups, for example

---

[22]Tuan completed his doctoral dissertation under the direction of Brauer; the dissertation was submitted to Princeton University, and was published in 1944 ([273]). Upon his return to China, Tuan founded a school for research in modular representation theory and the structure of finite groups at Peking University. Under his leadership, it became a center for research, and is flourishing today (see [274]).

[23]The *principal p-block* of a finite group is the p-block containing the principal character.

[24]Local subgroups of a finite group are normalizers of subgroups of prime-power order, and local group-theoretic analysis is the study of these subgroups in a given finite group. Gorenstein's article [162] contains an explanation of the method as it applies to this particular classification problem.

John G. Thompson

by doubly transitive simple permutation groups of degree $p + 1$. The proof of the theorem, along with other related results, was published in [31], (1943), following a year spent at the Institute for Advanced Study in Princeton with the support of a Guggenheim Fellowship. The fellowship gave him badly needed time to write up the, sometimes lengthy, proofs of the results that flowed from [29], and also an opportunity to visit Emil Artin at Indiana University. His conversation with Artin was the starting point of another chapter of his research, this time on algebraic number theory, and in particular his proof of Artin's conjecture concerning L-series with general group characters (see §4).

The theorems discussed above concerning blocks of characters in a finite group G whose order contains the prime p to the first power only, and their application to problems about finite simple groups, led Brauer to a systematic investigation of blocks in a general finite group. The results known today as *Brauer's First and Second Main Theorems* on blocks were obtained between 1944 and 1946, and were announced in three notes [32], [34], and [35]. In the first of these notes, he said:

> There can be little doubt that we are far from knowing all important properties of group characters. In particular, we are interested in further results which connect the group characters directly with the properties of the abstract group G. Any result of this kind means, in the last analysis, a result concerning the structure of the general group of finite order.

The First and Second Main Theorems imply new connections between p-blocks of characters and the structure of a finite group, and are based on the important notion of the *defect group* of a p-block. He began with the concept of the *defect d* of a p-block of type α; this was defined, in [32], by the formula

$$d = a - \alpha,$$

where p^a is the highest power of p dividing the order of the group. The p-blocks of highest type became blocks of defect 0. The defect of a p-block was also characterized as the exponent of the highest power of p dividing all the integers $|G|/f_i$, as the f_i run through the degrees of the ordinary irreducible characters belonging to the given p-block.

The *defect group* of a given p-block, defined in [34], was a p-subgroup of G, uniquely determined up to conjugacy, of order p^d, where d is the defect of the block. For example, the defect group of the principal p-block (containing the principal character) is a Sylow p-subgroup of G, by one of the characterizations of the defect given above. At the other extreme, the defect group of a p-block of defect 0 is the trivial subgroup, containing the identity element alone.

We summarize some of the properties of the defect group of a block in the following statements; a definition is given later, following Theorem 3.5. The properties listed here, and some others, were stated by Brauer in the notes [34] and [35]. An interesting application of defect groups appeared shortly afterwards, in Brauer's part of the proof of a conjecture of Nakayama [225] concerning p-blocks of ordinary irreducible characters of the symmetric group. Recall that the irreducible characters of the symmetric group S_n are parametrized by partitions of n (Chapter II, §5). Nakayama's conjecture gave a condition on two partitions of n for the corresponding irreducible characters to belong to the same p-block. It was proved in 1947 by Brauer and G. de B. Robinson ([36], [240]) at the University of Toronto.

Some of the properties of defect groups stated below involve the representations of the center of the group algebra $\omega^{(i)}$, corresponding to the ordinary irreducible characters $\chi^{(i)}$ of a finite group G. These were reviewed at the end of §1, and the notation introduced there will be carried over to the present discussion.

(i) The defect group D of a p-block B is a p-subgroup of G of order p^d, where d is the defect of the block.

(ii) There exists a p-regular element $v \in G$ such that D is a Sylow p-subgroup of the centralizer $C_G(v)$. Moreover, let c_k be the sum of the elements in the conjugacy class containing v; then $\omega_k^{(i)} \notin \mathfrak{p}$ for all the representations $\omega^{(i)}$ corresponding to characters in the block B.

(iii) If the class C_ℓ containing an element $h \in G$ does not contain elements of the centralizer $C_G(D)$, then $\omega_\ell^{(i)} \equiv 0 \pmod{\mathfrak{p}}$ for all $\omega^{(i)}$ belonging to characters in the block B.

The properties (i)-(iii) characterize the defect group D of the block B, up to conjugacy in G.

(iv) If u is an element of order $p^b, b \geq 1$, which is not conjugate to an element of D, and if v is a p-regular element commuting with u, then

$$\chi^{(i)}(uv) = 0,$$

for all characters $\chi^{(i)}$ belonging to the block B.

The last property (iv) is a generalization, to an arbitrary block, of the part of Theorem 3.1 which states that a character belonging to a p-block of defect zero vanishes on elements which are not p-regular.

The proofs of properties (i)-(iv) and a deep result concerning defect groups—the first of Brauer's Main Theorems on Blocks—appeared 10 years later, in [42], (1956), after he had moved to Harvard University. In the meantime, he had proved Artin's conjecture on L-series, the result known today as the *Brauer Induction Theorem*, and solved the splitting field problem (see §4 for these results), so it is not surprising that there was a delay. But there was more to it than that. In [117], Feit said, "The ten year delay in the publication of the First and Second Main Theorems of block theory was only partly due to the fact that Brauer was working on other things. It was also partly due to the fact that he felt that no one was interested in his work on modular representations. This was not really the case. For instance, I. M. Gelfand [in Moscow] was interested in some of his work and quite recently, he and some of his collaborators have proved results about the representations of Lie algebras which are analogous to some of Brauer's results in modular representation theory (see [13]). Of course Brauer could not know of Gelfand's interest. There were also some Japanese mathematicians who had become interested in the theory of modular representations." In particular, M. Osima had obtained independently some of the results about the theory of blocks, and had published a proof of the First Main Theorem ([235], 1955) before the publication of [42].

The following conceptual description (see Theorem 3.5) of the defect group of a block B in terms of the central idempotent in the modular group algebra KG associated with B can be obtained from properties (i)-(iii); a simplified proof was obtained later by Rosenberg ([244], Proposition 3.2). At the end of §1, it was shown that the representations $\overline{\omega^{(i)}}$ of the center of the group algebra KG corresponding to the ordinary irreducible characters in a p-block B all coincide, so that

$$\overline{\omega^{(i)}} = \overline{\omega^{(j)}},$$

for all characters $\chi^{(i)}$ and $\chi^{(j)}$ belonging to B. This means that the block B is associated with an irreducible representation η of the center of the group algebra KG, whose value at a basis element c_k of the center of the group algebra is given by

$$\eta(c_k) = \overline{\omega_k^{(i)}},$$

for $k = 1, 2, \ldots$, and any character $\chi^{(i)}$ belonging to B. The representation η has the property that $\eta(\varepsilon) = 1$, for the idempotent ε in the center of the group algebra KG such that $KG\varepsilon$ is the block ideal associated with the block B, and $\eta(\varepsilon') = 0$ for all other central block idempotents ε'.

THEOREM 3.5. *Let ε be the generator of the block ideal in KG of a p-block B, and let η be the representation of the center of the group algebra associated with B. Then $\varepsilon = \sum \xi_\ell c_\ell$, with coefficients $\xi_\ell \in K$. The set of indices $\{\ell\}$ such that $\xi_\ell \neq 0$ can be divided into disjoint subsets L and L', with the following properties:*

(i) the Sylow p-subgroups of the centralizers $C_G(x)$ of elements x belonging to any one of the conjugacy classes $C_\ell, \ell \in L$, are all conjugate in G;

(ii) the Sylow p-subgroups of the centralizers $C_G(x)$ of elements x belonging to any one of the conjugacy classes $C_{\ell'}, \ell' \in L'$, are all conjugate to proper subgroups of the p-subgroup defined in part (i); and

(iii) $\eta(c_\ell) \neq 0$ for some $\ell \in L$, while $\eta(c_{\ell'}) = 0$ for all $\ell' \in L'$.

The p-subgroup of G, uniquely determined up to conjugacy by part (i) of the theorem, is the *defect group* of the p-block B.

Brauer's First Main Theorem gave a correspondence between p-blocks of G and p-blocks of certain subgroups of G.

THEOREM 3.6 (Brauer's First Main Theorem). *Let G be a finite group, and let p be a prime dividing the order of G. Let D be a subgroup of G of order p^d, and let H be the normalizer $N_G(D)$. Then there is a bijection from the set of p-blocks of G with defect group D (or a conjugate of D) to the set of p-blocks of defect d of the subgroup H. All p-blocks of defect d of H have the same defect group D.*

The bijection is defined by means of a homomorphism σ from the center $Z(KG)$ of the group algebra KG to the center $Z(KH)$ of the group algebra of the normalizer H of D, known today as the *Brauer homomorphism*, and defined as follows. Let C_i be a conjugacy class of G, and let c_i be the corresponding basis element of $Z(KG)$, namely the sum of the elements of C_i. Let D be a subgroup of order a power of p. Put $\sigma(c_i)$ equal to the sum of the elements in $C_i \cap C_G(D)$, or zero, if the intersection is empty. The intersection is a union of conjugacy classes in $H = N_G(D)$, so that $\sigma(c_i) \in Z(KH)$ for each basis element c_i. In [**42**], Brauer proved:

LEMMA 3.1 (The Brauer Homomorphism). *The map $\sigma : Z(KG) \to Z(KH)$, obtained by extending the map defined above by linearity, is a homomorphism of algebras.*

Using the lemma, a correspondence from p-blocks of H to p-blocks of G is obtained. A p-block of H is associated with a representation $\psi : Z(KH) \to K$ of $Z(KH)$. The composition $\psi \circ \sigma$ is a representation of the center of the group algebra KG, and is associated with a unique p-block of G. Brauer's proof of the First Main

Theorem was obtained by a close study of the correspondence from blocks of H to blocks of G defined by σ, and was simplified by Rosenberg [**244**].

The decomposition matrix D of a finite group G gave a formula for the values of ordinary irreducible characters $\chi^{(i)}$ of G in terms of values of the irreducible Brauer characters $\varphi^{(j)}$, on the set of p-regular elements $v \in G$:

$$\chi^{(i)}(v) = \sum_{j=1}^{t} d_{ij}\varphi^{(j)}(v),$$

for $i = 1, \ldots, s$. Brauer's Second Main Theorem, published in a sequel [**43**] to [**42**], was a culmination of the whole general development of block theory he had achieved in the two decades following his first papers with Nesbitt at Toronto. It extended the preceding formula to give precise information about the values of ordinary characters at arbitrary, not necessarily p-regular, elements, in terms of Brauer characters of subgroups. Using it he was able to recover portions of the character table of a finite group from scanty information, such as the structure of the defect group of a p-block, extending the results he had obtained for characters of finite group whose orders were divisible by p to the first power.

In the statement of the theorem, reference is made to the fact that, for a prime p, an arbitrary element of G can be expressed uniquely in the form uv, with u an element of order a power of p, and v a p-regular element belonging to $C_G(u)$.

THEOREM 3.7 (Brauer's Second Main Theorem). *Let u be an element of order a power of p, and let $\varphi_j^{(u)}, j = 1, \ldots, t_u$ be the irreducible Brauer characters of the centralizer $H = C_G(u)$ of u. Then there exist algebraic integers,*[25] *$d_{ij}^{(u)}$, for $i = 1, \ldots, s$ and $j = 1, \ldots, t_u$, such that*

$$\chi^{(i)}(uv) = \sum_{j} d_{ij}^{(u)}\varphi_j^{(u)}(v),$$

for all p-regular elements $v \in H$. Moreover, if the ordinary character $\chi^{(i)}$ belongs to a p-block B of G, then $d_{ij}^{(u)} = 0$ unless $\varphi_j^{(u)}$ belongs to a p-block of H which corresponds to B through the Brauer homomorphism $\sigma : Z(KG) \to Z(KH)$.

An elegant, and simpler, proof of the theorem was obtained about ten years later by H. Nagao [**220**], and published in a volume of the *Nagoya Mathematical Journal* dedicated to Brauer on the occasion of his sixtieth birthday.

Each advance in the theory of blocks led Brauer to a better understanding of the theory of characters of a finite group, and to new applications in finite group theory. We conclude this section with some comments on three more results about finite groups obtained by these methods.

The first was a theorem that had been conjectured by M. Suzuki, and was proved, independently, by Brauer, Suzuki, and G. E. Wall [**56**].

THEOREM 3.8 (Brauer-Suzuki-Wall Theorem). *Let G be a finite group of even order, which coincides with its commutator group, and has the property that any two cyclic subgroups A and B of G with a nontrivial intersection are both contained in a cyclic subgroup of G. Then $G \cong PSL_2(q)$ for some prime power $q \geq 4$.*

The theorem was another application of character theory to the classification of finite simple groups. The proof started by showing that the Sylow 2-subgroups of

---

[25]Brauer called them *generalized decomposition numbers*.

Michio Suzuki, Donald G. Higman, and Walter Feit, July, 1997

a finite group satisfying the hypotheses of the theorem were either dihedral groups or elementary abelian groups. The next step was to obtain the structure of the centralizer H of an involution[26] in G, and a description of the conjugacy classes in G which contain elements of H. Using a theory of *exceptional characters*[27] first developed by Suzuki in his thesis [**267**], it was possible to obtain the values of the irreducible characters of G at elements of H not equal to 1, and congruences for the degrees of the irreducible characters. When this information was combined with formulas for multiplication of certain class sums in the center of the group algebra, the degrees of the irreducible characters and the order of the group were calculated. The group G was identified finally by applying a result of Zassenhaus [**293**] on certain doubly-transitive permutation groups.

While the published proof of the theorem involved only the theory of ordinary characters, Brauer had first obtained a lengthy proof using the block theory of groups with dihedral Sylow 2-subgroups. The history of this important paper, and the separate contributions of the authors, is interesting, and was assembled by Green [**164**].

The focus on the centralizer of an involution in the proof of Theorem 3.8 was an illustration of a general principle stated by Brauer in his address at the International

---

[26]An *involution* in a finite group is an element of order two.

[27]Suzuki explained his theory of exceptional characters in a general setting in [**269**]. For groups of even order, further information, obtained independently by Suzuki and by Brauer and Fowler [**54**], could be sometimes combined with the theory of exceptional characters to obtain formulas for the degrees of the irreducible characters and the order of a finite group, starting from information about the subgroup structure, exactly as in the proof of Theorem 3.8.

Congress of Mathematicians in 1954 at Amsterdam [**41**]. Brauer and his former
Ph.D. student, K. A. Fowler, had proved [**54**]:

THEOREM 3.9. *There exist only a finite number of simple groups G which contain an involution j such that the centralizer of j is isomorphic to a given finite
group.*

In his address, he described his plan as follows:

> The last theorem suggests the following problem: Given a group N
> containing an involution j in its center. What are the groups G
> containing N as a subgroup such that N is the normalizer of j in G?

Brauer had also announced Theorem 3.8 in the address, and was able to state
a further result, giving a characterization of the finite groups $PSL_3(q)$:

THEOREM 3.10. *Let G be a finite simple group containing an involution j whose
centralizer is isomorphic to $GL_2(q)$, for an odd prime power q. Then either $G \cong
PSL_3(q)$, or $q = 3$ and G is isomorphic to the simple Mathieu group of order 7912.*

After giving an outline of the proof of the theorem, Brauer continued:

> The problem treated here of characterizing special simple groups
> is of interest in connection with the important unsolved problem
> of determining all simple groups of finite order. In order to be able
> to recognize the known simple groups, we need workable charac-
> terizations of these groups. Though we are certainly very far from
> a solution of the general problem of the simple groups of finite
> order, at least some sort of a plan seems to evolve according to
> which the problem might be attacked. One hope would be that
> the only non-cyclic simple groups are the alternating groups, the
> finite analogues of the simple Lie groups and some distorted ver-
> sions of a few of the latter groups [the sporadic groups]. It would
> be necessary to see where these groups fit into our program and
> to show that no other cases can arise. It is not possible to say
> whether this plan is workable, since some of the most important
> links are missing.
>
> As a matter of fact, it is necessary to mention in this connection
> again the old conjecture that all groups of odd order are soluble. Of
> course, this would now obtain an added significance. No progress
> has been made on this question since it was first formulated.

The situation with regard to the problem of classifying the finite simple groups
was very different 15 years later; by then, the odd-order theorem had been proved
by Feit and Thompson (see Chapter III, §2), and Michael Aschbacher was able to
state, in [**11**], "The classification of finite simple groups now seems close at hand."
In his article on the subdivisions of the classification, *The classification of finite
simple groups I: Simple groups and local analysis* [**162**], published the year after
Brauer's death and dedicated to his memory, Daniel Gorenstein said, with reference
to Theorem 3.10:

> This last result, which Brauer announced in his address at the In-
> ternational Congress of Mathematicians in Amsterdam in 1954, rep-
> resented the starting point for the classification of simple groups in

terms of the structure of the centralizers of their involutions. Moreover, it foreshadowed the basic fact that conclusions of general classification theorems would necessarily include sporadic simple groups as exceptional cases ([the Mathieu group M_{11} of order 7912] being the sporadic group of least order).

Another result of a different nature, obtained independently by Brauer and Suzuki, and published jointly ([**57**], 1959) also played a part in the classification.

THEOREM 3.11 (The Brauer-Suzuki Theorem). *Let G be a group of finite even order. If the Sylow 2-subgroup P is a quaternion group (ordinary or generalized), then G is not simple.*

The proof was based on an application of Brauer's Second Main Theorem 3.7 to the 2-blocks of G. The case where P was an ordinary quaternion group of order 8 turned out to be the most difficult one.

The Brauer-Suzuki Theorem 3.11, and a generalization of it known as the Z^*-Theorem (proved by Glauberman [**160**]), were important tools in *local analysis* (see [**162**], Chapter IV). Concerning them, Gorenstein said ([**162**], p. 138), "We should also mention that the Z^*-Theorem [and the Brauer-Suzuki Theorem used in its proof] represents the single result of general local analysis that requires Brauer's theory of modular characters."

4. The Brauer Induction Theorem and its Applications

In his conversations with Emil Artin at Indiana University during his year as a Guggenheim fellow in 1941, Brauer learned that a result concerning characters of finite groups, if it could be proved, would settle a conjecture of Artin concerning L-functions with general group characters. He had recently proved that the determinant of the matrix of Cartan integers, in the modular representation theory of a finite group G over an algebraically closed field of characteristic $p > 0$, was a power of p ([**27**], 1941). In the next few years he realized that some of the work he had done in connection with the theorem on the determinant of the Cartan matrix could be used to prove the result on character theory needed to establish Artin's conjecture. This idea led him not only to a proof of the conjecture and other contributions to algebraic number theory, but also to new developments in the ordinary character theory of finite groups.

We begin with some of the ideas leading up to a statement of Artin's conjecture. Let F be an algebraic number field, and let E denote a finite Galois extension of F with Galois group $\mathcal{G}$. In [**9**], Artin defined an L-function $L(s, \chi, E/F)$ for each complex valued character χ of the finite group $\mathcal{G}$. As in the classical case (see Chapter I, §7 for the case of the rational field), L-functions were introduced in order to investigate the distribution of prime ideals in algebraic number fields. Artin's L-functions, defined below, introduced these powerful analytic tools, for the first time, in the setting of extensions of algebraic number fields with nonabelian Galois groups.

For each unramified finite prime $\mathfrak{p}$ of F, and a prime $\mathfrak{P}$ of E dividing $\mathfrak{p}$, it is possible to define a Frobenius automorphism, denoted by

$$\sigma(\mathfrak{P}) = \left(\frac{E/F}{\mathfrak{P}} \right),$$

belonging to the Galois group $\mathcal{G}$ (see Chapter II, §2, Proposition 2.3, for the case $F = \mathbb{Q}$, and [**191**], Chapter III, §2, for the general case). The Frobenius automorphisms corresponding to other prime ideals $\mathfrak{P}'$ in E dividing $\mathfrak{p}$ are conjugates of $\sigma(\mathfrak{P})$ in $\mathcal{G}$, so that $\mathfrak{p}$ is associated with a unique conjugacy class in $\mathcal{G}$. Let $\sigma \to M(\sigma)$ be a representation of $\mathcal{G}$ affording the given character χ. The L-function $L(s, \chi, E/F)$ is defined in terms of its Euler product factorization[28] by

$$L(s, \chi, E/F) = \prod_{\mathfrak{p}} |I - N(\mathfrak{p})^{-s} M_{\mathfrak{p}}|^{-1},$$

where $|\cdot|$ is the determinant, I is the identity matrix, and $N(\mathfrak{p})$ is the norm map from F to $\mathbb{Q}$ applied to $\mathfrak{p}$. The product is taken over all primes $\mathfrak{p}$ of F, unramified or not, with $M_{\mathfrak{p}} = M(\sigma(\mathfrak{P}))$ for an unramified prime $\mathfrak{p}$, and $M_{\mathfrak{p}}$ interpreted in a suitable way for ramified primes (see [**9**], §2). This extended an investigation carried out in an earlier paper [**7**] of an L-function defined by the preceding formula, with the product taken only over the set of unramified primes $\mathfrak{p}$ of F. The factors of the L-function defined above depend only on the character χ, and not on the choice of the representation M affording χ, because of the remarks made above in the case of an unramified prime $\mathfrak{p}$, and by Artin's interpretation of the factor in the case of ramified primes.

The product expansion of $L(s, \chi, E/F)$ is absolutely convergent in the half-plane $Re(s) > 1$, and the function $L(s, \chi, E/F)$ can be expanded in a Dirichlet series there.

The following results concerning the L-functions $L(s, \chi, E/F)$ were obtained by Artin [**9**].

THEOREM 4.1. (*i*) *If χ is a linear combination $\sum_i c_i \varphi_i$ of characters φ_i with rational coefficients c_i, then*

$$L(s, \chi, E/F) = \prod_i L(s, \varphi_i, E/F)^{c_i}.$$

(*ii*) *Let Ω be a subfield of E containing F, and let $\mathcal{H}$ be the subgroup of $\mathcal{G}$ corresponding to Ω according to the fundamental theorem of Galois theory.[29] Let ψ be a character of $\mathcal{H}$, and let $\psi^{\mathcal{G}}$ be the induced character.[30] Then*

$$L(s, \psi^{\mathcal{G}}, E/F) = L(s, \psi, E/\Omega).$$

(*iii*) *If the representation of $\mathcal{G}$ belonging to the character χ has the kernel $\mathcal{N}$, and if N is the corresponding subfield of E, then*

$$L(s, \chi, E/F) = L(s, \chi, N/F).$$

In case E/F is an extension with an abelian Galois group $\mathcal{G}$, and if χ is an irreducible character of $\mathcal{G}$, Artin proved, using the reciprocity law, that $L(s, \chi, E/F)$ coincides with a so-called abelian L-function, where χ is now interpreted as a character on a certain finite abelian group obtained as a finite subquotient of the ideal class group of E (see [**176**], §2, or [**191**], Chapter IV, §4). Hecke had proved, in a series of articles on L-functions and their application to number theory ([**175**], 1917-1920), that abelian L-functions satisfy a certain functional equation, and can

---

[28]See Chapter I, §7, Lemma 7.1, for the Euler product factorization of an L-series in the classical case.

[29]See Chapter I, Theorem 1.3.

[30]See Chapter II, §4.

be analytically continued over the complex plane, so that they are meromorphic functions on $\mathbb{C}$.

Artin brought all these ideas together with a new result about characters of finite groups ([**9**], §1):

THEOREM 4.2 (Artin Induction Theorem). *Every character χ of a finite group G can be expressed as a linear combination $\sum_i c_i \varphi_i$, where the φ_i are characters of G which are induced from characters of cyclic subgroups of G, and the coefficients c_i are rational numbers.*

Upon combining this theorem with the previous one, Artin deduced that each L-function $L(s, \chi, E/F)$ can be expressed as a product of rational powers of abelian L-functions:

$$L(s, \chi, E/F) = \prod_i L(s, \psi_i, E/\Omega_i)^{c_i},$$

where Ω_i are subfields of E belonging to cyclic subgroups $\mathcal{C}_i$ of the Galois group $\mathcal{G}$ of E over F, ψ_i is a character of $\mathcal{C}_i$, and the exponents c_i are rational numbers. By the result of Hecke stated above, the abelian L-functions occurring in the product are meromorphic functions on the complex plane. It follows that a suitable power $L(s, \chi, E/F)^m$, for some positive integer m, is a product of abelian L-functions, and consequently is a meromorphic function on $\mathbb{C}$. As m may be greater than 1, this does not imply that $L(s, \chi, E/F)$ is a single-valued function on $\mathbb{C}$. Artin conjectured that this is indeed the case. More precisely, Artin conjectured, and Brauer proved ([**37**], 1947), the following theorem.

THEOREM 4.3 (Brauer Induction Theorem). *Let G be a finite group, and let χ be a character of G. Then χ can be expressed as a linear combination, with coefficients in $\mathbb{Z}$, of induced characters of the form ψ_i^G, where each character ψ_i is a linear character[31] of a subgroup H_i of G, for $i = 1, 2, \ldots$.*

The theorem implies that $L(s, \chi, E/F)$ is a meromorphic function. Brauer pointed out, however, that there was a further unsettled question ([**37**], §1):

> A second, stronger conjecture remains open. If χ is a simple character of $\mathcal{G}$, different from the 1-character, then Artin surmises that $L(s, \chi, E/F)$ is an integral function.[32]

Brauer's proof of Theorem 4.3 was based on the following chain of reasoning. Let $C_1, \ldots, C_s$ be the conjugacy classes of the finite group G, and let x_i be a representative of the class C_i, for each i. Let Φ be an algebraic number field containing the character values of all characters of G and subgroups of G. He asserted that the theorem would follow if the following statement can be proved. If a congruence

(18) $$\sum_{i=1}^s r_i \omega(x_i) \equiv 0 \pmod{\mathfrak{q}^t},$$

modulo a power $\mathfrak{q}^t$ of a prime ideal $\mathfrak{q}$ of Φ, with $\mathfrak{q}$-integral coefficients r_i, holds for every induced character ω of the form $\omega = \lambda^G$, for a linear character λ of a

---

[31] A *linear character* of a finite group is an irreducible character of degree 1.

[32] Brauer, R., "On Artin's L-series with general group characters," *Annals of Mathematics* **48** (1947), pp. 502–514. Used with permission.

subgroup of G, then the corresponding congruence

$$(19) \qquad \sum r_i \chi(x_i) \equiv 0 \pmod{\mathfrak{q}^t}$$

holds for every character χ of G. Using the theory of modular characters, Brauer was able to show that the last sum can be split into partial sums, each of which must be congruent to 0 modulo $\mathfrak{q}^t$, if (19) holds. The partial sums are of the following type. Let q be the rational prime divisible by $\mathfrak{q}$, and let a be a q-regular element of G. Then (19), for all characters χ, implies

$$(20) \qquad \Sigma' r_i \chi(x_i) \equiv 0 \pmod{\mathfrak{q}^t},$$

where the sum is taken over representatives x_i of classes C_i which contain elements $x \in G$ having a as their q-regular factor. On the other hand, Brauer observed that (19) follows from (20). Therefore, with this insight from modular representation theory, Brauer was able to organize the proof in a different way.

Let a be a q-regular element, let Q be a Sylow q-subgroup of the centralizer of a, and let H be the subgroup generated by a and Q. Brauer proved that each irreducible character of H is induced from a linear character of a subgroup of H. Moreover, if (18) holds for all induced characters of the required form, then it holds for induced characters of the form ψ^G, for an irreducible character ψ of a subgroup H as above. It turned out that these latter congruences were sufficient to prove (20), and, after further argument, the theorem itself.

Brauer found his proof of the induction theorem at a time when he was immersed in research on the theory of blocks, defect groups, Cartan matrices, and other topics in modular representation theory. As he explained in connection with the reasoning above, "The theory of modular characters will not be used in the following. However, we mention the fact [about the partial sums] here, because it allows us to see the reason for the procedure used in this paper."

The subgroups H involved in the preceding discussion have the form

$$H = \langle a \rangle \times Q,$$

for a cyclic group $\langle a \rangle$ generated by a q-regular element a, and a q-subgroup Q, for some prime number q. Brauer called groups of this type *elementary*, and made the observation that the subgroups H_i occurring in the statement of Theorem 4.3 can all be taken to be elementary subgroups of G. This was an immediate consequence of his proof of the theorem. The sharpened version of the theorem, and the idea used in the proof of studying congruences for character values of induced characters from elementary subgroups, turned out to be extraordinarily powerful tools for the further development of character theory. No one understood this better than Brauer himself, and in a series of papers published over the next few years ([**38**], [**39**], [**40**]) he gave several applications of these ideas to ordinary character theory of finite groups and the theory of the Schur index.

In his first application of the Induction Theorem to other problems in character theory, Brauer returned to the old question, considered earlier by Maschke, Burnside, and Schur (see Chapter III, §5 and Chapter IV, §4) as to whether an arbitrary irreducible representation of a finite group can be written in the field of $|G|$-roots of unity. The most important progress towards the solution of the problem, after the early work, was made by Hasse, in an appendix to the joint paper [**49**] with Brauer and Noether, where he proved that the irreducible representations of a finite group G can all be written in a cyclotomic field, and, more precisely, in the field of $|G|^h$-th roots of unity, for a sufficiently large integer h. There the matter stood until 1945, when Brauer settled the problem in full generality ([**33**]) with a proof based on the methods of Schur and Hasse, combined with the theory of modular representations.

THEOREM 4.4 (Brauer's Splitting Field Theorem). *Every irreducible representation* $x \to A(x)$ *of a finite group* G *(in the field of complex numbers) can be written in the field* $\mathbb{Q}(\omega)$, *where* ω *is a primitive* $|G|$-*th root of unity. In other words, the representation* $x \to A(x)$ *is equivalent to a representation with matrix coefficients in* $\mathbb{Q}(\omega)$.

In 1947, in a paper [**37**] submitted only a few months after he sent in his proof of Artin's conjecture to the *Annals of Mathematics*, he gave a new proof of the splitting field theorem, this time making full use of the Induction Theorem 4.3. Both of Brauer's proofs of the theorem were based on the theory of the Schur index (Chapter IV, §4). We give here an outline of a direct proof, also based on the Brauer Induction Theorem, that was obtained later by Feit. First of all, a linear representation ψ of a subgroup H of G can be written in the field $\mathbb{Q}(\omega)$. This follows because the order of H divides $|G|$, and a linear representation of a finite group H is a representation of an abelian quotient group of H, so it can be written in the field of $|H|$-th roots of unity. Moreover, the induced representation ψ^G is also rational in $\mathbb{Q}(\omega)$, by the definition of induced representations (see Chapter II, §4). Now let χ be the character of the representation given in the statement of the theorem. By Theorem 4.3, the character χ is a linear combination, with coefficients in $\mathbb{Z}$, of characters of the form ψ^G, for linear characters ψ of subgroups of G. If all the coefficients are nonnegative, then it follows from what has been said that the representation A can be written in the field $\mathbb{Q}(\omega)$. If some of the coefficients are positive and some are negative, then it follows that there exist representations $x \to B(x)$ and $x \to C(x)$, both equivalent to representations with matrix coefficients in the field $\mathbb{Q}(\omega)$, such that the representation $x \to A(x) \oplus B(x)$ is equivalent to the representation $x \to C(x)$. Using Maschke's Theorem, it can be shown that common irreducible direct summands of B and C can be cancelled, and, with further discussion, that it may be assumed that the representations $x \to B(x)$ and $x \to C(x)$ have no irreducible direct summands in common. As the representation $x \to A(x)$ is irreducible, it follows readily that the given representation A is equivalent to C, and $B = 0$, completing the proof.

Brauer realized that the sharpened version of the induction theorem gave the elementary subgroups special significance for the calculation of the characters of a finite group. In [**37**], he obtained two more theorems elaborating this point, one for ordinary characters and one for Brauer characters. Here is the result for ordinary characters.

THEOREM 4.5. *Assume that the number* s *of conjugacy classes* $C_1, \ldots, C_s$ *of a finite group and the number of elements in each class are known. Suppose that a complete system of elementary subgroups* H *is given, that is, a set of representatives of the conjugacy classes of elementary subgroups. Assume further that for each* H *the number* ℓ *of conjugacy classes* $L_1, \ldots, L_\ell$ *of* H *is known, and that it is known to which class* C_i *the elements of a class* L_j *belong. Finally, assume the values of the linear characters of each elementary subgroup are known. Then the irreducible characters of* G *are completely determined.*

While this theorem was a good illustration of the power of the induction theorem, an application of it required more information than one would ordinarily have. A few years later, he found a criterion for a complex-valued class function to be a *generalized character*, that is, a linear combination of the irreducible characters with integer coefficients. This result, also proved using the induction theorem, has had a large impact on the development of character theory in the last half of the 20-th century; it can be applied to the construction of irreducible characters and used for other purposes as well. The statement is disarmingly simple (from [**40**], 1953).

THEOREM 4.6 (Brauer's Characterization of Characters). *A   complex-valued function θ defined on a finite group G is a generalized character if and only if the following conditions are satisfied:*

(i) θ is a class function, that is θ is constant on each conjugacy class of G.

(ii) For every elementary subgroup H of G, the restriction of θ to H is a generalized character of H.

The function θ defined on G is an irreducible character of G if and only if it satisfies conditions (i) and (ii) above, and, in addition:

(iii) The average value of $|\theta|$ on G is 1,

$$\sum_{x \in G} |\theta(x)|^2 = |G|,$$

and

(iv) The number $\theta(1)$ is positive.

The proof is a direct application of the sharpened version of the Induction Theorem 4.3 and the Frobenius Reciprocity Theorem (Chapter II, §4). Let us assume that θ satisfies conditions (i) and (ii) in the statement of the theorem; we shall prove that θ is a generalized character. As the irreducible characters $\chi^{(1)}, \chi^{(2)}, \ldots$ form a basis for the vector space of class functions, one has

$$\theta = a_1 \chi^{(1)} + a_2 \chi^{(2)} + \cdots,$$

with coefficients $a_i \in \mathbb{C}$. We have to prove that the coefficients a_i are in $\mathbb{Z}$. By the orthogonality relations, the coefficients are given by

$$a_i = \frac{1}{|G|} \sum_{x \in G} \theta(x) \chi^{(i)}(x^{-1}),$$

for $i = 1, 2, \ldots$. By Theorem 4.3, each character $\chi^{(i)}$ is a linear combination, with coefficients in $\mathbb{Z}$, of induced characters ψ^G, for linear characters ψ of elementary subgroups H. By a version of the Frobenius Reciprocity Theorem, one has

$$\frac{1}{|G|} \sum_{x \in G} \theta(x) \psi^G(x^{-1}) = \frac{1}{|H|} \sum_{h \in H} \theta(h) \psi(h^{-1}),$$

for each induced character ψ^G from an elementary subgroup H. By the assumption (ii), the restriction of θ to an elementary subgroup H is a linear combination, with coefficients in $\mathbb{Z}$, of irreducible characters of H. Combining all this information, and using the orthogonality relations for the irreducible characters of H, it follows that $a_i \in \mathbb{Z}$, as required. The rest of the proof of the theorem is not difficult, and is omitted.

In his obituary article about Brauer [**117**], Feit wrote,

Many of his most important results have been given alternative and better proofs, usually by others, sometimes by himself. This is at least partly due to the fact that he was always primarily interested in results and did not spend a great deal of time looking for the "best" proof.

The Brauer Induction Theorem 4.3 provides a striking illustration of the preceding remarks. The applications of the theorem to the proof of Artin's conjecture on L-series, the new proof of the splitting field theorem, and a reduction theorem for the computation of the Schur index [39], were all published immediately following the proof of the Induction Theorem, and established beyond any doubt that it was a fundamental new discovery. But the original proof of Theorem 4.3 in [37] was difficult to follow. Brauer realized this, and in his paper [40] containing Theorem 4.6, he pointed out that, while the main conclusion of the theorem is equivalent to the Induction Theorem, "... since the proof of the earlier theorem can be simplified considerably, I will prove Theorem 1 [Theorem 4.6] without reference to the preceding paper." In the same year, the proof of the Induction Theorem was greatly simplified by Peter Roquette [242]. Shortly afterwards, Brauer and John Tate gave another proof of the Induction Theorem [55], which is the standard proof given today, remarking that their proof was a further simplification, using Roquette's ideas.

The work reported on up to now, in which Brauer initiated the lines of research associated with his name today, by no means signaled the end, or even a slowing down, of his mathematical work. There followed a period of research in his later life, after he had reached the age of 60, when he published more than 50 papers. These included some on applications of character theory and the theory of blocks to the classification of finite simple groups and other topics in group theory, by himself, and in collaboration with H. S. Leonard [58], J. L. Alperin and D. Gorenstein [4], [5], P. Fong [59], [61], and W. Wong [60]. There were also refinements of the theory of blocks, and survey articles based on talks he gave at meetings. His last paper [45], entitled *Blocks of characters and the structure of finite groups*, was based on a talk he gave at a meeting of the American Mathematical Society at Storrs, Connecticut on October 30, 1976. It ended as follows:

An epilogue: Work not yet done.

Apart from the many questions in the preceding pages, there are many open problems. Since there are now so many mathematicians working in the theory of finite groups, perhaps some of them may try to solve them.

Most of the recent work on simple groups uses purely combinatorial arguments. A number of new concepts have been introduced which are certainly highly significant. It will be of interest to see if any of them play a role in problems of the representation theory of finite groups. ...

Even if the classification problem for simple groups can be solved in the most satisfactory form, we still do not know all about finite groups that we want to know, e.g. about the structure of the modular representations, about the invariants of finite linear groups, and so on.

I, for one, hope that finite group theory will keep mathematicians busy for a long time to come.

Bibliography

[1] Albert, A. A., *Structure of algebras*, Amer. Math. Society, New York, 1939.

[2] Albert, A. A. and Hasse, H., *A determination of all normal division algebras over an algebraic number field*, Trans. Amer. Math. Soc. **34** (1932), 722-726.

[3] Alexandrov, P. S., *In memory of Emmy Noether*, in *Emmy Noether, Collected Papers*, Springer-Verlag, Berlin (1983), 1-11.

[4] Alperin, J. L., Brauer, R., and Gorenstein, D., *Finite groups with quasi-dihedral and wreathed Sylow 2-subgroups*, Trans. Amer. Math. Soc. **151** (1970), 1-261.

[5] Alperin, J. L., Brauer, R., and Gorenstein, D., *Finite simple groups of 2-rank two*, Scripta Math. **29** (1973), 191-214; *Richard Brauer, Collected Papers*, M.I.T. Press, Cambridge, (1980), III, 120-143.

[6] Amitsur, S., *Finite subgroups of division rings*, Trans. Amer. Math. Soc. **80** (1955), 361-386.

[7] Artin, E., *Über eine neue Art von L-Reihen*, Hamb. Abh. **3** (1924), 89-108; Collected Papers, Addison-Wesley, Reading (1965), 105-124.

[8] Artin, E., *Zur Theorie der hyperkomplexen Zahlen*, Hamb. Abh. **5** (1928), 251-260; Collected Papers, 307-316.

[9] Artin, E., *Zur Theorie der L-Reihen mit allgemeinen Gruppencharakteren*, Hamb. Abh. **8** (1930), 292-306; Collected Papers, 165-179.

[10] Artin, E., *Galois Theory*, Notre Dame Mathematical Lectures, Number 2, Notre Dame, Indiana, 1946.

[11] Aschbacher, M., *An introduction to the classification of finite simple groups*, Bull. Amer. Math. Soc. (N.S.) **1** (1979), 39-41.

[12] Bender, H., *A group-theoretic proof of the $p^a q^b$-Theorem*, Math. Zeitschrift **126** (1972), 327-338.

[13] Bernstein, I. N., Gelfand, I. M., and Gelfand, S. I., *Category of $\mathfrak{g}$-modules*, Functional Analysis and its Applications **10** (1976), 87-92.

[14] Biermann, K. R., *Die Mathematik und ihre Dozenten an der Berliner Universität 1810-1920*, Akademie-Verlag, Berlin, 1973.

[15] Biermann, K. R., *Die Ära Schmidt-Schur-Bieberbach-von Mises (1920-1933)*, in *Die Mathematik und ihre Dozenten an der Berliner Universität 1810-1933*, Akademie-Verlag, Berlin, 1988.

[16] Bourbaki, N., *Modules et Anneaux Semi-simples*, Actualités Scientifiques et Industrielles **1261**, Hermann, Paris, 1958.

[17] Brauer, A., *Gedenkrede auf Issai Schur*, pp. V-XIV in *Issai Schur, Gesammelte Abhandlungen*, Bd. I, Springer-Verlag, Berlin, 1973.

[18] Brauer, R., *Über Zusammenhänge zwischen arithmetischen und invariantentheoretischen Eigenschaften von Gruppen linearer Substitutionen*, S'ber Akad. Wiss. Berlin (1926), 410-416; Collected Papers, M.I.T. Press, Cambridge, (1980), I, 5-11.

[19] Brauer, R., *Untersuchungen über die arithmetischen Eigenschaften von Gruppen linearer Substitutionen I*, Math. Zeitschrift **28** (1928), 677-696; Collected Papers I, 20-39.

[20] Brauer, R., *Über Systeme hyperkomplexer Zahlen*, Math. Zeitschrift **30** (1929), 79-107; Collected Papers I, 40-68.

[21] Brauer, R., *Untersuchungen über die arithmetischen Eigenschaften von Gruppen linearer Substitutionen II*, Math. Zeitschrift **31** (1930), 733-747; Collected Papers I, 88-102.

[22] Brauer, R., *Über die algebraische Struktur der Schiefkörpern*, J. reine angew. Math. **166** (1932), 241-252; Collected Papers I, 103-114.

[23] Brauer, R., *Über die Darstellung von Gruppen in Galoisschen Feldern*, Actualités Sci. Indust. **195** (1935), 15 pp.; Collected Papers I, 323-335.

[24] Brauer, R., *On modular and p-adic representations of algebras*, Proc. Nat. Acad. Sci. U.S.A. **25** (1939), 252-258; Collected Papers I, 199-205.

[25] Brauer, R., *On algebras which are connected with the semisimple continuous groups*, Ann. of Math. (1937), 857-872; Collected Papers III, 446-461.

[26] Brauer, R., *On the representations of groups of finite order*, Proc. Nat. Acad. Sci. U.S.A. **25** (1939), 290-295; Collected Papers I, 355-360.

[27] Brauer, R., *On the Cartan invariants of groups of finite order*, Ann. of Math. **42** (1941), 53-61; Collected Papers I, 361-369.

[28] Brauer, R., *On the connection between the ordinary and modular characters of groups of finite order*, Ann. of Math. **42** (1941), 926-935; Collected Papers I, 405-414.

[29] Brauer, R., *On groups whose order contains a prime to the first power I*, Amer. J. Math. **64** (1942), 401-420; Collected Papers I, 438-457.

[30] Brauer, R., *On groups whose order contains a prime to the first power II*, Amer. J. Math. **64** (1942), 421-440; Collected Papers I, 458-477.

[31] Brauer, R., *On permutation groups of prime degree and related classes of groups*, Ann. of Math. **44** (1943), 57-79; Collected Papers I, 478-500.

[32] Brauer, R., *On the arithmetic in a group ring*, Proc. Nat. Acad. Sci. U.S.A. **30** (1944), 109-114; Collected Papers I, 501-506.

[33] Brauer, R., *On the representation of a group of order g in the field of g-th roots of unity*, Amer. J. Math. **67** (1945), 461-471; Collected Papers I, 518-528.

[34] Brauer, R., *On blocks of characters of groups of finite order I*, Proc. Nat. Acad. Sci. U.S.A. **32** (1946), 182-186; Collected Papers I, 529-533.

[35] Brauer, R., *On blocks of characters of groups of finite order II*, Proc. Nat. Acad. Sci. U.S.A. **32** (1946), 215-219; Collected Papers I, 534-538.

[36] Brauer, R., *On a conjecture of Nakayama*, Trans. Roy. Soc. Canada. Sect. III. (3) **41**, 11-19; Collected Papers I, 560-568.

[37] Brauer, R., *On Artin's L-series with general group characters*, Ann. of Math. **48** (1947), 502-514; Collected Papers I, 539-551.

[38] Brauer, R., *Applications of induced characters*, Amer. J. Math. **69** (1947), 709-716; Collected Papers I, 552-559.

[39] Brauer, R., *On the algebraic structure of group rings*, J. Math. Soc. Japan **3** (1951), 237-251; Collected Papers I, 569-583.

[40] Brauer, R., *A characterization of the characters of groups of finite order*, Ann. of Math. **57** (1953), 357-377; Collected Papers I, 588-608.

[41] Brauer, R., *On the structure of groups of finite order*, Proc. Internat. Congr. Math., Amsterdam (1954), Vol. 1, Noordhoff, Groningen; North Holland, Amsterdam (1957), 209-217; Collected Papers II, 69-77.

[42] Brauer, R., *Zur Darstellungstheorie der Gruppen endlicher Ordnung*, Math. Zeitschrift **63** (1956), 406-444; Collected Papers II, 22-60.

[43] Brauer, R., *Zur Darstellungstheorie der Gruppen endlicher Ordnung II*, Math. Zeitschrift **72** (1959), 25-46; Collected Papers II, 119-140.

[44] Brauer, R. *Emil Artin*, Bull. Amer. Math. Soc. **73** (1967), 27-43; Collected Papers III, 671-687.

[45] Brauer, R., *Blocks of characters and structure of finite groups*. Bull. Amer. Math. Soc. **1** (1979), 21-38; Collected Papers III, 267-284.

[46] Brauer, R., *Preface*, in *Richard Brauer, Collected Papers*, M.I.T. Press, Cambridge, (1980), I, xv-xix.

[47] Brauer, R. and Noether, E., *Über minimale Zerfällungskörper irreduzibler Darstellungen*, S'ber. Akad. Wiss. Berlin (1927), 221-228; *Richard Brauer, Collected Papers*, M.I.T. Press, Cambridge, (1980), I, 12-19.

[48] Brauer, R. and Schur, I., *Zum Irreduzibilitätsbegriff in der Theorie der Gruppen linearer homogener Substitutionen*, S'ber. Akad. Wiss. Berlin (1930), 209-226; *Richard Brauer, Collected Papers*, M.I.T. Press, Cambridge, (1980), I, 69-87.

[49] Brauer, R., Hasse, H., and Noether, E., *Beweis eines Hauptsatzes in der Theorie der Algebren*, J. reine angew. Math. **167** (1931), 399-404; *Richard Brauer, Collected Papers*, M.I.T. Press, Cambridge, (1980), I, 115-120.

[50] Brauer, R. and Nesbitt, C., *On the modular representations of groups of finite order I*, Univ. Toronto Studies no. 4 (1937), 21 pp.; *Richard Brauer, Collected Papers*, M.I.T. Press, Cambridge, (1980), I, 336-354.

[51] Brauer, R. and Nesbitt, C., *On the regular representations of algebras*, Proc. Nat. Acad. Sci. U. S. A. **23** (1937), 236-240; *Richard Brauer, Collected Papers*, M.I.T. Press, Cambridge, (1980), I, 190-194.

[52] Brauer, R. and Nesbitt, C., *On the modular characters of groups*, Ann. of Math. **42** (1941), 556-590; *Richard Brauer, Collected Papers*, M.I.T. Press, Cambridge, (1980), I, 370-404.

[53] Brauer, R. and Tuan, H.-F., *On simple groups of finite order. I*, Bull. Amer. Math. Soc. **51** (1945), 756-766; *Richard Brauer, Collected Papers*, M.I.T. Press, Cambridge, (1980), I, 507-517.

[54] Brauer, R. and Fowler, K. A., *On groups of even order*, Ann. of Math. **62** (1955), 565-583; *Richard Brauer, Collected Papers*, M.I.T. Press, Cambridge, (1980), II, 3-21.

[55] Brauer, R. and Tate, J., *On the characters of finite groups*, Ann. of Math. **62** (1955), 1-7; *Richard Brauer, Collected Papers*, M.I.T. Press, Cambridge, (1980), I, 609-615.

[56] Brauer, R., Suzuki, M., and Wall, G. E., *A characterization of the one-dimensional unimodular projective groups over finite fields*, Illinois J. Math. **2** (1958), 718-745; *Richard Brauer, Collected Papers*, M.I.T. Press, Cambridge, (1980), II, 78-105.

[57] Brauer, R. and Suzuki, M., *On finite groups of even order whose 2-Sylow group is a quaternion group*, Proc. Nat. Acad. Sci. U.S.A. **45** (1959), 1757-1759; *Richard Brauer, Collected Papers*, M.I.T. Press, Cambridge, (1980), II, 141-143.

[58] Brauer, R. and Leonard, H. S., *On finite groups with an abelian Sylow group*, Canad. J. Math. **14** (1962), 436-450; *Richard Brauer, Collected Papers*, M.I.T. Press, Cambridge, (1980), II, 150-164.

[59] Brauer, R. and Fong, P., *A characterization of the Mathieu group M_{12}*, Trans. Amer. Math. Soc. **122** (1966), 18-47; *Richard Brauer, Collected Papers*, M.I.T. Press, Cambridge, (1980), II, 359-388.

[60] Brauer, R. and Wong, W. J., *Some properties of finite groups with wreathed Sylow 2-subgroups*, J. Algebra **19** (1971), 263-273; *Richard Brauer, Collected Papers*, M.I.T. Press, Cambridge, (1980), III, 58-68.

[61] Brauer, R. and Fong, P., *On the centralizers of p-elements in finite groups*, Bull. London Math. Soc. **6** (1974), 319-324; *Richard Brauer, Collected Papers*, M.I.T. Press, Cambridge, (1980), III, 227-232.

[62] Bühler, W. K., *Gauss*, Springer-Verlag, New York, 1981.

[63] Burnside, W., *On a class of automorphic functions*, Proc. London Math. Soc. **23** (1891), 49-88.

[64] Burnside, W., *Further note on automorphic functions*, Proc. London Math. Soc. **23** (1892), 281-285.

[65] Burnside, W., *Notes on the theory of groups of finite order*, Proc. London Math. Soc. **25** (1893), 9-18.

[66] Burnside, W., *On a class of groups defined by congruences*, Proc. London Math. Soc. **25** (1894), 113-139.

[67] Burnside, W., *Notes on the theory of groups of finite order*, Proc. London Math. Soc. **26** (1895), 191-214.

[68] Burnside, W., *Notes on the theory of groups of finite order (continued)*, Proc. London Math. Soc. **26** (1895), 325-338.

[69] Burnside, W., *Theory of groups of finite order* (first edition), Cambridge Univ. Press, Cambridge, 1897.

[70] Burnside, W., *On the continuous group that is defined by any given group of finite order*, Proc. London Math. Soc. **29** (1898), 207-224.

[71] Burnside, W., *On the continuous group that is defined by any given group of finite order*, (*Second paper*), Proc. London Math. Soc. **29** (1898), 546-565.

[72] Burnside, W., *On group-characteristics*, Proc. London Math. Soc. **33** (1900), 146-162.

[73] Burnside, W., *On some properties of groups of odd order*, Proc. London Math. Soc. **33** (1900), 162-185.

[74] Burnside, W., *On transitive groups of degree n and class $n-1$*, Proc. London Math. Soc. **32** (1900), 240-246.

[75] Burnside, W., *On a class of groups of finite order*, Trans. Cambridge Philos. Soc. **18** (1900), 269-276.

[76] Burnside, W., *On an unsettled question in the theory of discontinuous groups*, Quarterly J. Math. **33** (1902), 230-238.

[77] Burnside, W., *On the representation of a group of finite order as an irreducible group of linear substitutions and the direct establishment of the relations between the group characteristics*, Proc. London Math. Soc. (2) **1** (1903), 117-123.

[78] Burnside, W., *On the reduction of a group of homogeneous linear substitutions of finite order*, Acta Mathematica **28** (1904), 369-387.

[79] Burnside, W., *On groups of order $p^\alpha q^\beta$*, Proc. London Math. Soc. (2) **1** (1904), 388-392.

[80] Burnside, W., *On the condition of reducibility of any group of linear substitutions*, Proc. London Math. Soc. (2) **3** (1905), 430-434.

[81] Burnside, W., *On criteria for the finiteness of the order of a group of linear substitutions*, Proc. London Math. Soc. (2) **3** (1905), 435-440.

[82] Burnside, W., *On the complete reduction of any transitive permutation-group; and on the arithmetical nature of the coefficients in its irreducible components*, Proc. London Math. Soc. (2) **3** (1905), 239-252.

[83] Burnside, W., *On a general property of finite irreducible groups of linear substitutions*, Messenger of Mathematics **35** (1905), 51-55.

[84] Burnside, W., *On simply transitive groups of prime degree*, Quart. J. Math. **37** (1906), 215-221.

[85] Burnside, W., *On the theory of groups of finite order* (Presidential address), Proc. London Math. Soc. (2) **7** (1908), 1-7.

[86] Burnside, W., *On the group of the twenty-seven lines of a cubic surface*, Quart. J. Math. **40** (1909), 246-250.

[87] Burnside, W., *Theory of groups of finite order* (second edition), Cambridge University Press, Cambridge, 1911.

[88] Burnside, W., *On a group of order 25920 and the projective transformations of a cubic surface*, Proc. Camb. Phil. Soc. **23** (1926), 103-108.

[89] Cartan, É., *Sur la structure des groupes de transformations finis et continus*, Thèse, Paris, Nony, 1894; *Oeuvres complètes*, Gauthier-Villars, Paris, 1952-1955; I, vol. 1, 137-288.

[90] Cartan, É., *Les groupes bilinéaires et les systèmes de nombres complexes*, Ann. Fac. Sci. Toulouse **12** (1898), 1-99; *Oeuvres complètes*, Gauthier-Villars, Paris, 1952-1955, II, vol. 1, 7-106.

[91] Cartan, H. and Eilenberg, S., *Homological algebra*, Princeton, 1956.

[92] Carter, R. W., *Finite groups of Lie type: conjugacy classes and complex characters*, John Wiley and Sons, Chichester, 1985.

[93] Cayley, A., *On the Schwarzian derivative, and the polyhedral functions*, Trans. Camb. Philosophical Soc. **13** (1881), 5-68; Math. Papers XI, 148-216.

[94] Coleman, A. J., *The greatest mathematical paper of all time*, The Mathematical Intelligencer **11** 1989, 29-38.

[95] Conway, J. H., *Three lectures on exceptional groups*, in *Finite simple groups*, M. B. Powell and G. Higman, eds., Academic Press, London, 1971, 215-247.

[96] Conway, J. H., Curtis, R. T., Norton, S. P., Parker, R. A., and Wilson, R. A., *Atlas of finite groups*, Oxford University Press, Oxford, 1985.

[97] Curtis, C. W., *Linear algebra: an introductory approach*, Springer-Verlag, New York, 1984.

[98] Curtis, C. W., *Representation theory of finite groups: from Frobenius to Brauer*, The Mathematical Intelligencer **14** (1992), 48-57.

[99]   Curtis, C. W. and Reiner, I., *Methods of representation theory, Volume* I, John
       Wiley and Sons, New York, 1981.

[100] Curtis, C. W. and Reiner, I., *Methods of representation theory, Volume* II, John
       Wiley and Sons, New York, 1986.

[101] Davenport, H., *The higher arithmetic*, Harper, New York, 1960.

[102] Dedekind, R., *Zur Theorie der Ideale*, Göttinger Nachrichten (1894), 272-277;
       Gesammelte mathematische Werke II, Braunschweig, 1931.

[103] Dedekind, R., *Zur Theorie der aus n Haupteinheiten gebildeten complexen Grössen*,
       Göttinger Nachrichten (1885), 141-159; *Gesammelte mathematische Werke* II,
       Braunschweig, 1931.

[104] Dedekind, R., *Gesammelte mathematische Werke* II, Braunschweig, 1931.

[105] Deruyts, J., *Essai d'une théorie générale des formes algébriques*, Mém. Soc. Roy.
       Sci. Liège **17** (1892), 1-156.

[106] Deuring, M., *Algebren*, Erg. der Math. v. 4, Springer-Verlag, Berlin, 1935.

[107] Dickson, L. E., *Linear groups*, Leipzig, 1901.

[108] Dickson, L. E., *On the group defined for any given field by the multiplication table
       of any given finite group*, Trans. Amer. Math. Soc. **3** (1902), 285-301.

[109] Dickson, L. E., *Modular theory of group matrices*, Trans. Amer. Math. Soc. **8** (1907),
       389-398.

[110] Dickson, L. E., *Modular theory of group characters*, Bull. Amer. Math. Soc. **13**
       (1907), 477-488.

[111] Dickson, L. E., *Algebras and their arithmetics*, Chicago, 1923; German translation,
       Algebren und ihre Zahlentheorie, Zürich, 1927.

[112] Dickson, L. E., *Introduction to the theory of numbers*, Dover, New York, 1957.

[113] Dirichlet, P. G. Lejeune, *Beweis des Satzes, dass jede unbegrenzte arithmetische
       Progression, deren erstes Glied und Differenz ganze Zahlen ohne gemeinschaftlichen
       Factor sind, unendlich viele Primzahlen enthält*, Abh. König. Preuss. Akad. Wiss.
       (1837), 45-81; Werke I, 313-342.

[114] Dirichlet, P. G. Lejeune, *Vorlesungen über Zahlentheorie*, 4th ed., Published and
       supplemented by R. Dedekind, Vieweg, Braunschweig, 1894.

[115] Dyck, W., *Gruppentheoretische Studien*, Math. Annalen **20** (1882), 1-44.

[116] Feit, W., *Characters of finite groups*, W. A. Benjamin, New York, 1967.

[117] Feit, W., *Richard D. Brauer*, Bull. Amer. Math. Soc. (N.S.) **1** (1979), 1-20.

[118] Feit, W., *The representation theory of finite groups*, North-Holland, Amsterdam,
       1982.

[119] Feit, W., *Some computations of Schur indices*, Israel J. Math. **46** (1983), 274-300.

[120] Feit, W., Hall, M., and Thompson, J. G., *Finite groups in which the centralizer of
       any nonidentity element is nilpotent*, Math. Zeitschrift. **74** (1960), 1-17.

[121] Feit, W. and Thompson, J. G., *Groups which have a faithful representation of degree
       less than (p - 1)/2*, Pacific J. Math. **11** (1961), 1257-1262.

[122] Feit, W. and Thompson, J. G., *Solvability of groups of odd order*, Pacific J. Math.
       13 (1963), 775-1029.

[123] Formanek, E. and Sibley, D., *The group determinant determines the group*, Proc.
       Amer. Math. Soc. **112** (1991), 649-656.

[124] Forsyth, A. R., *William Burnside*, J. London Math. Soc. **3** (1928), 64-80.

[125] Frobenius, F. G., *Über Thetafunctionen mehrerer Variabeln*, J. reine angew. Math.
       96 (1884), 100-122; Ges. Abh. II, Springer-Verlag, Berlin, 1968, 149-171.

[126] Frobenius, F. G., *Neuer Beweis des Sylowschen Satzes*, J. reine angew. Math. **100**
       (1887), 179-181; Ges. Abh. II, 301-303.

[127] Frobenius, F. G., *Über die Congruenz nach einem aus zwei endlichen Gruppen
       gebildeten Doppelmodul*, J. reine angew. Math. **101** (1887), 273-299; Ges. Abh. II,
       304-330.

[128] Frobenius, F. G., *Über endliche Gruppen*, S'ber. Akad. Wiss. Berlin (1895), 81-112;
       Ges. Abh. II, 632-663.

[129] Frobenius, F. G., *Über vertauschbare Matrizen*, S'ber. Akad. Wiss. Berlin (1896),
       601-614; Ges. Abh. II, 705-718.

[130] Frobenius, F. G., *Über Beziehungen zwischen den Primidealen eines algebraischen Körpers und den Substitutionen seiner Gruppe*, S'ber. Akad. Wiss. Berlin (1896), 689-703; Ges. Abh. II, 719-733.

[131] Frobenius, F. G., *Über Gruppencharaktere*, S'ber. Akad. Wiss. Berlin (1896), 985-1021; Ges. Abh. III, 1-37.

[132] Frobenius, F. G., *Über die Primfactoren der Gruppendeterminante*, S'ber. Akad. Wiss. Berlin (1896), 1343-1382; Ges. Abh. III, 38-77.

[133] Frobenius, F. G., *Über die Darstellung der endlichen Gruppen durch lineare Substitutionen*, S'ber. Akad. Wiss. Berlin (1897), 944-1015; Ges. Abh. III, 82-103.

[134] Frobenius, F. G., *Über Relationen zwischen den Charakteren einer Gruppe und denen ihrer Untergruppen*, S'ber. Akad. Wiss. Berlin (1898), 501-515; Ges. Abh. III, 104-118.

[135] Frobenius, F. G., *Über die Composition der Charaktere einer Gruppe*, S'ber. Akad. Wiss. Berlin (1899), 330-339; Ges. Abh. III, 119-128.

[136] Frobenius, F. G., *Über die Darstellung der endlichen Gruppen durch lineare Substitutionen II*, S'ber. Akad. Wiss. Berlin (1899), 482-500; Ges. Abh. III, 129-147.

[137] Frobenius, F. G., *Über die Charaktere der symmetrischen Gruppe*, S'ber Akad. Wiss. Berlin (1900), 516-534; Ges. Abh. III, 148-166.

[138] Frobenius, F. G., *Über die Charaktere der alternirenden Gruppe*, S'ber. Akad. Wiss. Berlin (1901), 303-315; Ges. Abh. III, 167-179.

[139] Frobenius, F. G., *Über auflösbare Gruppen III*, S'ber. Akad. Wiss. Berlin (1901), 849-875; Ges. Abh. III, 180-188.

[140] Frobenius, F. G., *Über auflösbare Gruppen IV*, S'ber. Akad. Wiss. Berlin (1901), 1216-1230; Ges. Abh. III, 189-203.

[141] Frobenius, F. G., *Über Gruppen des Grades p oder $p+1$*, S'ber. Akad. Wiss. Berlin (1902), 351-369; Ges. Abh. III, 220-238.

[142] Frobenius, F. G., *Über Gruppen der Ordnung $p^\alpha q^\beta$*, Acta Mathematica **26** (1902), 189-198; Ges. Abh. III, 210-219.

[143] Frobenius, F. G., *Über die charakteristischen Einheiten der symmetrischen Gruppe*, S'ber. Akad. Wiss. Berlin (1903), 328-358; Ges. Abh. III, 244-274.

[144] Frobenius, F. G., *Theorie der hyperkomplexen Grössen*, S'ber. Akad. Wiss. Berlin (1903), 504-537; Ges. Abh. III, 284-317.

[145] Frobenius, F. G., *Theorie der hyperkomplexen Grössen II*, S'ber. Akad. Wiss. Berlin (1903), 634-645; Ges. Abh. III, 318-329.

[146] Frobenius, F. G., *Über einen Fundamentalsatz der Gruppentheorie II*, S'ber Akad. Wiss. Berlin (1907), 428-437; Ges. Abh. III, 394-403.

[147] Frobenius, F. G., *Über Matrizen aus positiven Elementen*, S'ber Akad. Wiss. Berlin (1908), 471-476; Ges. Abh. III, 404-409.

[148] Frobenius, F. G., *Über Matrizen aus positiven Elementen II*, S'ber Akad. Wiss. Berlin (1909) 514-518; Ges. Abh. III, 410-414.

[149] Frobenius, F. G., *Gruppentheoretische Ableitung der 32 Kristallklassen*, S'ber Akad. Wiss. Berlin (1911), 681-691; Ges. Abh. III, 519-529.

[150] Frobenius, F. G., *Über Matrizen aus nicht negativen Elementen*, S'ber Akad. Wiss. Berlin (1912), 456-477; Ges. Abh. III, 546-567.

[151] Frobenius, F. G. und Schur, I., *Über die reellen Darstellungen der endlichen Gruppen*, S'ber. Akad. Wiss. Berlin (1906), 186-208; Ges. Abh. III, 355-377.

[152] Frobenius, F. G. und Schur, I., *Über die Äquivalenz der Gruppen linearer Substitutionen*, S'ber. Akad. Wiss. Berlin (1906), 209-217; Ges. Abh. III, 378-386.

[153] Frobenius, F. G. und Stickelberger, L., *Über Gruppen von vertauschbaren Elementen*, J. reine angew. Math. **86** (1879) 217-262; Ges. Abh. I, 545-590.

[154] Galois, E., *Oeuvres Mathématiques*, Gauthier-Villars, Paris, 1897.

[155] Gauss, C. F., *Disquisitiones Arithmeticae*, Leipzig, 1801; Werke, vol. I; English translation by A. A. Clarke, Yale University Press, New Haven, 1966.

[156] Gauss, C. F., *Theorematis fundamentalis in doctrina de residuis quadraticis demonstrationes et ampliationes novae*, 1818; Werke, vol. II, 47-64.

[157] Gauss, C. F., *Theoria residuorum biquadraticorum*, Commentatio prima, 1828; Werke, vol. II, p. 65-92.

[158] Gauss, C. F., *Mathematisches Tagebuch, 1796-1814*, Edited by K.-R. Biermann, Ostwalds Klassiker 256.

[159] Gierster, J., *Die Untergruppen der Galois'schen Gruppe der Modulargleichungen für den Fall eines primzahligen Transformationsgrades*, Math. Ann. **18** (1881), 319-365.

[160] Glauberman, G., *Central elements in core-free groups*, J. Algebra **4** (1966), 403-420.

[161] Godeaux, L., *Notice sur Jacques Deruyts*, Annuaire de l'Academie Royale de Belgique **115** (1949), 21-43.

[162] Gorenstein, D., *The classification of finite simple groups I: Simple groups and local analysis*, Bull. Amer. Math. Soc. (N.S.) **1** (1979), 43-199.

[163] Gorenstein, D., *Finite simple groups: an introduction to their classification*, Plenum Press, New York, 1982.

[164] Green, J. A., *Richard Dagobert Brauer*, Bull. London Math. Soc. **10** (1978), 317-342.

[165] Green, J. A., *Polynomial representations of GL_n*, Lecture Notes in Mathematics **830**, Springer-Verlag, Berlin Heidelberg, 1980.

[166] Green, J. A., *Classical invariants and the general linear group*, Progress in Mathematics, **95** (1991), 247-272.

[167] Green, J. A., *On a formula of J. Deruyts*, J. Pure and Applied Algebra **107** (1996), 219-232.

[168] Hall, M., *The theory of groups*, Macmillan, New York, 1959.

[169] Hardy, G. H., *The J-type and the S-type among mathematicians*, Nature **134** (1934), 250; Collected papers of G. H. Hardy, London Mathematical Society, vol. 7, 610-611.

[170] Hasse, H., *Theory of cyclic algebras over an algebraic number field*, Trans. Amer. Math. Soc. **34** (1932), 171-214.

[171] Hasse, H., *History of class field theory*, Chapter XI in *Algebraic Number Theory*, ed. by J. W. S. Cassels and A. Frölich, Thompson Book Co., Washington, D.C., 1967.

[172] Hawkins, T., *The origins of the theory of group characters*, Archive for History of Exact Sciences **7** (1971), 142-170.

[173] Hawkins, T., *Hypercomplex numbers, Lie groups, and the creation of group representation theory*, Archive for History of Exact Sciences **8** (1972), 243-287.

[174] Hawkins, T., *New light on Frobenius' creation of the theory of group characters*, Archive for History of Exact Sciences **12** (1974), 217-243.

[175] Hecke, E., *Über die L-Funktionen und der Dirichletschen Primzahlsatz für einen beliebigen Zahlkörper*, Nachrichten der Gesellschaft der Wissenschaften zu Göttingen, Math.-phys. Klasse (1917), 299-318; *Eine neue Art von Zetafunktionen und ihre Beziehungen zur Verteilungen der Primzahlen, Erste Mitteilung*, Math. Zeitschrift **1** (1918), 357-376; *Eine neue Art von Zetafunktionen und ihre Beziehungen zur Verteilungen der Primzahlen, Zweite Mitteilung*, Math. Zeitschrift **6** (1920), 11-51; Mathematische Werke, Vandenhoeck and Ruprecht, Göttingen, (1959), 178-197, 215-234, 249-289.

[176] Heilbronn, H., *Zeta functions and L-functions*, in *Algebraic Number Theory*, edited by J. W. S. Cassels and A. Fröhlich, Thompson Book Company, Washington, D.C. (1967), Chapter VIII.

[177] Hilbert, D., *Die Theorie der algebraischen Zahlkörper*, Jahresbericht der Deutschen Mathematikervereinigung **4** (1897), 175-546; Ges. Abh. I, 63-363.

[178] Hilbert, D., *Grundzüge einer allgemeinen Theorie der linearen Integralgleichungen, Erste Mitteilung*, Göttinger Nachrichten 1904, 49-91.

[179] Hodge, W. V. D., *H. F. Baker (1866-1956)*, J. London Math. Soc. **32** (1957), 112-128.

[180] Hölder, O., *Die Gruppen der Ordnungen p^3, pq^2, pqr, p^4*, Math. Ann. **43** (1893), 301-412.

[181] Hölder, O., *Die Gruppen mit quadratfreier Ordnungzahl*, Göttinger Nachrichten (1895), 211-229.

[182] Humphreys, J., *Reflection groups and Coxeter groups*, Cambridge Studies in Advanced Mathematics, Cambridge University Press, 1990.

[183] Huppert, B., *Endliche Gruppen*, Springer-Verlag, Berlin, 1967.

[184] Hurwitz, A., *Über Riemann'sche Flächen mit gegebenen Verzweigungspunkten*, Math. Annalen **39** (1891), 1-61; Mathematische Werke I, 321-383.

[185] Hurwitz, A., *Zur Invariantentheorie*, Math. Ann. **45** (1894), 381-404; Mathematische Werke II, 508-532.

[186] Hurwitz, A. *Über die Anzahl der Riemann'sche Flächen mit gegebenen Verzweigungspunkten*, Math. Annalen **55** (1902), 53-66; Mathematische Werke I, 492-505.

[187] Hurwitz, A., *Über Beziehungen zwischen den Primidealen eines algebraischen Körpers und den Substitutionen seiner Gruppe*, Math. Zeitschrift **25** (1926), 661-665; Mathematische Werke II, 733-739.

[188] Ireland, K. and Rosen, M., *A classical introduction to modern number theory*, Springer-Verlag, New York, 1982.

[189] James, G. and Kerber, A., The representation theory of the symmetric group, Encyclopedia of Mathematics and its Applications, Vol. 16, Addison Wesley Publishing Co., Reading, 1981.

[190] Janusz, G., *Primitive idempotents in group algebras*, Proc. Amer. Math. Soc. **17** (1966), 520-523.

[191] Janusz, G., *Algebraic Number Fields*, Academic Press, New York, 1973; second edition, American Mathematical Society, Providence, 1996.

[192] Jordan, C., *Traité des substitutions et des équations algébriques*, Gauthier-Villars, Paris, 1870.

[193] Jordan, C., *Sur les groupes d'ordre $p^m q^2$*, Liouville's Journal (5) **4** (1898), 21-26.

[194] Kanigel, R., *The man who knew infinity: a life of the genius Ramanujan*, Charles Scribner and Sons, New York, 1991.

[195] Kimberling, C., *Emmy Noether and her Influence*, in *Emmy Noether: a Tribute to her Life and Work*, Marcel Dekker, New York, 1981, pp. 3-61.

[196] Klein, F., *Vorlesungen über das Ikosaeder und die Auflösung der Gleichungen vom fünften Grade*, B. G. Teubner, Leipzig, 1884.

[197] Knapp, W., *On Burnside's method*, J. Algebra **175** (1995), 644-660.

[198] Kronecker, L., *Auseinandersetzung einiger Eigenschaften der Klassenzahl idealer complexer Zahlen*, Monatsb. König. Preuss. Akad. Wiss. Berlin (1870); Werke I, Leipzig, 1895, 271-282.

[199] Kronecker, L., *Über die Irreductibilität von Gleichungen*, Monatsb. König. Preuss. Akad. Wiss. Berlin (1880), 155-162; Werke II, 83-93.

[200] Krull, W., *Über verallgemeinerte endliche Abelsche Gruppen*, Math. Zeitschrift **23** (1925), 161-196.

[201] Lagrange, J. L., *Réflexions sur la résolution algébrique des équations*, Nouveaux Mém. de l'Acad. R. des Sc. et B.-L. de Berlin, 1770-71; Oeuvres, vol. III, p. 332.

[202] Lagrange, J. L., *Traité de la résolution numérique des équations*. 2nd ed., Paris, 1808, Notes XIII-XIV; Oeuvres, vol. VIII, pp. 295-367.

[203] Lam, T. Y., *Noether's mathematics. Representation theory*. pp. 145-156, in *Emmy Noether: a tribute to her life and work*, Marcel Dekker, New York, 1981.

[204] Lasker, E., *Zur Theorie der Moduln und Ideale*, Math. Ann. **60** (1905) 20-116.

[205] Ledermann, W., *Issai Schur and his school in Berlin*, Bull. London Math. Soc. **15** (1983), 97-106.

[206] Loewy, A., *Sur les formes quadratiques définies à indéterminées conjuguées de M. Hermite*, Comptes Rendus, Académie des Sciences, Paris, **123** (1896), 168-171.

[207] Loewy, A., *Über die Reducibilität der Gruppen linearer homogener Substitutionen*, Trans. Amer. Math. Soc. 4 (1903), 44-64.

[208] Loewy, A., *Über die Reducibilität der reellen Gruppen linearer homogener Substitutionen*, Trans. Amer. Math. Soc. 4 (1903), 171-177.

[209] Macaulay, F. S., *On the resolution of a given modular system into primary systems including some properties of Hilbert numbers*, Math. Ann. **74** (1913), 66-121.

[210] Macdonald, I. G., *Symmetric functions and Hall polynomials*, Oxford University Press, Oxford, 1979.

[211] MacLane, S., *Homology*, Springer-Verlag, Berlin, 1963.

[212] Maschke, H., *Über den arithmetischen Charakter der Coefficienten der Substitutionen endlicher linearer Substitutionsgruppen*, Math. Ann. **50** (1898), 492-498.

[213] Maschke, H., *Beweis des Satzes, dass diejenigen endlichen linearen Substitutionsgruppen, in welchen einige durchgehends verschwindende Coefficienten auftreten, intransitiv sind*, Math. Ann. **52** (1899), 363-368.

[214] Miyake, K., *A note on the arithmetic background to Frobenius' theory of group characters*, Expo. Math. **7** (1989), 347-358.

[215] Molien, T., *Über Systeme höherer complexer Zahlen*, Math. Ann. **41** (1893), 83-156.

[216] Molien, T., *Über die Invarianten der linearen Substitutionsgruppen*, S'ber Akad. Wiss. Berlin (1897), 1152-1156.

[217] Moore, E. H., *A doubly-infinite system of simple groups*, in Mathematical Papers read at the International Mathematics Congress held in Connection with the World's Columbian Exposition: Chicago, 1893, ed. E. H. Moore et al., Macmillan, New York, 1896. 208-242.

[218] Moore, E. H., *An universal invariant for finite groups of linear transformations: with applications in the theory of the canonical form of a linear substitution of finite period*, Math. Ann. **50** (1898), 213-219.

[219] Mosenthal, V. and Wagner, A., *A bibliography of William Burnside* (1852-1927), Historia Mathematica **5** (1978), 307-312.

[220] Nagao, H., *A proof of Brauer's theorem on generalized decomposition numbers*, Nagoya Math. J. **22** (1963), 73-77.

[221] Nagao, H. and Nakayama, T., *On the structure of M_o- and M_u-modules*, Math. Zeitschrift **59** (1953), 164-170.

[222] Nagao, H. and Tsushima, Y., *Representations of finite groups*, Academic Press, San Diego, 1989.

[223] Nakayama, T., *Some studies on regular representations, induced representations and modular representations*, Ann. of Math. **39** (1938), 361-369.

[224] Nakayama, T., *On Frobeniusean algebras I*, Ann. of Math. **40** (1939), 611-633.

[225] Nakayama, T., *On some modular properties of irreducible representations of symmetric groups. II*, Jap. J. Math. **17** (1941), 411-423.

[226] Nesbitt, C., *On the regular representations of algebras*, Ann. of Math. **39** (1938), 634-658.

[227] Neumann, P. M., *Helmut Wielandt on permutation groups*, in *Helmut Wielandt, Mathematische Werke*, Walter de Gruyter, Berlin, 1994; I, 3-30.

[228] Noether, E., *Körper und Systeme rationaler Funktionen*, Math. Ann. **76** (1915), 161-196; Ges. Abh., Springer-Verlag, Berlin, 1983, 145-180.

[229] Noether, E., *Der Endlichkeitssatz der Invarianten endlicher Gruppen*, Math. Ann. **77** (1916), 89-92; Ges. Abh. 181-184.

[230] Noether, E., *Gleichungen mit vorgeschriebener Gruppe*, Math. Ann. **78** (1918), 221-229; Ges. Abh. 231-239.

[231] Noether, E., *Idealtheorie in Ringbereichen*, Math. Ann. **83** (1921), 24-66; Ges. Abh. 354-396.

[232] Noether, E., *Abstrakter Aufbau der Idealtheorie in algebraischen Zahl- und Funktionenkörpern*, Math. Ann. **96** (1927), 26-61; Ges. Abh. 493-528.

[233] Noether, E., *Hyperkomplexe Grössen und Darstellungstheorie*, Math. Zeitschrift. **30** (1929), 641-692; Ges. Abh. 563-592.

[234] Noether, E., *Nichtkommutative Algebren*, Math. Zeitschrift **37** (1933), 514-541; Ges. Abh. 642-669.

[235] Osima, M., *Notes on blocks of group characters*, Math. J. Okayama Univ. **4** (1955), 175-188.

[236] Parshall, K. H. and Rowe, D. E., *The emergence of the American mathematical research community, 1876-1900: J. J. Sylvester, Felix Klein, and E. H. Moore*, American Mathematical Society, London Mathematical Society, Providence, 1994.

[237] Peirce, B., *Linear associative algebra*, Amer. J. Math. *4* (1881), 97-221.

[238] Reid, C., *Hilbert*, Springer-Verlag, New York, 1970.

[239] Remak, R., *Über die Zerlegung der endlichen Gruppen in direkte unzerlegbare Faktoren*, J. reine angew Math. **139** (1911), 293-308.

[240] Robinson, G. de B., *On a conjecture by Nakayama*, Trans. Roy. Soc. Canada. Sect. III. (3) **41** (1947), 20-25.

[241] Roggenkamp, K., *On Dedekind's group determinant and Frobenius' higher characters*, Darstellungstheorietage (Erfurt 1992), 21-42, Sitzungsber. Math.-Naturwiss. Kl. 4, Akad. Gemein. Wiss. Erfurt, Erfurt, 1992.

[242] Roquette, P., *Arithmetische Untersuchung des Charakterringes einer endlichen Gruppe*, J. Reine Angew. Math. **190** (1952), 148-168.

[243] Roseblade, J. E., *Philip Hall*, Bull. London Math. Soc. **16** (1984), 603-626.

[244] Rosenberg, A., *Blocks and centres of group algebras*, Math. Zeitschrift **76** (1961), 209-216.

[245] Schmidt, E., *Zur Theorie der linearen und nichtlinearen Integralgleichungen I. Teil: Entwicklung willkürlicher Functionen nach Systemen vorgeschriebener*, Math. Ann. **63**, 433-472.

[246] Schmidt, O., *Über unendliche Gruppen mit endlicher Kette*, Math. Zeitschrift **29** (1928), 34-41.

[247] Schur, I., *Über eine Klasse von Matrizen, die sich einer gegebenen Matrix zuordnen lassen*, Dissertation, Berlin, 1901; Ges. Abh., Springer-Verlag, Berlin, 1973, I, 1-73.

[248] Schur, I., *Über die Darstellung der endlichen Gruppen durch gebrochene lineare Substitutionen*, J. reine angew. Math. **127** (1904), 20-50; Ges. Abh. I, 86-116.

[249] Schur, I., *Neue Begründung der Theorie der Gruppencharaktere*, S'ber. Akad. Wiss. Berlin (1905), 406-432; Ges. Abh. I, 143-169.

[250] Schur, I., *Arithmetische Untersuchungen über endliche Gruppen linearer Substitutionen*, S'ber Akad. Wiss. Berlin (1906), 164-184; Ges. Abh. I, 177-197.

[251] Schur, I., *Untersuchungen über die Darstellung der endlichen Gruppen durch gebrochene lineare Substitutionen*, J. reine angew. Math. **132** (1907), 85-137; Ges. Abh. I, 198-250.

[252] Schur, I., *Neuer Beweis eines Satz von W. Burnside*, Jahresbericht der Deutschen Mathematiker-Vereinigung **17** (1908), 171-176; Ges. Abh. I, 266-271.

[253] Schur, I., *Zur Theorie der linearen homogenen Integralgleichungen*, Math. Ann. **67** (1909), 306-339; Ges. Abh. I, 312-345.

[254] Schur, I., *Bemerkungen zur Theorie der beschränkten Bilinearformen mit unendlich vielen Veränderlichen*, J. reine angew. Math. **140** (1911), 1-28; Ges. Abh. I, 464-491.

[255] Schur, I., *Über die Darstellung der symmetrischen und der alternierenden Gruppe durch gebrochene lineare Substitutionen*, J. reine angew. Math. **139** (1911), 155-250; Ges. Abh. I, 346-441.

[256] Schur, I., *Bemerkungen zu der vorstehenden Arbeit des Herrn Speiser*, Math. Zeitschrift **5** (1919), 7-10.

[257] Schur, I., *Antrittsrede*, Ges. Abh. II, 413-415.

[258] Schur, I., *Neue Anwendungen der Integralrechnung auf Probleme der Invariantentheorie*, S'ber. Akad. Wiss. Berlin (1924), I. 189-208; II. 297-321; III. 346-355; Ges. Abh. II, 440-459, 460-484, 485-494.

[259] Schur, I., *Über die rationalen Darstellungen der allgemeinen linearen Gruppe*, S'ber. Akad. Wiss. Berlin (1927), 58-75; Ges. Abh. III, 68-85.

[260] Schur, I., *Über die stetigen Darstellungen der allgemeinen linearen Gruppe*, S'ber. Akad. Wiss. Berlin (1928), 100-124; Ges. Abh. III, 89-113.

[261] Schur, I., *Zur Theorie der einfach transitiven Permutationsgruppen*, S'ber. Akad. Wiss. Berlin (1933), 598-623; Ges. Abh. III, 266-291.

[262] Selberg, A., *An elementary proof of the prime number theorem*, Ann. of Math. **50** (1949), 305-319.

[263] Serre, J.-P., *A course in arithmetic*, Springer-Verlag, New York, 1973.

[264] Slodowy, P., *The early development of the representation theory of semisimple Lie groups: A. Hurwitz, I. Schur, H. Weyl*, RIMS-1190, Research Institute for Mathematical Sciences, Kyoto University, Kyoto, Japan, 1998.

[265] Speiser, A., *Zahlentheoretische Sätze aus der Gruppentheorie*, Math. Zeitschrift **5** (1919), 1-6.

[266] Study, E., *Complexe Zahlen und Transformationsgruppen*, Ber. k. sächs. Ges. d. Wiss. Leipzig, (1889), 177-227.

[267] Suzuki, M., *A characterization of simple groups LF(2,p)*, J. Fac. Sci. Univ. Tokyo. Sect. I. **6** (1951), 259-293.

[268] Suzuki, M., *The nonexistence of a certain type of simple group of odd order*, Proc. Amer. Math. Soc. **8** (1957), 686-695.

[269] Suzuki, M., *Applications of group characters*, Proc. Sympos. Pure Math., Vol. 1, Amer. Math. Soc., Providence, R. I., (1959), 88-99.

[270] Sylow, M. L., *Théorèmes sur les groupes de substitutions*, Math. Ann. **5** (1872), 584-594.

[271] Tchebotarev, N., *Die Bestimmung der Dichtigkeit einer Menge von Primzahlen, welche zu einer gegebenen Substitutionsklasse gehören*, Math. Ann. **95** (1926), 191-228.

[272] Thompson, J. G., *Nonsolvable finite groups all of whose local subgroups are solvable, I-VI*, Bull. Amer. Math. Soc. **74** (1968), 383-437; Pacific J. Math. **33** (1970), 451-536; **39** (1971), 483-534; **48** (1973), 511-592; **50** (1974), 215-297; **51** (1974), 573-630.

[273] Tuan, H.-F., *On groups whose orders contain a prime to the first power*, Ann. of Math. **45** (1944), 110-140.

[274] Tuan, H.-F., *Some problems in the block theory of modular representations*, Contemp. Math. **82** (1989), 181-190.

[275] Turnbull, H. W., *Alfred Young, 1873-1940*, Journal London Math. Soc. **16** (1941), 194-207.

[276] Van der Waerden, B. L., *Moderne Algebra*, Springer-Verlag, Berlin, 1931; Second Edition, 1937.

[277] Von Neumann, J., *Zur Theorie der Darstellungen kontinuierlicher Gruppen*, S'ber. Akad. Wiss. Berlin (1927), 76-90; Collected Works I, 134-148.

[278] Weber, H., *Lehrbuch der Algebra, Bd. 2*, Vieweg, Braunschweig, 1896.

[279] Wedderburn, J. H. M., *On hypercomplex numbers*, Proc. London Math. Soc. Ser. 2, **6** (1908), 77-118.

[280] Weierstrass, K., *Zur Theorie der aus n Haupteinheiten gebildeten complexen Grössen*, Göttinger Nachrichten (1884), 395-414; Werke II, 311-332.

[281] Weil, A., *Number of solutions of equations in a finite field*, Bull. Am. Math. Soc. **55** (1949), 497-508; Collected Papers, vol. I, 399-410.

[282] Weil, A., *La cyclotomie jadis et naguère*, Sem. Bourbaki, juin, 1974; Collected Papers, vol. III, 311-327.

[283] Weil, A., *Number theory: an approach through history; from Hammurapi to Legendre*, Birkhäuser, Boston, 1983.

[284] Weyl, H., *Theorie der Darstellung kontinuierlicher halb-einfacher Gruppen durch lineare Transformationen* I, Math. Zeitschrift **23** (1925), 271-309; II, Math. Zeitschrift **24** (1926), 328-376; III, Math. Zeitschrift **24** (1926), 377-395; Ges. Abh. II, 543-647.

[285] Weyl, H., *Gruppentheorie und Quantenmechanik*, Zürich, 1928; English translation by H. Robertson, *Theory of Groups and Quantum Mechanics*, E. P. Dutton, New York, 1931.

[286] Weyl, H., *Emmy Noether*, Scripta Mathematica **3** (1935), 201-220; Ges. Abh., Springer-Verlag, New York, 1968, vol. 3, 425-444.

[287] Weyl, H., *The classical groups*, Princeton University Press, Princeton, 1939.

[288] Wielandt, H., *Finite permutation groups*, translated from the German by R. Bercov, Academic Press, New York, 1964.

[289] Wigner, E., *Gruppentheorie und ihre Anwendung auf der Quantenmechanik der Atomspektren*, Vieweg, Braunschweig, 1931; English translation by J. Griffin, *Group theory and its applications to the quantum mechanics of atomic spectra*, Academic Press, New York, 1959.

[290] Wussing, H., *Die Genesis des abstrakten Gruppenbegriffes*, Berlin, 1969.

[291] Young, A., *On quantitative substitutional analysis*, Proc. London Math. Soc. **33** (1901), 97-146.

[292] Young, A., *On quantitative substitutional analysis (second paper)*, Proc. London Math. Soc. **34** (1902), 361-397.

[293] Zassenhaus, H., *Kennzeichnung endlicher linearer Gruppen als Permutationsgruppen*, Abh. Math. Sem. Univ. Hamburg **11** (1936), 17-40.

Index